Functional Polymer Solutions and Gels

Functional Polymer Solutions and Gels—Physics and Novel Applications

Special Issue Editors

Florian J. Stadler
Bing Du

MDPI • Basel • Beijing • Wuhan • Barcelona • Belgrade • Manchester • Tokyo • Cluj • Tianjin

Special Issue Editors
Florian J. Stadler
Shenzhen University
China

Bing Du
Shenzhen University
China

Editorial Office
MDPI
St. Alban-Anlage 66
4052 Basel, Switzerland

This is a reprint of articles from the Special Issue published online in the open access journal *Polymers* (ISSN 2073-4360) (available at: https://www.mdpi.com/journal/polymers/special_issues/Polymer_Solutions_Gels).

For citation purposes, cite each article independently as indicated on the article page online and as indicated below:

LastName, A.A.; LastName, B.B.; LastName, C.C. Article Title. *Journal Name* **Year**, *Article Number*, Page Range.

ISBN 978-3-03936-230-1 (Pbk)
ISBN 978-3-03936-231-8 (PDF)

Contents

About the Special Issue Editors

Florian J. Stadler, Prof. Dr.-Ing. is distinguished professor at Shenzhen University, which he joined in 2014 after studies at the Friedrich-Alexander University Erlangen-Nuremberg (FAU), Germany and further stations in Japan, Belgium, and South Korea. His focus is on the research in polymer physics, especially rheology, polymer chemistry of functional polymers, and mostly metal oxide-based nanoparticles for environmental applications. He published more than 250 papers in a large variety of scientific journals.

Bing Du, Dr. rer. nat., is lecturer of College of Materials Science and Engineering, Shenzhen University. She worked at Institute of Polymer Research from Helmholtz-Zentrum Geesthacht and received her Ph.D. degree in 2013. Her current research interests are development of polymer/carbon nano-composites, electrochemical sensors for detecting biomolecules, catalytic membrane reactor.

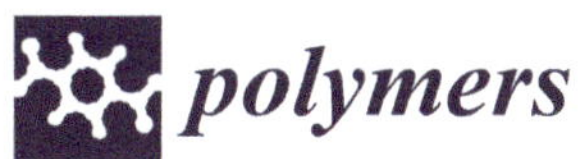 

Editorial

Functional Polymer Solutions and Gels—Physics and Novel Applications

Bing Du and Florian J. Stadler *

College of Materials Science and Engineering, Shenzhen Key Laboratory of Polymer Science and Technology, Guangdong Research Center for Interfacial Engineering of Functional Materials, Nanshan District Key Lab for Biopolymers and Safety Evaluation, Shenzhen University, Shenzhen 518055, China; dubing@szu.edu.cn
* Correspondence: fjstadler@szu.edu.cn

Received: 3 March 2020; Accepted: 16 March 2020; Published: 18 March 2020

Recent years have seen significant improvements in the understanding of functional soft matter. Advances in organic chemistry have identified a wide variety of possible interactions, ranging from hydrophobic interactions, e.g., based on cyclodextrin or hydrophobic end-capping, host–guest interactions, and multiple hydrogen bonding to metal–ligand bonding, such as for terpyridines and catechols. Those functional moieties can link polymer chains, usually in aqueous solution, and thus assemble these solutions into gels. These discoveries of the last ca. 30 years by chemists worldwide have made it possible to understand biological soft matter significantly better, as well as to create our own soft materials with tailored properties, such as a pH-controlled behavior. Their current and future applications are often focused on the biomedical field, particularly on drug release and tissue engineering. However, functional polymer solutions and gels can also be envisioned for many other applications. For example, functional nanofibers electrospun from solution can also be used for advanced applications. The resulting nanofibers combine an incredible highly specific surface area with an excellent performance as membranes with high flux, good separator selectivity, as well as extraordinary selective absorption for both functional nanoparticles and pollutants in water.

This Special Issue of *Polymers* attracted contributions from several diverse fields of polymers which can only exemplarily illustrate the broadness of the topic of polymer solutions and gels.

Polymers with special moieties, leading to thermoresponsive and stimuli-responsive behavior, was one of the main topics.

Yan et al. [1] studied the interactions between tacticity of poly (N-iso-propylacrylamide) PNIPAM and 1-butyl-3-methylimidazolium bis(trifluoromethanesulfonyl) imide ([BMIM][TFSI]) ionic liquid, in which a higher isotaxy of PNIPAM leads to a higher amount of interaction and, consequently, to more temperature stable gels, i.e., to an increased upper critical solution temperature. Interestingly, the results show that it is possible to have a decoupling of gelation and turbidity, which is counterintuitive and expands upon earlier knowledge on the physicochemical behavior of PNIPAM in ionic liquids [2]. These results show the relevance of ionic liquids for the solubilization of polymers, which has revolutionized the unwrapping of tightly packed crystalline cellulose [3–6].

Polymer solutions with catechol functionalities were shown to significantly influence thermoresponsive behavior as well as the end groups, which could be modeled statistically, demonstrating the combined effects of the end groups derived from the rather hydrophobic RAFT agents and catechol groups [7], which has led to further elucidation as well as confirmed previous observations on the properties of thermoresponsive polymers containing catechol moieties [8,9]. This work also extends on previous reports on PNIPAM-based polymer with other structures [10–12]. Similarly to catechols, terpyridine groups—as one species of metal–ligand complexing functionalities—are used as a terminal group of self-assembling di-block copolymers which exhibit an unusually fast self-healing behavior as well as a highly concentration and ion-type-dependent rheology [13].

Pure polymer physics was also represented in the theoretical study of Wang [14], which studied the drag reduction effect in polymer solutions, an effect which has been a focal point of interest for both experimental and theoretical polymer physicists for a long time [15–17], as drag reduction promises to be a technology that can contribute to the international climate crisis by reducing energy consumption for pumping fluids.

The realm of environmental technology was represented by Luo et al. [18], who introduced a way to selectively adsorb highly environmentally polluting chromium (VI) ions that are often produced during leather production [19]. This work introduces a nonwoven able to adsorb chromium but also to release it and, thus, allowing for regeneration of the adsorbent. In comparison to many adsorption approaches based on nanoparticles, nonwovens have the advantage of macroscopic size and, thus, easier handling as a filter-like material.

The biomedical applications in the Special Issue were represented by several contributions. Guan et al. [20] demonstrated the usability of alginate–acrylamide hydrogels for the production of vascular grafts—hydrogels that can be used for blood vessel repair, leading to similarly good mechanical performance as their natural porcine counterparts. Steffi et al. [21] developed a new way to improve bone-forming cell differentiation by using a strontium-doped titania tannic acid polyphenol layer on titanium for implants. Such a coating of implants could improve the interface between the implant and the bone, especially in the case of osteoporosis. Skwira et al. [22] contributed to the Special Issue in the broad field of drug delivery with their paper investigating the release properties of ciprofloxacin antibiotic from special silica-based composites with ethylcellulose and polydimethylsiloxane.

Li et al. [23] showed a biomedical-related use of hydrogels—they demonstrated how gallic acid (GA) could accumulate from red ginseng by adsorption on chitosan-based hydrogels. GA is a natural polyphenolic molecule, interesting for its antioxidant, anti-inflammatory, anticancer, and antiviral activities [24].

Khan et al. [25] showed one example of the use of polymer gels as dosimeters, providing a high-resolution equivalent of tissue in quality control of radiation therapy. Warman et al. [26] reported the development of a device for the tomographic imaging of radio-fluorogenic gels (RFGs), which allows for scanning the 3D distribution of radiation in a gel.

Asphalt binders are another type of gel-like material covered in this Special Issue. Yu et al. [27] showed how efficiently styrene–butadiene–styrene (SBS) tri-block copolymers are sufficiently UV-stable as an asphalt binder for permanently improving the properties of asphalt. Gong et al. [28] contributed to the same stream of research by investigating the rheological behavior of these asphalt–SBS mixtures, which complements a research line of polymer–asphalt mixtures for improving rheological properties [29,30].

Electrospinning technology is one of the most efficient and straightforward approaches to fabricate the fibrous membrane or scaffold [31]. Through the powerful tension, various solutions based on polymers or their composite are drawn and instantaneously solidified into micro-/nanoscale fibers. Owing to the nanoscale of the fibers, the electrospun products possess an extremely high ratio surface, conjunct network structure, and quite large porosity, providing plenty of possibilities to expand versatile materials. In particular, the fibrous structure can remarkably reduce the density of the materials; thus, it can effectively "soften" the substance with a lightweight feature. Many researchers develop copious types of novel soft materials, which have been widely utilized for tissue scaffold [32,33], wound dressing [34], fibrous membrane for catalyst [35], energy cell [36], and electrochemical [37] and mechanical [38] sensors. These active developments profit from the fascinating matrix based on nanofibrous structure, which extensively adapts to many fields as supporting materials. On the other hand, the electrospun membrane can also play a crucial role as one relative independent structural component because of its excellent intrinsic properties. In this Special Issue, Monteserín et al. [39] reported on interesting layer-by-layer epoxy resin composites reinforced by electrospun nanofiber veils. They included electrospun material made of polyamide 6 modified with TiO_2 nanoparticles into carbon fiber/epoxy composite as a single structure. From this work, it was observed that the nanofibers could effectively improve the flexural stress at failure and fracture toughness of the composite. When the fibers were modified with TiO_2, the composite exhibited new antibacterial performance, which widened the application of the material.

The field of functional polymer solutions and gels has, so far, experienced a lot of attention from chemists, who are making numerous interesting, complex systems whose physical behavior is very complex. Hence, while polymer chemistry also including functional moieties has reached a certain level of maturity, main topics remain which are lacking a good understanding of the physics of such systems. The clear application direction for functional polymer solutions and gels is towards biomedical applications, which automatically means that the questions to be answered are rather complex as, in most cases, they involve interactions of complex functional polymer materials with even more complex—especially biological—systems. Classical polymer physics on melts and solutions without functional moieties has reached maturity after being intensively researched from ca. 1950 until the present date. Functional polymers were systematically introduced to research much later, and it will certainly take several decades until the same level of understanding is reached for the physics of functional polymers. The future of physics and applications of functional polymers will be highly diverse, as is shown in this Special Issue. We expect that special emphasis will be paid on tailoring the properties of the polymers to a particular application direction, as well as to gradually increase the fraction of modified natural polymeric systems in comparison to classical synthetic polymer systems.

Functional polymer solutions and gels are often targeted for high-value applications, especially in the biomedical field. The complexities of these fields, owing to the multidimensional property profile of these materials interacting with very complex systems, also means that classical characterization methods are no longer sufficient for totally capturing the essential parameters of the system. As a consequence, we expect to see the development of combined methods, where two classical methods are merged together to yield 2 or more kinds of measurements simultaneously. Examples of these which the editors are familiar with include the combination of rheology with spectroscopic methods (nuclear magnetic resonance (NMR), Raman spectroscopy, Fourier transform infrared spectroscopy (FTIR), and dielectric spectroscopy). Furthermore, further improvement of existing methods will become necessary—one recent example being non-vacuum scanning electron microscopy (SEM), useful for investigating living biological samples or fast chip calorimetry, improving traditional speed limitations of dynamic scanning calorimetry of ca. 1–2 K/s by several orders of magnitude. Furthermore, we will see the development of completely new methods as well, which will suit needs of the scientific community.

Lastly, as one of the primary areas of functional soft materials is the biomedical area, increasing collaborations with industrial partners will become necessary as the regulatory requirements for testing the safety of materials for in vivo use—i.e., applying the material to human or animal patients—are strongly regulated, requiring extensive and expensive tests that research institutions usually cannot pay for without strong support from industrial partners.

Conflicts of Interest: The authors declare no conflicts of interest.

References

1. Yan, Z.C.; Biswas, C.S.; Stadler, F.J. Rheological Study on the Thermoreversible Gelation of Stereo-Controlled Poly(N-Isopropylacrylamide) in an Imidazolium Ionic Liquid. *Polymers (Basel)* **2019**, *11*, 783. [CrossRef] [PubMed]
2. Biswas, C.S.; Stadler, F.J.; Yan, Z.-C. Tacticity effect on the upper critical solution temperature behavior of Poly(N-isopropylacrylamide) in an imidazolium ionic liquid. *Polymer* **2018**, *155*, 101–108. [CrossRef]
3. Zhang, H.; Wu, J.; Zhang, J.; He, J. 1-Allyl-3-methylimidazolium Chloride Room Temperature Ionic Liquid: A New and Powerful Nonderivatizing Solvent for Cellulose. *Macromolecules* **2005**, *38*, 8272–8277. [CrossRef]
4. Vitz, J.; Erdmenger, T.; Haensch, C.; Schubert, U.S. Extended dissolution studies of cellulose in imidazolium based ionic liquids. *Green Chem.* **2009**, *11*, 417–424. [CrossRef]

5. Chen, X.; Zhang, Y.; Wang, H.; Wang, S.-W.; Liang, S.; Colby, R.H. Solution rheology of cellulose in 1-butyl-3-methyl imidazolium chloride. *J. Rheol.* **2011**, *55*, 485–494. [CrossRef]

6. Lv, Y.; Wu, J.; Zhang, J.; Niu, Y.; Liu, C.-Y.; He, J.; Zhang, J. Rheological properties of cellulose/ionic liquid/dimethylsulfoxide (DMSO) solutions. *Polymer* **2012**, *53*, 2524–2531. [CrossRef]

7. Garcia-Penas, A.; Biswas, C.S.; Liang, W.; Wang, Y.; Yang, P.; Stadler, F.J. Effect of Hydrophobic Interactions on Lower Critical Solution Temperature for Poly(N-isopropylacrylamide-co-dopamine Methacrylamide) Copolymers. *Polymers (Basel)* **2019**, *11*, 991. [CrossRef]

8. Vatankhah-Varnoosfaderani, M.; Hashmi, S.; GhavamiNejad, A.; Stadler, F.J. Rapid self-healing and triple stimuli responsiveness of a supramolecular polymer gel based on boron–catechol interactions in a novel water-soluble mussel-inspired copolymer. *Polym. Chem. UK* **2014**, *5*, 512–523. [CrossRef]

9. Vatankhah-Varnoosfaderani, M.; GhavamiNejad, A.; Hashmi, S.; Stadler, F.J. Hydrogen bonding in aprotic solvents, a new strategy for gelation of bioinspired catecholic copolymers with N-isopropylamide. *Macromol. Rapid Commun.* **2015**, *36*, 447–452. [CrossRef]

10. García-Peñas, A.; Biswas, C.S.; Liang, W.; Wang, Y.; Stadler, F.J. Lower Critical Solution Temperature in Poly(N-Isopropylacrylamide): Comparison of Detection Methods and Molar Mass Distribution Influence. *Macromol. Chem. Phys.* **2019**, *220*, 1900129. [CrossRef]

11. García-Peñas, A.; Sharma, G.; Kumar, A.; Galluzzi, M.; Du, L.; Stadler, F.J. Effect of Cross-Linker in Poly(N-Isopropyl Acrylamide)-Grafted-Gelatin Gels Prepared by Microwave-Assisted Synthesis. *Chem. Select* **2019**, *4*, 10346–10351. [CrossRef]

12. García-Peñas, A.; Wang, Y.; Muñoz-Bonilla, A.; Fernández-García, M.; Stadler, F.J. Lower critical solution temperature sensitivity to structural changes in poly(N -isopropyl acrylamide) homopolymers. *J. Sci. Part B Polym. Phys.* **2019**, *57*, 1386–1393. [CrossRef]

13. Yan, Z.C.; Stadler, F.J.; Guillet, P.; Mugemana, C.; Fustin, C.A.; Gohy, J.F.; Bailly, C. Linear and Nonlinear Dynamic Behavior of Polymer Micellar Assemblies Connected by Metallo-Supramolecular Interactions. *Polymers (Basel)* **2019**, *11*, 1532. [CrossRef]

14. Wang, Y. Reynolds Stress Model for Viscoelastic Drag-Reducing Flow Induced by Polymer Solution. *Polymers* **2019**, *11*, 1659. [CrossRef] [PubMed]

15. Morgan, S.E.; Mccormick, C.L. Water-Soluble Copolymers 32—Macromolecular Drag Reduction—A Review of Predictive Theories and the Effects of Polymer Structure. *Prog. Polym. Sci.* **1990**, *15*, 507–549. [CrossRef]

16. Zhu, Y.X.; Granick, S. Apparent slip of Newtonian fluids past adsorbed polymer layers. *Macromolecules* **2002**, *35*, 4658–4663. [CrossRef]

17. Escudier, M.P.; Nickson, A.K.; Poole, R.J. Turbulent flow of viscoelastic shear-thinning liquids through a rectangular duct: Quantification of turbulence anisotropy. *J. Non Newton Fluid* **2009**, *160*, 2–10. [CrossRef]

18. Luo, Z.; Xu, J.; Zhu, D.; Wang, D.; Xu, J.; Jiang, H.; Geng, W.; Wei, W.; Lian, Z. Ion-Imprinted Polypropylene Fibers Fabricated by the Plasma-Mediated Grafting Strategy for Efficient and Selective Adsorption of Cr(VI). *Polymers* **2019**, *11*, 1508. [CrossRef]

19. Kahlon, S.K.; Sharma, G.; Julka, J.M.; Kumar, A.; Sharma, S.; Stadler, F.J. Impact of heavy metals and nanoparticles on aquatic biota. *Environ. Chem. Lett.* **2018**, *16*, 919–946. [CrossRef]

20. Guan, G.; Yu, C.; Xing, M.; Wu, Y.; Hu, X.; Wang, H.; Wang, L. Hydrogel Small-Diameter Vascular Graft Reinforced with a Braided Fiber Strut with Improved Mechanical Properties. *Polymers (Basel)* **2019**, *11*, 810. [CrossRef]

21. Steffi, C.; Shi, Z.; Kong, C.H.; Chong, S.W.; Wang, D.; Wang, W. Use of Polyphenol Tannic Acid to Functionalize Titanium with Strontium for Enhancement of Osteoblast Differentiation and Reduction of Osteoclast Activity. *Polymers* **2019**, *11*, 1256. [CrossRef]

22. Skwira, A.; Szewczyk, A.; Prokopowicz, M. The Effect of Polydimethylsiloxane-Ethylcellulose Coating Blends on the Surface Characterization and Drug Release of Ciprofloxacin-Loaded Mesoporous Silica. *Polymers* **2019**, *11*, 1450. [CrossRef] [PubMed]

23. Li, G.; Rwo, K.H. Hydrophilic Molecularly Imprinted Chitosan Based on Deep Eutectic Solvents for the Enrichment of Gallic Acid in Red Ginseng Tea. *Polymers* **2019**, *11*, 1434. [CrossRef]

24. Dludla, P.V.; Nkambule, B.B.; Jack, B.; Mkandla, Z.; Mutize, T.; Silvestri, S.; Orlando, P.; Tiano, L.; Louw, J.; Mazibuko-Mbeje, S.E. Inflammation and Oxidative Stress in an Obese State and the Protective Effects of Gallic Acid. *Nutrients* **2018**, *11*, 23. [CrossRef] [PubMed]

25. Khan, M.; Heilemann, G.; Lechner, W.; Georg, D.; Berg, A.G. Basic Properties of a New Polymer Gel for 3D-Dosimetry at High Dose-Rates Typical for FFF Irradiation Based on Dithiothreitol and Methacrylic Acid (MAGADIT): Sensitivity, Range, Reproducibility, Accuracy, Dose Rate Effect and Impact of Oxygen Scavenger. *Polymers* **2019**, *11*, 1717. [CrossRef] [PubMed]

26. Warman, J.M.; de Haas, M.P.; Luthjens, L.H.; Yao, T.; Navarro-Campos, J.; Yuksel, S.; Aarts, J.; Thiele, S.; Houter, J.; in het Zandt, W. FluoroTome 1: An Apparatus for Tomographic Imaging of Radio-Fluorogenic (RFG) Gels. *Polymers* **2019**, *11*, 1729. [CrossRef] [PubMed]

27. Yu, H.; Bai, X.; Qian, G.; Wei, H.; Gong, X.; Jin, J.; Li, Z. Impact of Ultraviolet Radiation on the Aging Properties of SBS-Modified Asphalt Binders. *Polymers (Basel)* **2019**, *11*, 1111. [CrossRef]

28. Gong, X.; Dong, Z.; Liu, Z.; Yu, H.; Hu, K. Examination of Poly (Styrene-Butadiene-Styrene)-Modified Asphalt Performance in Bonding Modified Aggregates Using Parallel Plates Method. *Polymers* **2019**, *11*, 2100. [CrossRef] [PubMed]

29. Rojas, J.M.; Hernandez, N.A.; Manero, O.; Revilla, J. Rheology and Microstructure of Functionalized Polymer-Modified Asphalt. *J. Appl. Polym. Sci.* **2010**, *115*, 15–25. [CrossRef]

30. Samaniuk, J.R.; Hermans, E.; Verwijlen, T.; Pauchard, V.; Vermant, J. Soft-Glassy Rheology of Asphaltenes at Liquid Interfaces. *J. Disper. Sci. Technol.* **2015**, *36*, 1444–1451. [CrossRef]

31. Persano, L.; Camposeo, A.; Tekmen, C.; Pisignano, D. Industrial Upscaling of Electrospinning and Applications of Polymer Nanofibers: A Review. *Macromol. Mater. Eng.* **2013**, *298*, 504–520. [CrossRef]

32. Scaffaro, R.; Maio, A.; Lopresti, F.; Botta, L. Nanocarbons in Electrospun Polymeric Nanomats for Tissue Engineering: A Review. *Polymers* **2017**, *9*, 76. [CrossRef] [PubMed]

33. Kishan, A.P.; Cosgriff-Hernandez, E.M. Recent advancements in electrospinning design for tissue engineering applications: A review. *J. Biomed. Mater. Res. Part A* **2017**, *105*, 2892–2905. [CrossRef] [PubMed]

34. Dong, R.-H.; Jia, Y.-X.; Qin, C.-C.; Zhan, L.; Yan, X.; Cui, L.; Zhou, Y.; Jiang, X.; Long, Y.-Z. In situ deposition of a personalized nanofibrous dressing via a handy electrospinning device for skin wound care. *Nanoscale* **2016**, *8*, 3482–3488. [CrossRef] [PubMed]

35. Lu, P.; Murray, S.; Zhu, M. Chapter 23—Electrospun Nanofibers for Catalysts. In *Electrospinning: Nanofabrication and Applications*; Ding, B., Wang, X., Yu, J., Eds.; William Andrew Publishing; Elsevier: Amsterdam, The Netherlands, 2019; pp. 695–717. [CrossRef]

36. Sun, N.; Wen, Z.; Zhao, F.; Yang, Y.; Shao, H.; Zhou, C.; Shen, Q.; Feng, K.; Peng, M.; Li, Y.; et al. All flexible electrospun papers based self-charging power system. *Nano Energy* **2017**, *38*, 210–217. [CrossRef]

37. Ganesh, V.A.; Dinachali, S.S.; Nair, A.S.; Ramakrishna, S. Robust Superamphiphobic Film from Electrospun TiO2 Nanostructures. *ACS Appl. Mater. Interfaces* **2013**, *5*, 1527–1532. [CrossRef]

38. Ma, S.; Ye, T.; Zhang, T.; Wang, Z.; Li, K.; Chen, M.; Zhang, J.; Wang, Z.; Ramakrishna, S.; Wei, L. Highly Oriented Electrospun P(VDF-TrFE) Fibers via Mechanical Stretching for Wearable Motion Sensing. *Adv. Mater. Technol.* **2018**, *3*, 1800033. [CrossRef]

39. Monteserín, C.; Blanco, M.; Murillo, N.; Pérez-Márquez, A.; Maudes, J.; Gayoso, J.; Laza, J.M.; Hernáez, E.; Aranzabe, E.; Vilas, J.L. Novel Antibacterial and Toughened Carbon-Fibre/Epoxy Composites by the Incorporation of TiO$_2$ Nanoparticles Modified Electrospun Nanofibre Veils. *Polymers* **2019**, *11*, 1524. [CrossRef]

Article

Linear and Nonlinear Dynamic Behavior of Polymer Micellar Assemblies Connected by Metallo-Supramolecular Interactions

Zhi-Chao Yan [1], Florian J. Stadler [1,2,*], Pierre Guillet [2,3], Clément Mugemana [2,4], Charles-André Fustin [2], Jean-François Gohy [2] and Christian Bailly [5]

[1] Shenzhen Key Laboratory of Polymer Science and Technology, Guangdong Research Center for Interfacial Engineering of Functional Materials, College of Materials Science and Engineering, Shenzhen University, Shenzhen 518060, China; yanzhch@szu.edu.cn

[2] Institute of Condensed Matter and Nanosciences (IMCN), Bio and Soft Matter Division (BSMA), Université Catholique de Louvain, Place Pasteur 1, B-1348 Louvain-la-Neuve, Belgium; pierre.guillet@univ-avignon.fr (P.G.); clement.mugemana@list.lu (C.M.); charles-andre.fustin@uclouvain.be (C.-A.F.); jean-francois.gohy@uclouvain.be (J.-F.G.)

[3] Equipe Chimie Bioorganique et Systèmes Amphiphiles, Institut des Biomolécules Max Mousseron (UMR 5247 UM-CNRS-ENSCM) & Avignon University, 301 rue Baruch de Spinoza, 84916 Avignon CEDEX 9, France

[4] Luxembourg Institute of Science and Technology (LIST), 5 Avenue des Hauts-Fourneaux, 4362 Esch-sur-Alzette, Luxembourg

[5] Institute of Condensed Matter and Nanosciences (IMCN), Bio and Soft Matter Division (BSMA), Université Catholique de Louvain, Place Croix du Sud 1, B-1348 Louvain-la-Neuve, Belgium; christian.bailly@uclouvain.be

* Correspondence: fjstadler@szu.edu.cn

Received: 13 August 2019; Accepted: 15 September 2019; Published: 20 September 2019

Abstract: The linear and nonlinear rheology of associative colloidal polymer assemblies with metallo-supramolecular interactions is herein studied. Polystyrene-*b*-poly(*tert*-butylacrylate) with a terpyridine ligand at the end of the acrylate block is self-assembled into micelles in ethanol, a selective solvent for the latter block, and supramolecularly connected by complexation to divalent metal ions. The dependence of the system elasticity on polymer concentration can be semi-quantitatively understood by a geometrical packing model. For strongly associated (Ni^{2+}, Fe^{2+}) and sufficiently concentrated systems (15 w/v%), any given ligand end-group has a virtually 100% probability of being located in an overlapping hairy region between two micelles. By assuming a 50% probability of intermicellar crosslinks being formed, an excellent prediction of the plateau modulus was achieved and compared with the experimental results. For strongly associated but somewhat more dilute systems (12 w/v%) that still have significant overlap between hairy regions, the experimental modulus was lower than the predicted value, as the effective number of crosslinkers was further reduced along with possible density heterogeneities. The reversible destruction of the network by shear forces can be observed from the strain dependence of the storage and loss moduli. The storage moduli of the Ni^{2+} and Zn^{2+} systems at a lower concentration (12 w/v%) showed a rarely observed feature (i.e., a peak at the transition from linear to nonlinear regime). This peak disappeared at a higher concentration (15 w/v%). This behavior can be rationalized based on concentration-dependent network stretchability.

Keywords: associative polymer colloids; micellar assemblies; rheology

---

1. Introduction

Supramolecular self-assembly is one of the most fascinating topics in soft matter. It gives synthetic materials some of the merits of biological systems, including self-healing properties, response to external stimuli, and the ability to reversibly modify stiffness as a reaction to external stress levels [1–18]. Supramolecular self-assembly strategy has been shown to work excellently for constructing dynamically bonded gels [19,20]. In particular, supramolecular micellar gels can be created with somewhat slower dynamic behavior, according to a binary self-assembly strategy that combines both phase-separation-induced block aggregation and the bonding from supramolecular intermicellar interactions [7]. As regards micellar gels made by block copolymers, a kinetically controlled and even reversible self-assembly is possible, making these gels good candidates for self-repairing and injectable soft materials.

So far, the research of supramolecular micellar gels is still mainly focused on chemically creating new functionality and designing new structures. The material's physical properties have not yet been thoroughly evaluated, especially the relationship between rheological behavior and structural information like micelle size, intermicellar distance, relative volume of core and corona, and strength of supramolecular interactions. In an earlier study, the structure of a wormlike micelle supramolecularly bridged by telechelic polymers was explored by Ligoure et al. [6]. A similar approach was also conducted by Lodge et al. [21,22], who found evidence for establishing crosslinks between wormlike micelles by a poly(ethylene oxide) with alkyl stickers installed at the chain ends. However, their samples were mixtures of micelles and linear chains, which cannot answer the question in pure micellar gels of how supramolecular networks are formed simply by intermicellar bonding among arms in coronae.

Among the different non-covalent interactions forming supramolecular bonds, the metal–ligand bond is particularly interesting because of its great directionality, broad selection of ligands, and easily tuned interaction strength that can be selected by choosing the appropriate metal ion. Guillet et al. [7,15] combined colloidal self-assembly and metallo-supramolecular interactions by exploring the behavior of polystyrene-*b*-poly(*tert*-butylacrylate) (PS-*b*-PtBA) with a terpyridine ligand at the end of the acrylate block, which is self-assembled in ethanolic solution. The addition of divalent ions such as Zn^2, Ni^{2+}, and Fe^{2+} yield flower-like micelles in diluted solutions and temporary networks in concentrated and semi-concentrated systems. The connections formed by such micelles are shown in Scheme 1. The rheological behavior of such gels can be extensively tuned by choice of metal ion. Besides metal–terpyridine coordination, we have also reported interactions between dopamine and metal ions in thermosensitive copolymers for self-assembly into switchable polymer gels. This process depends on a variety of factors, including pH [23,24], temperature [23], solvent type [19,24,25], ion type [26], and nanoparticles [27,28]. In most studies, the rheological properties have been restricted to a few test setups. In these preliminary studies, a frequency sweep was often carried out to prove gelation at an almost constant phase angle δ (or a constant slope in storage and loss modulus) [20,29], while the absence of such a constant slope has indicated that the material is sufficiently elastic (as in the case of a typical entangled polymer) [30]. However, a relationship between these rheological results (e.g., the plateau modulus and structural and connecting situations of micelles) have not been theoretically established. In addition, the nonlinear rheological behavior of micellar gels has not been thoroughly explored.

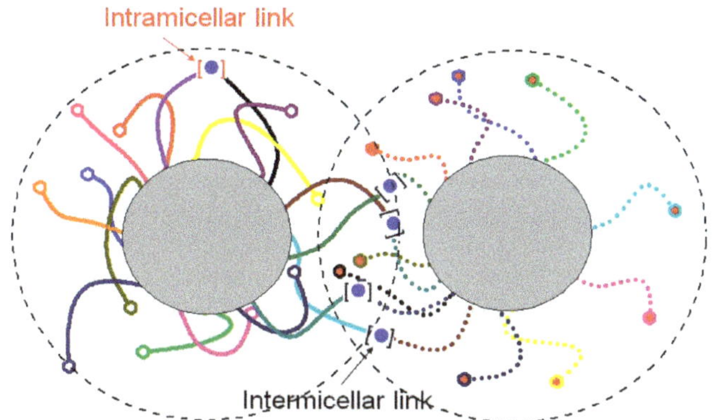

Scheme 1. Cartoon for connection formed by self-assembly of PS-*b*-PtBA with a terpyridine ligand. The core and coronae are PS and PtBA, respectively. ● Represents the metal ion.

This paper expands on an understanding of the concentrated systems. It focuses on the surprising finding that even small concentration differences can completely change rheological characteristics. Based on simple geometric considerations as well as structural and chemical information about this system, we first calculate a key structural parameter—the intermicellar linking fraction—then consider implications on gelling behavior. Predictions are compared with linear rheological measurements for two concentrations (12 and 15 w/v%) and three ions (zinc(II), iron(II), and nickel(II)). We further explore the transition from linear to nonlinear rheological behavior.

2. Experiments

Preliminary results with these systems have previously been published [7]. The polymer is a polystyrene-*b*-poly(*tert*-butyl acrylate) (PS-*b*-PtBA) block copolymer with average polymerization degrees of 80 ($M_n \approx 8300$ g/mol) and 200 ($M_n \approx 25{,}600$ g/mol) for the PS and PtBA blocks, respectively, and a polydispersity index of 1.11 for the whole copolymer. The end of the PtBA block is functionalized with a terpyridine ligand. The diblock copolymer forms micelles in ethanol, with a glassy PS core and PtBA corona. As determined by static light scattering, these micelles have a radius of gyration of approximately 15 nm and contain 28 chains on average. Metal ions can be added to this suspension of hairy colloids following a procedure described elsewhere [7]. Up to two terpyridine ligands form complexes with the divalent metal ions used in this article (Fe^{2+}, Ni^{2+}, and Zn^{2+}), leading to supramolecular bonds either between chains from different micelles (forming crosslinks), or from the same micelle (forming loops) [7]. Three metal ions (Zn^{2+}, Fe^{2+}, and Ni^{2+}) are added as chloride salts. The quantity of ions is half an equivalent with respect to the terpyridine [7].

Small-angle x-ray scattering (SAXS) experiments were performed on station DUBBLE at the European Synchrotron Radiation Facility, Grenoble, France. The X-ray wavelength was $\lambda = 1.55$ Å. SAXS patterns were collected with a two-dimensional multiwire gas-filled detector. The wavenumber scale ($q = 4\pi \times \sin\theta/\lambda$, where 2θ is the scattering angle) was calibrated using a sample of wet collagen (rat tail tendon).

The rheological characterization was performed with a TA Instruments ARES (New Castle, DE, USA) using either a 25 mm/0.02 rad cone and plate geometry, a Malvern Kinexus (Malvern, Cambridge, UK) using a 25 mm/0.5° cone, or a TA Instruments AR-G2 (New Castle, DE, USA) using a 40 mm/2 or a 20 mm/1° cone. The measurements were carried out at room temperature in an ethanol-saturated atmosphere to minimize evaporation of the solvent. All measurements were carried out at 20 °C. Before measurement, an equilibration time of around 10 min was applied, which was found to be sufficient for all conditions. Dynamic frequency sweeps and nonlinear dynamic strain sweeps were also performed.

3. Results

3.1. Structure of the Micellar Network

To explain the rheological data, the results of the analytical characterization are essential. Figure 1 shows the SAXS data collected for the investigated samples at a concentration of 12 w/v%. For all ions, a clear peak around $q = 0.03$ Å^{-1} is observed, which corresponds to a long period of 27 nm and has been ascribed to intermicellar distance (i.e., the core-to-core distance d_{cc}). Figure 1 also shows that no other order is present, which indicates that the micelles are organized in a liquid-like fashion. The value of d_{cc} for the 15-w/v% sample is calculated. Considering the relatively low concentration and poorly ordered properties, we assume micelle spheres are homogeneously dispersed in solutions by simple cubic packing. Thus, d_{cc} can be calculated as

$$d_{cc} = d_{cc,\exp} \sqrt[3]{\frac{c_{\exp}}{c}} \tag{1}$$

where $d_{cc,\exp}$ and $c_{\exp}$ are experimentally obtained values (i.e., $d_{cc,\exp} = 27$ nm for $c_{\exp} = 12$ w/v%). For the sample with $c = 15$ w/v%, d_{cc} is calculated as 25 nm.

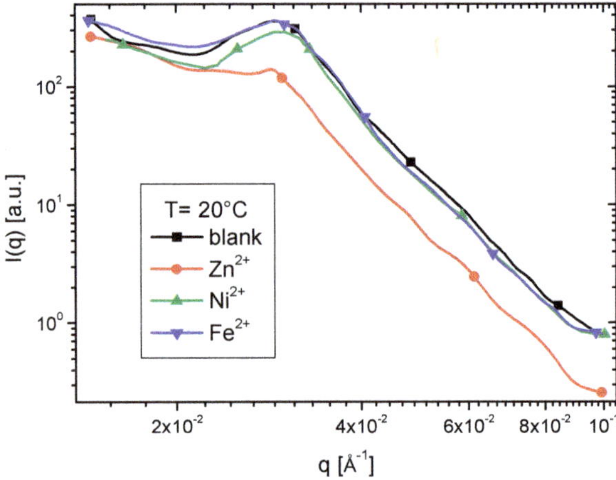

Figure 1. SAXS data collected on the investigated PS-b-PtBA micellar solutions in ethanol at a concentration of 12 w/v%.

It is known from dynamic light scattering that the hydrodynamic radius R_h of an isolated micelle is around 20 nm [7]. Static light scattering has been used to measure the radius of gyration of these micelles, determining a value of 15 nm [7]. The ratio between the radius of gyration and the hydrodynamic radius is, therefore, 0.75—a value close to the one expected for ideal spherical micelles (0.78). An ideal chain packing for the glassy PS cores leads to a core diameter of 9.1 nm. Considering packing errors and the slight swelling of the cores due to very small fractions of incorporated PtBA and ethanol, we assume that the "real" diameter is rather around 10 nm (core radius $R_c = 5$ nm). The number of chains per micelle on average is 28 (measured by light scattering) [7].

We assume that the micelles are equal in diameter and spherical. This is confirmed by TEM [15] and is logical considering the polymer concentration and block molecular weight (M_w) ratio. The overlapping volume V_0 of two equally sized spheres is given by [31]

$$V_0 = \frac{1}{12}\pi(4R_h + d_{cc})(2R_h - d_{cc})^2 \tag{2}$$

Using the volume of the corona $V_c = 33,000$ nm^3, which is calculated as the difference of micelle and core volumes ($V_{\mathrm{micelle}} - V_{\mathrm{core}}$), the overlapping fraction of one micelle with one neighbor f_1 can be calculated as

$$f_1 = \frac{V_0}{V_c} \tag{3}$$

The relation of f_1 and concentration (c) are determined by combining Equations (1)–(3).

Two coronae totally overlap when the micellar core of one micelle just touches the corona of the neighboring one (i.e., when $d_{cc} \leq 25$ nm (Scheme 2a)), which occurs around our concentration of 15 w/v%. For the 12-w/v% sample, coronae partially overlap (Scheme 2b). However, with dilution the overlap concentration at which the two coronae start to touch (Scheme 2c) reduces to as low as 3.7 w/v%, calculated from Equation (1). Moreover, the weak scattering signal indicates the system is very disordered (Figure 1).

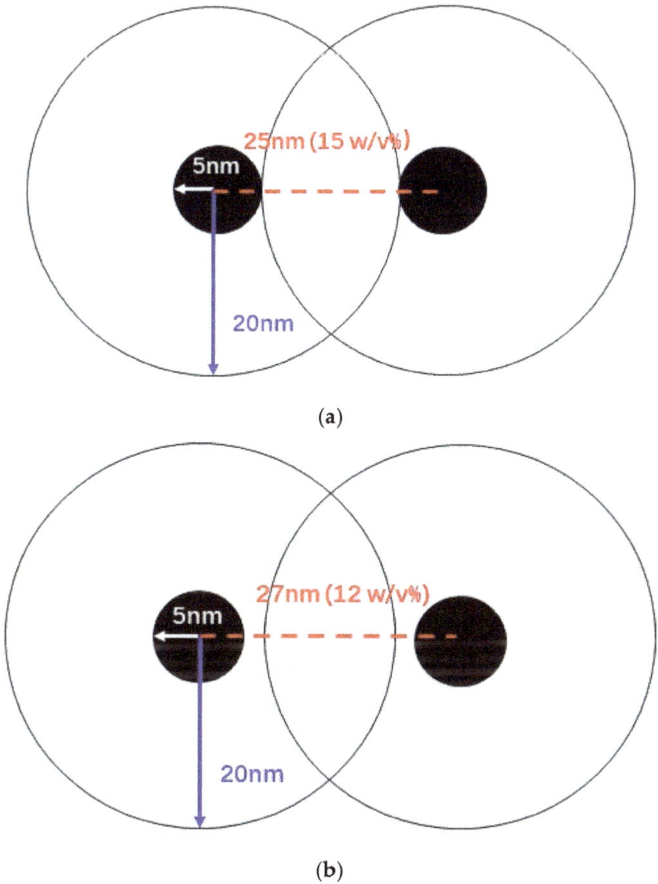

(a)

(b)

Scheme 2. *Cont.*

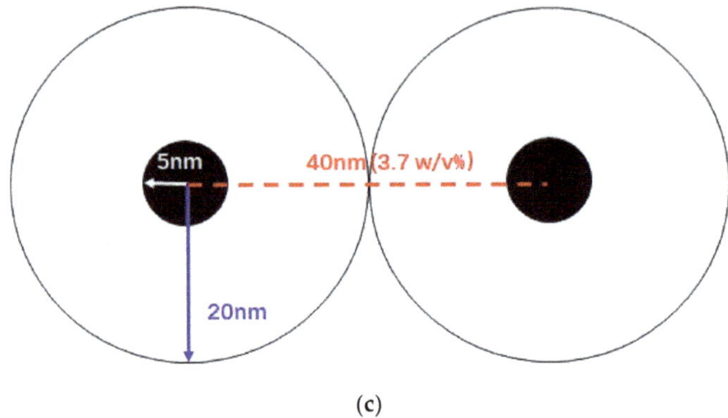

(c)

Scheme 2. Cartoon of the size and distance of micelles (solid small circle representing core and unfilled large circle representing corona) at (**a**) concentration for total overlap of corona (15 w/v%); (**b**) concentration for partial overlap (12 w/v%); and (**c**) overlap concentration (3.7 w/v%).

The overlapping fraction f_1 is plotted as a function of concentration in Figure 2 (left *y*-axis). The value of f_1 is null until 3.7 w/v%, then increases smoothly beyond that, suggesting that below a certain threshold no network can form because the micelles do not overlap and, hence, cannot form any intermicellar connection [32–34].

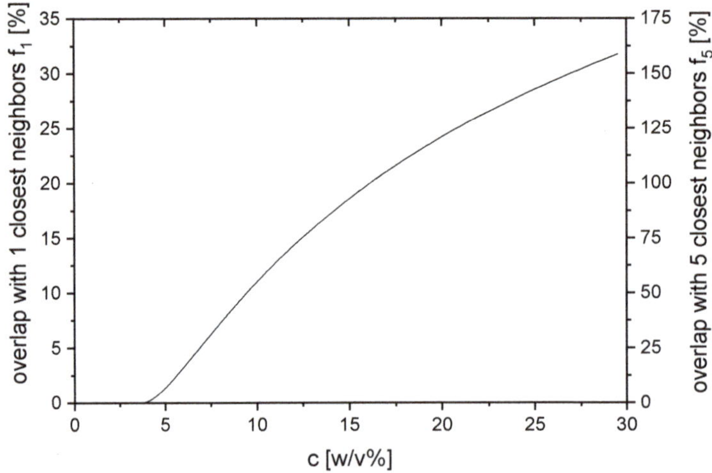

Figure 2. Overlapping fractions f_1 and f_5 as a function of concentration.

The disordered nature of the packing observed from the SAXS data (Figure 1) indicates that close packing is impossible and even cubic structures can hardly be reached. This puts the maximum number of closest neighbors below six (simple cubic packing). We reasonably assume that each micelle has about five closest neighbors, an assumption that has the best agreement with experimental data (see below).

Under the assumption of five closest neighbors, the combined overlapping fraction f_5 (calculated from additive behavior of five neighboring micelles ($f_5 = 5 \times f_1$)) is 100% at a 16-w/v% concentration and reaches 72% at 12 w/v% (Figure 2, right *y*-axis). Of course, the additive overlap assumption is too simple, as multiple overlaps in the same area can occur and complicate the picture, but it is a reasonable guideline and a fully quantitative analysis would require detailed knowledge of the packing, which cannot be gathered without 3D simulations of the structure. Based on the overlapping volume fraction, the volume fractions involving chains originating from one, two, and three different coronae (defined as n_1, n_2, and n_3, respectively) can be calculated as

$$n_1 = 1 - n_2 - n_3$$
$$n_2 = \begin{cases} f_5 & \text{when } c < 16 \text{ w/v\%} \\ 1 - n_3 & \text{when } c > 16 \text{ w/v\%} \end{cases}$$
$$n_3 = \begin{cases} 0 & \text{when } c < 16 \text{ w/v\%} \\ f_5 - 1 & \text{when } c > 16 \text{ w/v\%} \end{cases} \tag{4}$$

Their results are plotted in Figure 3. Up to about a 16-w/v% concentration, non-overlapping volume is present (n_1), while above 16 w/v%, n_1 becomes null but an overlap of three coronae (n_3) appears. The value of n_2 increases first with concentration but decreases beyond 16 w/v% due to the occurrence of a triple overlap.

Although a quantitative estimation has been made above, it is, regrettably, not possible to calculate the exact overlap, as we do not know how to account for the concentration difference between overlapping and non-overlapping regions of the coronae. There are two effects that cannot be disseminated without making a number of assumptions. First, it is clear that when two coronae overlap, the concentration is twice as high inside this overlapping area as outside. Because of this, there is a weak driving force, which will drive the chains out of the overlapping area. However, it is expected that the contribution will be small. Second, the exact geometry of the assembly is unknown and underlies statistical fluctuations. We know from the calculations that five closest neighbors are the most likely packing. However, it is not a disordered crystal but a random structure with average core-to-core distances underlying statistical fluctuations. As a result, based on the assumptions being put into Figure 3, we can estimate that a probable 100% overlap (i.e., the disappearance of n_1) is reached close to a concentration of 20 w/v%. This number is somewhat higher than Figure 3 suggests, as we have to assume that the overlap does not occur with perfect homogeneity.

Please note that in this simple approach it is assumed that no overlap of three coronae can occur so long as non-overlapping coronae are still present. In reality, the situation evolves more progressively, which will smooth the curves somewhat. The effect of this simplification, however, should be small. For example, we have to assume that, in reality, the disappearance of n_1 and the appearance of n_3 does not occur at 16 w/v%, but rather at a certain interval where both areas have no overlap (n_1) and where with three overlapping coronae (n_3) coexist due to geometric reasons. This is also the reason why we have to assume that, as mentioned in the previous paragraph, 100% overlap is reached at slightly higher concentrations than Figure 3 indicates.

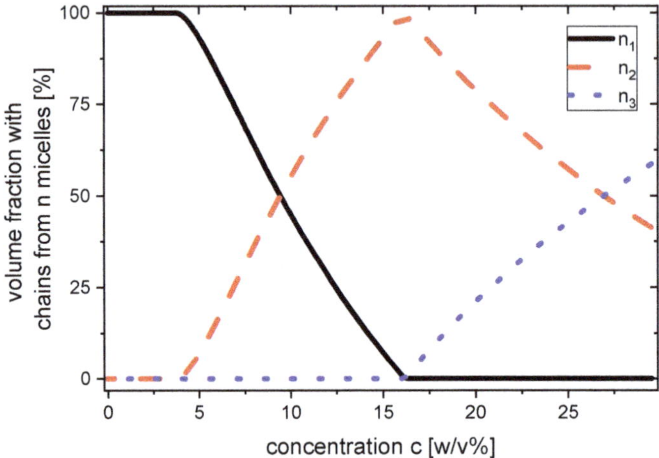

Figure 3. Volume fraction of non (n_1), binary (n_2), and triple (n_3) coronae overlap as a function of concentration.

One might argue that, according to the model of Daoud–Cotton [35], the density of chains close to the surface of the micellar core is higher than that in the outer areas of the micellar shell. The fact that the chains are only at 28% of their stretched length, which corresponds to the approximate value of free chains (not bound in a micelle) in solution is a clear indicator that the connection of the PtBA arms to the micellar core does not play a major role in coil size. Hence, we can also conclude that no distinct concentration gradient is present in micelle coronae. This sets the micellar gels in this article apart from other soft colloids (e.g., multi-arm stars) [35,36].

If we reasonably assume that the overlapping fraction based on five closest neighbors determines the amount of potential supramolecular crosslinkers in the system, it becomes possible to calculate the fraction of crosslinks that should exist between chains from different micelles (intermicellar links, f_{link}) with the help of Equation (5).

Here, we assume that two end-functionalized chains can 100% link considering the high bonding strength of terpyridine and Ni^{2+}/Fe^{2+}. Besides, the chains should not repel each other. These assumptions should work safely at relatively low concentrations, as we have here. Hence, there are two possibilities for the bonding partners, chains from other micelles (intermicellar bond) and chains from the same micelle (intramicellar bond). Under overlap, the intermicellar bonding probability is 0% when no overlap is present (n_1), 50% (1/2) when two micellar coronae overlap (n_2), and 66.67% (2/3) when three micellar coronae overlap (n_3). This simple probabilistic analysis neglects conformational free-energy differences between the intra- and intermicellar bonds. The outcome is the fraction of intermicellar links f_{link}:

$$f_{link} = \frac{\frac{1}{2}n_2 + \frac{2}{3}n_3}{n_1 + n_2 + n_3} \tag{5}$$

Figure 4 shows the fraction of intermicellar links f_{link} as a function of concentration. It becomes obvious that for smaller concentrations than 3.7 w/v%, only a negligible fraction of intermicellar links can exist (barely enough for a linear "train" of micelles), while for concentrations around 16 w/v%, a 50–50 share of inter- and intramicellar crosslinks is found. The reason for the sudden change in slope around 16 w/v% is the disappearance of non-overlapping volume. The fraction of intermicellar links for a given micelle is directly proportional to the total number of its intermicellar links (simply multiplied by the total number of ligand end-groups, $28 \times f_{link}$). This is also plotted in Figure 4.

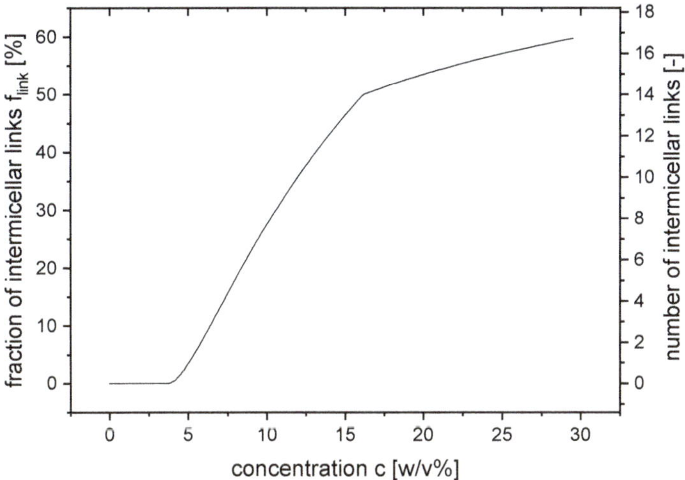

Figure 4. Average fraction and number of intermicellar links as a function of concentration.

Qualitative tube inversion tests (i.e., the sample sticks to the bottom of the tube even when the tube is turned upside down), have shown that for concentrations around 9 w/v%, a weak but stable gel is formed when using Fe^{2+} or Ni^{2+} ions. When comparing this value to Figure 4, we see that a concentration of 9 w/v% roughly corresponds to six intermicellar links per micelle, which is numerically marginal to form a network. This is a strong indicator that the assumptions and calculations given above provide a reasonable prediction.

3.2. Influence of the Type of the Ions and Polymer Concentration

The frequency-dependent storage modulus and dynamic viscosity of colloidal suspensions are given in Figure 5 and Figure S1. Very diverse behaviors are observed depending on ion or concentration. The blank sample has very low viscosity (~0.03 Pa·s, roughly 30 times the viscosity of water) and no measurable elasticity. The 12-w/v% sample containing Zn^{2+}-ions has a 40 times higher viscosity than the blank sample, showing slightly shear thinning and very weak but measurable elasticity. This sample thus shows only a minor departure from Newtonian behavior. In sharp contrast, the samples containing Ni^{2+} and with Fe^{2+} ions are characterized by much higher G', and are dominated by their elasticity, since their storage moduli are significantly larger than their loss moduli (see the supporting information for $G''(\omega)$). Furthermore, $G'(\omega)$ and $G''(\omega)$ are approximately constant over a wide range of frequencies. Such behavior is typical of a rubbery network. The complexes thus create (reversible) crosslinks between the micelles. As the elastic moduli are very low, these materials are best described as ultrasoft elastic gels.

When the concentration is increased to 15 w/v%, the behavior of the blank sample stays roughly unaffected, while a significantly higher modulus is found for samples with ions. The trend of the data remains very similar to the 12-w/v% samples, except that G' and $|\eta^*(\omega)|$ are higher by a factor of approximately 10. Although for Ni^{2+} and Fe^{2+} samples, the power law slope is close to 1 for $|\eta^*(\omega)|$ and close to 0 for $G'(\omega)$, reflecting almost perfect elastic behavior, it is still slightly below that limiting level. Hence, none of the samples can be considered as a truly ideal rubber state.

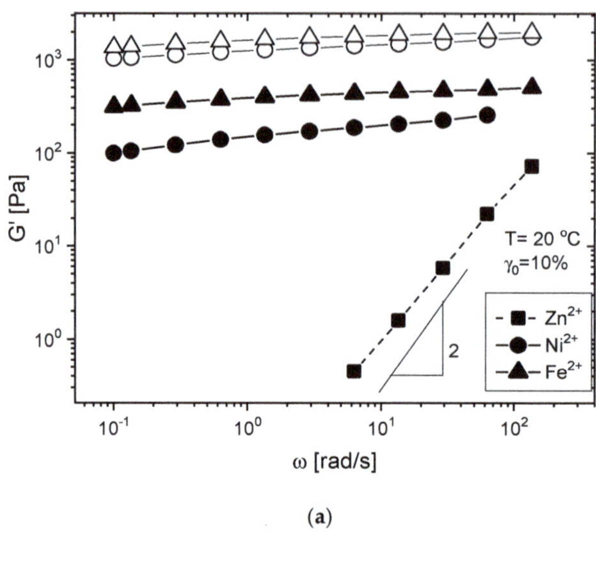

(a)

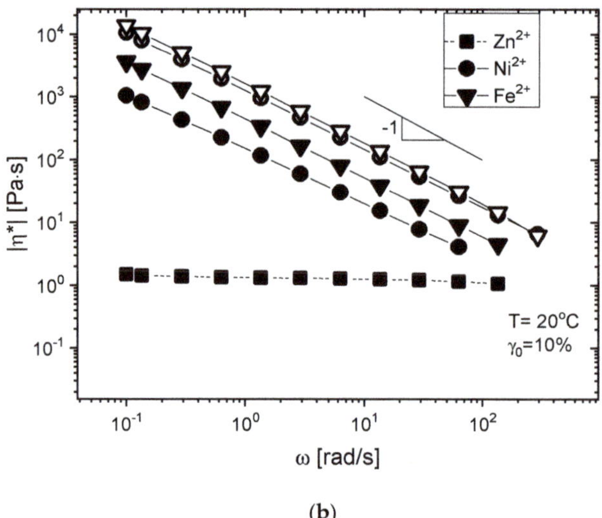

(b)

Figure 5. Frequency-dependent rheological data for (a) $G'(\omega)$ and (b) $|\eta^*(\omega)|$ at $T = 20\,°\mathrm{C}$. Filled symbols: 12 w/v% concentration, from ref [7]. Open symbols: 15 w/v% concentration.

The ion nature has a profound influence on the strength of the gel, as can be observed by a comparison of the storage and loss modulus at a given frequency ω (10 rad/s) and concentration (Figure 6). Although comparison under the same Deborah number ($\omega \times \tau_d$) is also widely used for samples, the terminal relaxation time τ_d is not experimentally reached here, and we mainly focus on the plateau moduli that reflects the network density. Therefore, the comparison at a given ω is fair, considering the G' plateau is broad enough to be approximately frequency independent. As reported above, Zn^{2+}, at a 12-w/v% concentration only leads to a 40-fold viscosity increase in respect to the blank. At the same concentration, and unlike the Zn^{2+} sample, the Ni^{2+} sample withstands the tube inversion test. Rheologically, this is understood as a gelation. By comparing with Zn^{2+}, Ni^{2+} leads to an increase of G' by a factor of 600 and makes G' larger than G'' (see Figure 6). Fe^{2+} leads to a further increase of G'. Thus, Ni^{2+} and Fe^{2+} samples both have a rubber-like behavior, but are much softer than a conventional elastomer due to the static dilution by solvent, leading to a low crosslinking density. Hence, this behavior is best described by the term "soft rubber". The different rheological behaviors among three ions are due to the strength of the complexes. Since Zn^{2+} forms relatively weak complexes with terpyridines, Fe^{2+} and Ni^{2+} form much stronger complexes that are difficult to disassociate [37].

Going from a 12- to 15-w/v% concentration essentially has negligible effect on the viscosity of the blank. The behavior of the 15-w/v% samples loaded with ions, however, is distinctly different from their 12% counterparts in terms of the moduli and different nonlinear behavior discussed later. For the Zn^{2+} containing the 12-w/v% sample, $G''(\omega)$ is higher than $G'(\omega)$, while $G''(\omega)$ is lower than $G'(\omega)$ for the Ni^{2+}/Fe^{2+} containing 12-w/v% samples. For the 15-w/v% Fe^{2+}/Ni^{2+} samples, $G'(\omega = 10$ rad/s$)$ is almost independent of the ion used, with all values around 1800 Pa. However, differences are still clear for $G''(\omega = 10$ rad/s$)$. These results indicate that a threshold is reached for the storage modulus at a 15-w/v% concentration. Unlike the 12-w/v% samples, the 15-w/v% samples behave like a solid as soon as ions are added. The stronger Ni^{2+}- and Fe^{2+} supramolecular bonds lead to a more pronounced elastic behavior characterized by a lower phase angle δ (tan $\delta = G''/G'$). The most likely mechanism is that the number of unlinked chain ends is lower because the equilibrium constant for complex formation is higher. Based on investigations on the lifetime and association constant of ion–terpyridine complexes by Dobrawa and Würthner [38], we can conclude that, at equilibrium, the vast majority of the terpyridine end groups are actually complexed by the ions in the Ni^{2+} and Fe^{2+} samples.

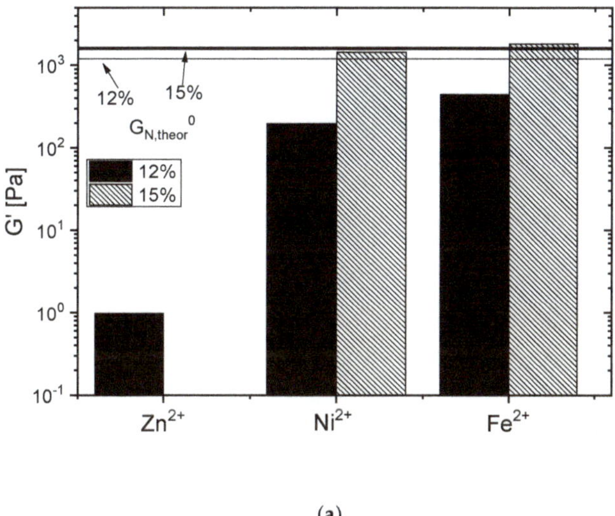

(a)

Figure 6. *Cont.*

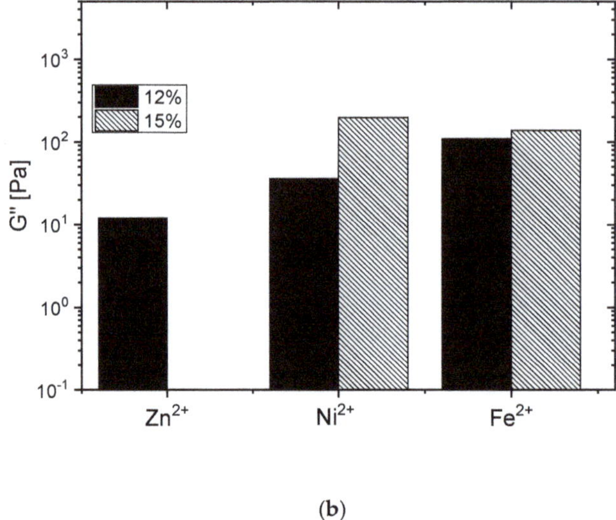

(**b**)

Figure 6. (a) Storage and (b) loss moduli, G' and G'', at $\omega = 10$ rad/s and $T = 20\ °C$. The thick and thin horizontal lines are theoretical plateau modulus $G_{N,theor}{}^0$ for the 12- and 15-w/v% samples, respectively.

3.3. Entanglement Molecular Weight and Plateau Modulus

The next interesting question is how many of the bonds are intermicellar or intramicellar? This can be estimated by comparing the experimental plateau modulus $G_{0,exp}$ of these samples to predictions obtained from the structural model outlined above.

Linear viscoelasticity rubber theory is only correct *stricto sensu* if we assume that we have a permanent homogeneous network without loops or dangling ends. Moreover, entanglements between network strands can complicate the picture. The first condition is not fully met by our supramolecular systems because of the transient nature of the bonds and existence of intramicellar links beside intermicellar ones. Hence, the elastic moduli must be lower than the theoretical values. Stable intermicellar bridges can be assumed for the Ni^{2+} and the Fe^{2+} samples but not for the 12-w/v% Zn^{2+} and blank sample. The second condition (no entanglements) is true if the dilution is so high that the chains are unentangled. An estimation of entanglement molecular weight (M_e) under dilution can be obtained from the following equation [39]:

$$M_e = M_{e,bulk}/\phi^{\,\alpha} \tag{6}$$

where ϕ is the polymer volume fraction in solution, $\alpha = 1.3$ is the dilution exponent, and $M_{e,bulk}$ is the entanglement molecular weight in a bulk state. According to Fetters et al. [40] and Tong et al. [41], $M_{e,bulk}$ of alkyl acrylate polymers is in the range between 7700–15,200 g/mol. However, the value of poly(*tert*-butylacrylate) is not given. Based on the available literature data [40,41], it is estimated (based on the bulkiness of the side chains and the established M_e-values) that $M_{e,bulk}$ (PtBA) is in the range of ~12,000 g/mol. Because the polystyrene chains are in the glassy rigid core, they do not play a role here. Using 0.79 g/cm^3 as density for ethanol and 1.07 g/cm^3 as density for the PS-PtBA copolymers, we find that a 12-w/v% fraction is equivalent to the volume fraction $\phi = 0.156$ and that, similarly, a 15-w/v% fraction corresponds to $\phi = 0.193$. As a result, the diluted M_e, as calculated from Equation (6), is 134,000 and 102,000 g/mol for 12 w/v% and 15 w/v%, respectively. Both values are much larger than a closed elastic strand (i.e., two supramolecularly connected arms ($2 \times M_{arm} \approx 51,200$ g/mol). Hence, physical entanglements among corona chains do not play a significant role for our samples and we will only discuss the quasi-permanent crosslinks—intermicellarly crosslinked PtBA-chains—in the following.

We can now estimate the theoretical elastic modulus $G_{0,\text{theor}}$ of our supramolecular systems, viewed as permanent networks, by using the equivalent crosslinking density predicted from Figure 4:

$$G_{0,\text{theor}} = nRT \tag{7}$$

where n is the elastic strand density comprising intermicellarly connected PtBA arms. The values of n are obtained by dividing the number of intermicellar bonds of a given micelle by the volume pervaded by arms (the volume of corona; i.e., $28 \times f_{link}/V_c$). The values of $G_{0,\text{theor}}$ are thus calculated based on n. Both n and $G_{0,\text{theor}}$ are shown in Table 1.

Table 1. Plateau moduli calculated at $T = 20\,^{\circ}\text{C}$.

	12 w/v%	15 w/v%
ϕ [-]	0.156	0.193
n [mol/m^3]	0.50	0.65
$G_{0,\text{theor}}$ (Equation (7)) [Pa] [a]	1200	1600
$G_{0,\text{exp.}}$ (experimental, Fe^{2+}-containing samples at 10 rad/s) [Pa]	450	1900

[a] Assuming permanent bonds with infinite lifetime.

A comparison between predicted and experimental values for the moduli at $\omega = 10$ rad/s (Table 1 and Figure 6) shows that for the 12-w/v% samples the experimental modulus is lower by a factor of about 2.7 versus the theoretical one for the Fe^{2+} sample, and even lower for the other samples. For the 15-w/v% samples, values between theoretical and experimental results are almost quantitatively consistent. At higher frequencies, undoubtedly higher moduli would be present, which, however, cannot be measured reliably, because of inertia effects.

Although the simplified geometric model only provides an approximation of the maximal crosslinking density, it is clear that the behavior of the 12-w/v% and 15-w/v% samples is different. The ratio of 2.7 for the 12-w/v% samples indicates that the majority of the supramolecular crosslinkers are either open (i.e., free chain ends) or do not contribute to the crosslinking network (i.e., closed loops, presumably corresponding to intramicellar supramolecular bonds). Isolated groups of micelles crosslinked with each other can also be responsible for the lower effective experimental crosslinking density. However, it is expected that for the samples discussed here, this effect is negligible.

G_N is proportional to the elastic strand density comprising intermicellarly connected arms. Hence, only about 37.5% (1/2.7) of the predicted intermicellar bonds in the 12-w/v% sample actually contribute to the network strength for the Fe^{2+} sample. For the other samples, this ratio is even lower. The phase angle $\delta \approx 11^{\circ}$ for the Ni^{2+} and the Fe^{2+} samples implies a relatively low amount of free chain ends. For the ion-treated 15-w/v% samples, almost all of the predicted intermicellar bonds contribute to the network. Actually, this result applies only to crosslinkers stable at the time scale of the experiments (measured for aqueous solution with a time of stability for a Fe^{2+} and Ni^{2+} ions in the order of 30 min, according to the literature data [38]). For shorter times (corresponding to higher frequencies), a higher modulus would presumably be found and, thus, a higher proportion of closed crosslinkers would be present.

Such a significant difference in elastic behavior for a minor concentration difference (from 12 to 15 w/v%) is rather unexpected and is indicative of a major modification of the supramolecular structure. For the 15-w/v% sample with sufficiently overlapped coronae, the consistency between predicted and experimental G_N^0 confirms our approximation that half (exactly 46% according to Figure 4) of terpyridine end groups link intermicellarly, while another half link intramicellarly. This also implies that the fraction of the open terpyridine end groups is negligibly small. For the 12% sample, the same argument does not yield such a nice agreement. Assuming a statistical distribution of the terpyridine end groups, 36% of the arms are predicted to intermicellarly link (see Figure 4), which is significantly higher than (although qualitatively consistent with) the value of 13.5% (=36% × 37.5%) suggested by the experiment.

This deviation of 12-w/v% samples can be explained by two factors. Firstly, the fact that some areas of the corona volume overlap, while others do not, means that the distribution of chains is not homogeneous and, hence, some micelles may not participate in the percolating elastic network. This effect can further be compounded by the following argument. The connected chains will be somewhat stretched out of their equilibrium configuration after making the bond, because the resulting supramolecular assembly will suddenly be tethered on two surfaces instead of one. Hence, they will pull the micelles together while trying to get back to equilibrium. This is likely to exacerbate local concentration heterogeneities, an effect that will be much more pronounced for the 12-w/v% sample as the overlapping fraction of neighboring micelles is only 72% (i.e., 28% of the micellar corona is not shared) when not taking the inhomogeneities into account, which will increase this fraction. The consequence of this is that the non-overlapped parts of the corona are not pulled by neighboring micelles, while overlapping coronae tend to influence each other and get closer to each other due to thermodynamic reasons. As the 15-w/v% sample practically does not contain any non-overlapping coronae fractions, this effect is much less pronounced, although it can be quite severe at lower concentrations. Secondly, the average distance measured by SAXS is a number average, but the average intersecting volume does not scale linearly with distance. If the distance between the spheres is polydisperse, the average overlapping volume will deviate locally from the value calculated from the average distance. For the current conditions, there is probably a somewhat smaller overlap than the one calculated from the average size.

3.4. Strain Dependence

The strain dependence of Fe^{2+}, Ni^{2+}, and Zn^{2+} samples at a given angular frequency ω is presented in Figure 7a–c. The behavior is qualitatively similar between Fe^{2+} and Ni^{2+} samples at the same polymer concentration. Below 100% deformation, $G'(\omega, \gamma_0)$, $G''(\omega, \gamma_0)$, and hence $\tan\delta$ remain constant, which means that the system is still in the linear viscoelastic regime. Moreover, the loss factor is distinctly below 1, indicating an elastic-dominated behavior. At a strain around 100%, an upturn in $G'(\omega, \gamma_0)$ is evident for the 12-w/v% sample, however, not for the 15-w/v% sample. At a higher strain γ_0 around 300%, an even stronger peak in $G''(\omega, \gamma_0)$ is found, which is very close to the G' and G'' crossover point $\omega_c(\gamma_0)$. This crossover point reflects the transition from elastic to plastic behavior [42–45]. We note that the dynamic moduli reflect average responses, and that the instantaneous responses could show dramatic differences and yet still have the same average value. At higher strains, $G'(\gamma_0)$ and $G''(\gamma_0)$ for the 12-w/v% sample decrease with slopes of 2 and 1, respectively. This corresponds to a sharp increase of $\tan\delta$ (γ_0).

This feature combining an upturn in both G' and G'' during the transition to nonlinear behavior has been rarely reported. It is described as strain sweep type IV by Sim et al. [46]. Most of the reports of a type IV strain sweep stem from characterization of semidilute supramolecular/associative solutions [47], including hydrophobic ethoxylated urethane (HEUR), hydrophobically modified alkali-swellable associative polymer (HASE), and similar systems [47,48]. Annable et al. [49] and Tanaka and Edwards [50–52] analyzed these findings from a theoretical point of view. In general, the behavior can be explained with the help of a transient network. The network is stretched until a certain stress is exerted on the chains. When that critical stress is reached, the supramolecular bond breaks and the behavior changes from elastic to plastic-dominated. For the 12-w/v% samples, the critical stress is around 70 Pa for the Ni^{2+} sample and 180 Pa for the Fe^{2+} sample. At higher strains, the stress remains approximately constant at this level (Figure S2), as can be inferred from the −1 slope of the $G''(\gamma_0)$ in the nonlinear region (see Figure 7a). Based on literature [46–49], the stretching (peak in $G'(\gamma_0)$) only occurs in a certain concentration regime. Therefore, a certain degree of dilution is needed to find the chain stretch. This stretching appears to only occur when the chains can be stretched to a certain degree before breaking the first bonds, which depends on the average distance between the micellar cores (as discussed later). It seems that this regime is reached for the 12-w/v% samples but not for the 15-w/v% samples.

For the 12-w/v% Zn^{2+} sample (Figure 7c), an almost perfectly linear behavior is found up to deformations of $\gamma_0 \approx 1000\%$. Above this threshold, a clear decrease in $G'(\omega)$ and a weak decrease in $G''(\omega)$ is found. This indicates that the Zn^{2+} sample has a very weak structure that can only be broken at a very high strain, suggesting that this structure exists in clusters of only very few micelles. This does not mean that the material consists of small stable clusters in a liquid. Instead, weak dynamic interactions exist among different micelles, which connect several micelles at a certain point in time. The cluster, however, is dynamic (i.e., the micelles involved in it change). This concept is similar to sticky reptation [2].

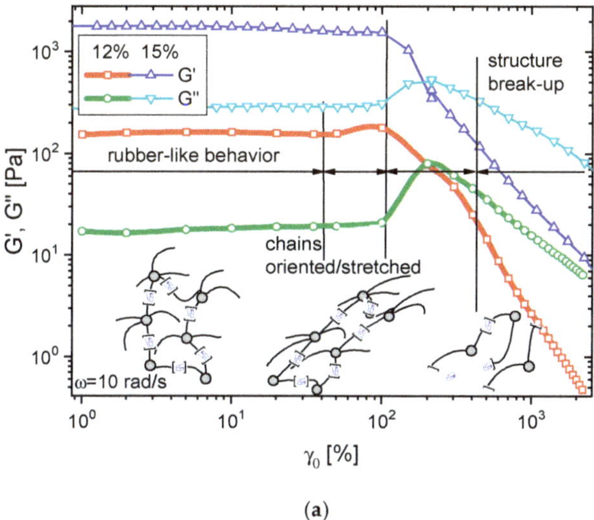

(a)

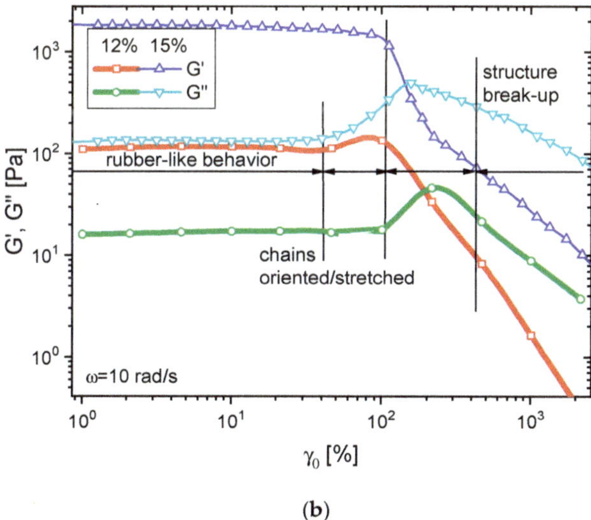

(b)

Figure 7. *Cont.*

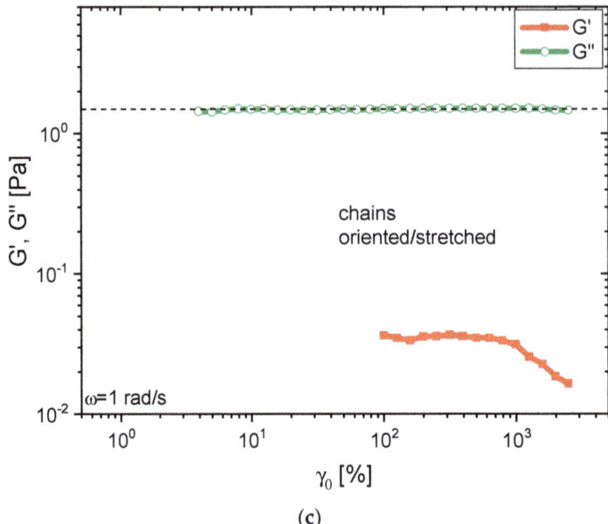

(c)

Figure 7. Strain sweep of (**a**) 12-w/v% Ni^{2+} (from ref [7]) and 15-w/v% Ni^{2+} samples; (**b**) 12-w/v% Fe^{2+} and 15-w/v% Fe^{2+} samples; and (**c**) the 12-w/v% Zn^{2+} sample.

The structure of the 12-w/v% samples can be understood from the schemes inserted in Figure 7a. At a low strain γ_0 the micelles are connected by the supramolecular bonds (indicated by ●), forming a network. At strains around 50%, the intermicellar links become oriented and stretched, which leads to an increase in the storage modulus $G'(\omega, \gamma_0)$. At a somewhat higher strain γ_0, $G''(\omega, \gamma_0)$ also increases distinctly, reflecting higher dissipation due to a massive breakup of the intermicellar links, which can also be concluded from the decrease of $G'(\omega, \gamma_0)$. At even higher deformations, the number of intermicellar links is reduced so much that the network breaks down, which leads to a behavior typical of polymer solutions and melts.

The transition to nonlinear behavior of the 15-w/v% samples is significantly different from that of the 12% samples, despite the minor difference in concentration. For the former, the distance between the cores is shorter. This leads to a larger overlap of the coronae and, thus, to a higher likeliness of intermicellar bonds. This in turn increases the likeliness of twists and contact of the intermicellar "hairs", which self-evidently leads to lower stretchability before breaking. A close look at Figure 7a,b reveals that the onset of non-linearity is around 40% for both samples. However, contrary to the 12-w/v% samples, there is no $G'(\gamma_0)$ upturn for the 15-w/v% samples (this corresponds to "type III" behavior in the nomenclature of Sim et al. [46]). This difference can be interpreted by assuming that the 15-w/v% samples already show the first signs of network breakdown before they get a chance to be stretched. At very high strains (γ_0 around 300%), a constant slope is reached for the loss factor (G''/G') of both samples, which is lower for the 15-w/v% sample than for the 12-w/v% sample. Hence, the former sample structure degrades faster. This can also be inferred from the different negative slopes observed for $G'(\gamma_0)$ and is logical considering the much higher micellar overlap of the 15-w/v% sample, leading to much higher network "healing" probability under flow.

From micellar inter-distance and the chain length of the acrylate blocks, it is possible to estimate the maximum stretching before breakage. The calculations were made assuming that the cores are rigid and that the chain is tethered perpendicular to the shear plane initially. The equilibrium length of connecting strands is the distance between surfaces of two neighboring micelles (i.e., 17 nm for the 12-w/v% sample and 15 nm for the 15-w/v% sample). If the strand is stretched to a maximum extent, the strand length is the contour length of 2 × 200 monomers. By approximating the length of a covalent bond as 0.154 nm, the maximum stretch leads to 2 × 200 × 0.154 nm = 61.6 nm. Hence, for a 12-w/v%

concentration, the theoretical stretching limit is at 61.6 nm/17 nm = 360%, while 61.6 nm/15 nm = 410% is possible for a 15-w/v% concentration. Comparison of this threshold with the data in Figure 7 suggests that there is a link with the microscopic nonlinear behavior dominated by dissipation (constant positive slope tanδ region).

As for the contribution of cores, although the type of large amplitude oscillatory shear (LAOS) behavior is affected by colloidal particles [53], PS core has a much smaller size than the coronae, and the concentrations of PS cores are rather low. Therefore, the PS core is similar in its function to a crosslinker and has only a limited effect on LAOS types at the studied concentrations. A study of the core's effect on LAOS could be realized by either increasing the concentration of polymers or tuning the relative length between core and coronae chains. This is beyond the present topic and prohibited by limited samples, but deserves systematic research in the future.

At last, we note that rheology combined with simultaneous scattering could be an effective tool to clarify the structural evolution under large-amplitude strain. This technique has been widely used to characterize the phase behavior of block copolymers during shear [54]. Specifically, for less ordered systems as ours (e.g., solutions, gels, and suspensions), a lot of literature focuses on in situ research of the alignment of block polymers under transient and steady shear [55–66]. However, the research on the associative colloidal polymer assemblies is still missing as far as we know. We performed time-resolved SAXS measurements to characterize the present system, but the scattering was too weak to get valuable data, probably because the electron contrast between PS cores and ethanolic solutions of PtBA was rather low. Fortunately, the change of network under shear can be theoretically interpreted based on the classical model of Tanaka and Edwards [50–52], which addresses the orientation and stretching of strands in a network—even without further experimental evidence from scattering, a discussion based on such a model can be used in analyzing the nonlinear rheological behavior of micellar gels.

4. Summary and Conclusions

Linear and nonlinear rheological characterization combined with simple geometric considerations about micellar overlap fractions provides a very good insight into the structure and dynamic behavior of the studied micellar assemblies in the absence or presence of metallo–supramolecular interactions, depending on ion type and polymer concentration.

In the linear regime, addition of Zn^{2+} only leads to viscosity increase for the 12-w/v% sample, but does not affect the Newtonian behavior. On the other hand, Ni^{2+} and Fe^{2+} induce a major shift towards an elastic dynamic network behavior because of semi-permanent intermicellar supramolecular interactions. Based on a comparison between theoretical and experimental elastic moduli combined with known concentrations of ligand-terminated corona chains, we conclude that only a 13.5% fraction of potential crosslinks are effective for network elasticity at a 12-w/v% polymer concentration, while 46% are effective at 15 w/v%. In the latter case, the experimental result is consistent with the theoretical prediction calculated by the intermicellar bond formation probability with the help of a simple geometric model using micellar overlap concepts. At a 12-w/v% polymer concentration, however, elastic strand density is much less than anticipated. Estimations of entanglement molecular weights of the solutions indicate that entanglements play no significant additional role in explaining the elasticity of the systems.

In the nonlinear rheological aspect, the Ni^{2+} and Fe^{2+} elastic networks show stretching at intermediate deformation (around 100%) and dynamic breaking of the supramolecular bonds at high deformation. The 15-w/v% samples start breaking down at a lower strain than the 12-w/v% ones, but demonstrate better reforming ability at large deformations because of higher micellar overlap.

Based on all results, the equilibrium structure of the gels can be pictured as soft interpenetrating spheres that are connected intermicellarly by less than 50% of the chains. The remaining chains either do not have a supramolecular bond or are linked intramicellarly and, thus, do not contribute to the network. The ratio between these two types of functional groups (inter- vs. intramicellarly bonded) depends on the polymer concentration. The ion used additionally determines how many of the potentially available groups are efficiently bonded. The low strength and lifetime of the Zn^{2+} ions [37] makes only a weak shear thinning fluid at a 12-w/v% concentration. The Ni^{2+}- and Fe^{2+}-based supramolecular bonds possess a much higher stability [37] and resulting micellar assemblies, and hence behave as very soft elastomers that could be self-healing [7]. One might argue that shear banding might be observed. However, as this would involve the collective and simultaneous break of all supramolecular bonds at the periphery, it is not a very possible process and even if it did occur the self-healing capability of the supramolecular bonds would quickly close it.

Samples at a 12-w/v% concentration show a small peak in storage modulus G' around 80–100% deformation. This can be explained by stretching of the supramolecular network, in agreement with some literature. The higher concentration 15-w/v% samples do not show this peak, which is explained by the fact that the network starts to break before it can be stretched, as the intensive twist and contact of intermicellar links make them less stretchable.

Supplementary Materials: The following are available online at http://www.mdpi.com/2073-4360/11/10/1532/s1.

Author Contributions: Conceptualization, F.J.S.; Data curation, Z.-C.Y.; Formal analysis, Z.-C.Y. and F.J.S.; Investigation, Z.-C.Y., F.J.S., P.G., C.M., C.-A.F., J.-F.G. and C.B.; Methodology, F.J.S.; Writing—original draft, Z.-C.Y. and F.J.S.; Writing—review & editing, P.G., C.M., C.-A.F., J.-F.G. and C.B. P.G., C.M.—polymer synthesis, SAXS, gel preparation, and analytical characterization.

Funding: The authors thank the Communauté française de Belgique for ARC SUPRATUNE and BELSPO for financial support in the frame of the network IAP 6/27. C.M. thanks the FRIA for financial support. F.J.S. thanks the National Natural Science Foundation of China (Grant No. 21574086) for financial support. Y.Z.C. thanks the National Natural Science Foundation of China (Grant No. 21803039) for financial support.

Conflicts of Interest: The authors declare no conflict of interest.

References

1. Afifi, H.; da Silva, M.A.; Nouvel, C.; Six, J.L.; Ligoure, C.; Dreiss, C.A. Associative networks of cholesterol-modified dextran with short and long micelles. *Soft Matter* **2011**, *7*, 4888–4899. [CrossRef]
2. Leibler, L.; Rubinstein, M.; Colby, R.H. Dynamics of reversible networks. *Macromolecules* **1991**, *24*, 4701–4707. [CrossRef]
3. Cordier, P.; Tournilhac, F.; Soulie-Ziakovic, C.; Leibler, L. Self-healing and thermoreversible rubber from supramolecular assembly. *Nature* **2008**, *451*, 977–980. [CrossRef] [PubMed]
4. Stephanou, P.S.; Baig, C.; Tsolou, G.; Mavrantzas, V.G.; Kroger, M. Quantifying chain reptation in entangled polymer melts: Topological and dynamical mapping of atomistic simulation results onto the tube model. *J. Chem. Phys.* **2010**, *132*, 124904. [CrossRef] [PubMed]
5. Van Ruymbeke, E.; Orfanou, K.; Kapnistos, M.; Iatrou, H.; Pitsikalis, M.; Hadjichristidis, N.; Lohse, D.J.; Vlassopoulos, D. Entangled dendritic polymers and beyond: Rheology of symmetric cayley-tree polymers and macromolecular self-assemblies. *Macromolecules* **2007**, *40*, 5941–5952. [CrossRef]
6. Ramos, L.; Ligoure, C. Structure of a new type of transient network: Entangled wormlike micelles bridged by telechelic polymers. *Macromolecules* **2007**, *40*, 1248–1251. [CrossRef]
7. Guillet, P.; Mugemana, C.; Stadler, F.J.; Schubert, U.S.; Fustin, C.-A.; Bailly, C.; Gohy, J.F. Connecting micelles by metallo-supramolecular interactions: Towards stimuli responsive hierarchical materials. *Soft Matter* **2009**, *5*, 3409–3411. [CrossRef]
8. Lehn, J.M. From supramolecular chemistry towards constitutional dynamic chemistry and adaptive chemistry. *Chem. Soc. Rev.* **2007**, *36*, 151–160. [CrossRef]
9. Whitesides, G.M.; Boncheva, M. Beyond molecules: Self-assembly of mesoscopic and macroscopic components. *Proc. Natl. Acad. Sci. USA* **2002**, *99*, 4769–4774. [CrossRef]
10. Lehn, J.M. Supramolecular chemistry: From molecular information towards self-organization and complex matter. *Rep. Prog. Phys.* **2004**, *67*, 249–265. [CrossRef]
11. Lehn, J.M. Toward complex matter: Supramolecular chemistry and self-organization. *Proc. Natl. Acad. Sci. USA* **2002**, *99*, 4763–4768. [CrossRef]
12. Lehn, J.M. Toward self-organization and complex matter. *Science* **2002**, *295*, 2400–2403. [CrossRef] [PubMed]

13. Mugemana, C.; Guillet, P.; Fustin, C.A.; Gohy, J.F. Metallo-supramolecular block copolymer micelles: Recent achievements. *Soft Matter* **2011**, *7*, 3673–3678. [CrossRef]

14. Tixier, T.; Tabuteau, H.; Carriere, A.; Ramos, L.; Ligoure, C. Transition from "brittle" to "ductile" rheological behavior by tuning the morphology of self-assembled networks. *Soft Matter* **2010**, *6*, 2699–2707. [CrossRef]

15. Guillet, P.; Fustin, C.A.; Mugemana, C.; Ott, C.; Schubert, U.S.; Gohy, J.F. Tuning block copolymer micelles by metal-ligand interactions. *Soft Matter* **2008**, *4*, 2278–2282. [CrossRef]

16. Brassinne, J.; Gohy, J.F.; Fustin, C.A. Controlling the cross-linking density of supramolecular hydrogels formed by heterotelechelic associating copolymers. *Macromolecules* **2014**, *47*, 4514–4524. [CrossRef]

17. Brassinne, J.; Fustin, C.A.; Gohy, J.F. Control over the assembly and rheology of supramolecular networks via multi-responsive double hydrophilic copolymers. *Polym. Chem.* **2017**, *8*, 1527–1539. [CrossRef]

18. Kotsuchibashi, Y.; Ebara, M.; Aoyagi, T.; Narain, R. Recent advances in dual temperature responsive block copolymers and their potential as biomedical applications. *Polymers* **2016**, *8*, 380. [CrossRef]

19. Vatankhah-Varnoosfaderani, M.; Hashmi, S.; Stadler, F.J.; GhavamiNejad, A. Mussel-inspired 3D networks with stiff-irreversible or soft-reversible characteristics—It's all a matter of solvent. *Polym. Test.* **2017**, *62*, 96–101. [CrossRef]

20. He, L.; Ran, X.; Li, J.X.; Gao, Q.Q.; Kuang, Y.M.; Guo, L.J. A highly transparent and autonomic self-healing organogel from solvent regulation based on hydrazide derivatives. *J. Mater. Chem. A* **2018**, *6*, 16600–16609. [CrossRef]

21. Yoshida, T.; Taribagil, R.; Hillmyer, M.A.; Lodge, T.P. Viscoelastic synergy in aqueous mixtures of wormlike micelles and model amphiphilic triblock copolymers. *Macromolecules* **2007**, *40*, 1615–1623. [CrossRef]

22. Lodge, T.P.; Taribagil, R.; Yoshida, T.; Hillmyer, M.A. SANS evidence for the cross-linking of wormlike micelles by a model hydrophobically modified polymer. *Macromolecules* **2007**, *40*, 4728–4731. [CrossRef]

23. Vatankhah-Varnoosfaderani, M.; Hashmi, S.; GhavamiNejad, A.; Stadler, F.J. Rapid self-healing and triple stimuli responsiveness of a supramolecular polymer gel based on boron–catechol interactions in a novel water-soluble mussel-inspired copolymer. *Polym. Chem.* **2014**, *5*, 512–523. [CrossRef]

24. Vatankhah-Varnoosfaderani, M.; Ghavaminejad, A.; Hashmi, S.; Stadler, F.J. Mussel-inspired pH-triggered reversible foamed multi-responsive gel—The surprising effect of water. *Chem. Commun.* **2013**, *49*, 4685–4687. [CrossRef]

25. Vatankhah-Varnoosfaderani, M.; GhavamiNejad, A.; Hashmi, S.; Stadler, F.J. Hydrogen bonding in aprotic solvents, a new strategy for gelation of bioinspired catecholic copolymers with *N*-isopropylamide. *Macromol. Rapid Commun.* **2015**, *36*, 447–452. [CrossRef] [PubMed]

26. Stadler, F.J.; Hashmi, S.; GhavamiNejad, A.; Vatankhah Varnoosfaderani, M. Rheology of dopamine containing polymers. *Ann. Trans. Nord. Rheol. Soc.* **2017**, *25*, 115–120.

27. GhavamiNejad, A.; Sasikala, A.R.K.; Unnithan, A.R.; Thomas, R.G.; Jeong, Y.Y.; Vatankhah-Varnoosfaderani, M.; Stadler, F.J.; Park, C.H.; Kim, C.S. Mussel-Inspired electrospun smart magnetic nanofibers for hyperthermic chemotherapy. *Adv. Funct. Mater.* **2015**, *25*, 2867–2875. [CrossRef]

28. GhavamiNejad, A.; Hashmi, S.; Vatankhah-Varnoosfaderani, M.; Stadler, F.J. Effect of H_2O and reduced graphene oxide on the structure and rheology of self-healing, stimuli responsive catecholic gels. *Rheol. Acta* **2016**, *55*, 163–176. [CrossRef]

29. Kawamoto, K.; Grindy, S.C.; Liu, J.; Holten-Andersen, N.; Johnson, J.A. Dual role for 1,2,4,5-tetrazines in polymer networks: Combining Diels-Alder reactions and metal coordination to generate functional supramolecular gels. *ACS Macro Lett.* **2015**, *4*, 458–461. [CrossRef]

30. Golkaram, M.; Fodor, C.; van Ruymbeke, E.; Loos, K. Linear viscoelasticity of weakly hydrogen-bonded polymers near and below the sol-gel transition. *Macromolecules* **2018**, *51*, 4910–4916. [CrossRef]

31. Weisstein, E.W. Sphere-Sphere Intersection. Available online: http://mathworld.wolfram.com/Sphere-SphereIntersection.html (accessed on 17 September 2019).

32. Stadler, F.J. Quantifying primary loops in polymer gels by linear viscoelasticity. *Proc. Natl. Acad. Sci. USA* **2013**, *110*, E1972. [CrossRef] [PubMed]

33. Zhou, H.; Woo, J.; Cok, A.M.; Wang, M.; Olsen, B.D.; Johnson, J.A. Counting primary loops in polymer gels. *Proc. Natl. Acad. Sci. USA* **2012**, *109*, 19119–19124. [CrossRef] [PubMed]

34. Tang, G.; Du, B.; Stadler, F.J. A novel approach to analyze the rheological properties of hydrogels with network structure simulation. *J. Polym. Res.* **2018**, *25*, 4. [CrossRef]

35. Daoud, M.; Cotton, J.P. Star shaped polymers: A model for the conformation and its concentration dependence. *J. Phys.* **1982**, *43*, 531–538. [CrossRef]

36. Vlassopoulos, D.; Pakula, T.; Fytas, G.; Roovers, J.; Karatasos, K.; Hadjichristidis, N. Ordering and viscoelastic relaxation in multiarm star polymer melts. *Polymer* **1997**, *39*, 617–622. [CrossRef]

37. Holyer, R.H.; Hubbard, C.D.; Kettle, S.F.A.; Wilkins, R.G. The Kinetics of Replacement Reactions of Complexes of the Transition Metals with 1,10-Phenanthroline and 2,2′-Bipyridine. *Inorg. Chem.* **1965**, *4*, 929–935. [CrossRef]

38. Dobrawa, R.; Wurthner, F. Metallosupramolecular approach toward functional coordination polymers. *J. Polym. Sci. Polym. Chem.* **2005**, *43*, 4981–4995. [CrossRef]

39. Osaki, K.; Nishimura, Y.; Kurata, M. Viscoelastic Properties of Semidilute Polystyrene Solutions. *Macromolecules* **1985**, *18*, 1153–1157. [CrossRef]

40. Fetters, L.J.; Lohse, D.J.; Colby, R.H. Chain dimensions and entanglement spacings. In *Physical Properties of Polymers*, 2nd ed.; Mark, J.E., Ed.; Springer: Heidelberg, Germany, 2007.

41. Tong, J.D.; Leclere, P.; Doneux, C.; Bredas, J.L.; Lazzaroni, R.; Jerome, R. Morphology and mechanical properties of poly(methylmethacrylate)-*b*-poly(alkylacrylate)-*b*-poly(methylmethacrylate). *Polymer* **2001**, *42*, 3503–3514. [CrossRef]

42. Zeng, F.; Han, Y.; Yan, Z.-C.; Liu, C.-Y.; Chen, C.-F. Supramolecular polymer gel with multi stimuli responsive, self-healing and erasable properties generated by host–guest interactions. *Polymer* **2013**, *54*, 6929–6935. [CrossRef]

43. Liu, J.; Feng, Y.; Liu, Z.-X.; Yan, Z.-C.; He, Y.-M.; Liu, C.-Y.; Fan, Q.-H. N-Boc-Protected 1,2-Diphenylethylenediamine-Based Dendritic Organogels with Multiple-Stimulus-Responsive Properties. *Chem. Asian J.* **2013**, *8*, 572–581. [CrossRef] [PubMed]

44. Liu, Z.-X.; Feng, Y.; Zhao, Z.-Y.; Yan, Z.-C.; He, Y.-M.; Luo, X.-J.; Liu, C.-Y.; Fan, Q.-H. A New Class of Dendritic Metallogels with Multiple Stimuli-Responsiveness and as Templates for the In Situ Synthesis of Silver Nanoparticles. *Chem. Eur. J.* **2014**, *20*, 533–541. [CrossRef]

45. Liu, Z.-X.; Feng, Y.; Yan, Z.-C.; He, Y.-M.; Liu, C.-Y.; Fan, Q.-H. Multistimuli Responsive Dendritic Organogels Based on Azobenzene-Containing Poly(aryl ether) Dendron. *Chem. Mater.* **2012**, *24*, 3751–3757. [CrossRef]

46. Sim, H.G.; Ahn, K.H.; Lee, S.J. Large amplitude oscillatory shear behavior of complex fluids investigated by a network model: A guideline for classification. *J. Non-Newton. Fluid* **2003**, *112*, 237–250. [CrossRef]

47. Berret, J.F.; Calvet, D.; Collet, A.; Viguier, M. Fluorocarbon associative polymers. *Curr. Opin. Colloid Interface Sci.* **2003**, *8*, 206–306. [CrossRef]

48. Tirtaatmadja, V.; Tam, K.C.; Jenkins, R.D. Superposition of oscillations on steady shear flow as a technique for investigating the structure of associative polymers. *Macromolecules* **1997**, *30*, 1426–1433. [CrossRef]

49. Annable, T.; Buscall, R.; Ettelaie, R.; Whittlestone, D. The rheology of solutions of associating polymers—Comparison of experimental behavior with transient network theory. *J. Rheol.* **1993**, *37*, 695–726. [CrossRef]

50. Tanaka, F.; Edwards, S.F. Viscoelastic properties of physically cross-linked networks. 1. Nonlinear stationary viscoelasticity. *J. Non-Newton. Fluid* **1992**, *43*, 247–271. [CrossRef]

51. Tanaka, F.; Edwards, S.F. Viscoelastic properties of physically cross-linked networks. 2. Dynamic mechanical moduli. *J. Non-Newton. Fluid* **1992**, *43*, 273–288. [CrossRef]

52. Tanaka, F.; Edwards, S.F. Viscoelastic properties of physically cross-linked networks. 3. Time-dependent phenomena. *J. Non-Newton. Fluid* **1992**, *43*, 289–309. [CrossRef]

53. Yziquel, F.; Carreau, P.J.; Moan, M.; Tanguy, P.A. Rheological modeling of concentrated colloidal suspensions. *J. Non-Newton. Fluid* **1999**, *86*, 133–155. [CrossRef]

54. Hamley, I.W.; Castelletto, V. Small-angle scattering of block copolymers in the melt, solution and crystal states. *Prog. Polym. Sci.* **2004**, *29*, 909–948.

55. Hamley, I.W.; Pople, J.A.; Booth, C.; Derici, L.; Imperor-Clerc, M.; Davidson, P. Shear-induced orientation of the body-centered-cubic phase in a diblock copolymer gel. *Phys. Rev. E* **1998**, *58*, 7620–7628. [CrossRef]

56. Pople, J.A.; Hamley, I.W.; Diakun, G.P. An integrated Couette system for in situ shearing of polymer and surfactant solutions and gels with simultaneous small angle x-ray scattering. *Rev. Sci. Instrum.* **1998**, *69*, 3015–3021. [CrossRef]

57. Sing, M.K.; Glassman, M.J.; Vronay-Ruggles, X.T.; Burghardt, W.R.; Olsen, B.D. Structure and rheology of dual-associative protein hydrogels under nonlinear shear flow. *Soft Matter* **2017**, *13*, 8511–8524. [CrossRef]

58. Caputo, F.E.; Ugaz, V.M.; Burghardt, W.R.; Berret, J.-F. Transient 1–2 plane small-angle x-ray scattering measurements of micellar orientation in aligning and tumbling nematic surfactant solutions. *J. Rheol.* **2002**, *46*, 927–946. [CrossRef]
59. Caputo, F.E.; Burghardt, W.R.; Krishnan, K.; Bates, F.S.; Lodge, T.P. Time-resolved small-angle X-ray scattering measurements of a polymer bicontinuous microemulsion structure factor under shear. *Phys. Rev. E* **2002**, *66*, 041401. [CrossRef]
60. Martin, H.P.; Brooks, N.J.; Seddon, J.M.; Luckham, P.F.; Terrill, N.J.; Kowalski, A.J.; Cabral, J.T. Microfluidic processing of concentrated surfactant mixtures: Online SAXS, microscopy and rheology. *Soft Matter* **2016**, *12*, 1750–1758. [CrossRef]
61. Kim, J.M.; Eberle, A.P.R.; Gurnon, A.K.; Porcar, L.; Wagner, N.J. The microstructure and rheology of a model, thixotropic nanoparticle gel under steady shear and large amplitude oscillatory shear (LAOS). *J. Rheol.* **2014**, *58*, 1301–1328. [CrossRef]
62. Calabrese, M.A.; Rogers, S.A.; Murphy, R.P.; Wagner, N.J. The rheology and microstructure of branched micelles under shear. *J. Rheol.* **2015**, *59*, 1299–1328. [CrossRef]
63. Rogers, S.; Kohlbrecher, J.; Lettinga, M.P. The molecular origin of stress generation in worm-like micelles, using a rheo-SANS LAOS approach. *Soft Matter* **2012**, *8*, 7831–7839. [CrossRef]
64. Lonetti, B.; Kohlbrecher, J.; Willner, L.; Dhont, J.K.G.; Lettinga, M.P. Dynamic response of block copolymer wormlike micelles to shear flow. *J. Phys. Condens. Matter* **2008**, *20*, 404207. [CrossRef]
65. Lopez-Barron, C.R.; Wagner, N.J.; Porcar, L. Layering, melting, and recrystallization of a close-packed micellar crystal under steady and large-amplitude oscillatory shear flows. *J. Rheol.* **2015**, *59*, 793–820. [CrossRef]
66. Westermeier, F.; Pennicard, D.; Hirsemann, H.; Wagner, U.H.; Rau, C.; Graafsma, H.; Schall, P.; Paul Lettinga, M.; Struth, B. Connecting structure, dynamics and viscosity in sheared soft colloidal liquids: A medley of anisotropic fluctuations. *Soft Matter* **2016**, *12*, 171–180. [CrossRef]

 *polymers*

Article

Rheological Study on the Thermoreversible Gelation of Stereo-Controlled Poly(*N*-Isopropylacrylamide) in an Imidazolium Ionic Liquid

Zhi-Chao Yan *, Chandra Sekhar Biswas and Florian J. Stadler

College of Materials Science and Engineering, Shenzhen Key Laboratory of Polymer Science and Technology, Guangdong Research Center for Interfacial Engineering of Functional Materials, Nanshan District Key Lab for Biopolymers and Safety Evaluation, Shenzhen University, Shenzhen 518055, China; chandra01234@gmail.com (C.S.B.); fjstadler@szu.edu.cn (F.J.S.)
* Correspondence: yanzhch@szu.edu.cn

Received: 19 March 2019; Accepted: 22 April 2019; Published: 2 May 2019

Abstract: The thermoreversible sol-gel transition for an ionic liquid (IL) solution of isotactic-rich poly (*N*-isopropylacrylamides) (PNIPAMs) is investigated by rheological technique. The meso-diad content of PNIPAMs ranges between 47% and 79%, and molecular weight (M_n) is ~35,000 and ~70,000 g/mol for two series of samples. PNIPAMs are soluble in 1-butyl-3-methylimidazolium bis(trifluoromethanesulfonyl) imide ([BMIM][TFSI]) at high temperatures but undergo a gelation with decreasing temperatures. The transition temperature determined from G'-G'' crossover increases with isotacticity, consistent with the previous cloud-point result at the same scanning rate, indicating imide groups along the same side of backbones are prone to be aggregated for formation of a gel. The transition point based on Winter-Chambon criterion is on average higher than that of the G'-G'' crossover method and is insensitive to tacticity and molecular weight, since it correlates with percolation of globules rather than the further formation of elastic network ($G' > G''$). For the first time, the phase diagram composed of both G'-G'' crossover points for gelation and cloud points is established in PNIPAM/IL mixtures. For low-M_n PNIPAMs, the crossover-point line intersects the cloud-point line. Hence, from solution to opaque gel, the sample will experience two different transitional phases, either clear gel or opaque sol. A clear gel is formed due to partial phase separation of isotactic segments that could act as junctions of network. However, when the partial phase separation is not faster than the formation of globules, an opaque sol will be formed. For high-M_n PNIPAMs, crossover points are below cloud points at all concentrations, so their gelation only follows the opaque sol route. Such phase diagram is attributed to the poorer solubility of high-M_n polymers for entropic reasons. The phase diagram composed of Winter-Chambon melting points, crossover points for melting, and clear points is similar with the gelation phase diagram, confirming the mechanism above.

Keywords: Poly(*N*-isopropylacrylamide); tacticity; ionic liquid; rheology

---

1. Introduction

Stimuli-responsive gels have received great attention owing to their scientific interest and potential applications [1,2]. The merit of these materials lies on their rapid and reversible response to environmental changes, such as temperature, pH, light, and stress. For thermosensitive gels, the response relies on the abrupt change in solubility upon cooling or heating. As a typical thermo-responsive polymer, poly (*N*-isopropylacrylamide) (PNIPAM) in aqueous solutions exhibits a lower critical solution temperature (LCST) phase behavior around ambient temperatures (≈33 °C) [3]. These notable features make PNIPAMs widely used in thermoreversible hydrogel for sensing and biomedical materials [4–7]. Despite the appealing applications, however, PNIPAM hydrogels suffer from the volatility of water, preventing their use at high temperatures and storage at open atmosphere. Moreover, as the major component, water cannot offer additional functionality to the gel besides polymers. In order to solve the above problems, non-volatile and functional solvents are desired to substitute water. Ionic liquids (ILs) are non-volatile room-temperature molten salts with thermal and chemical stability and conductivity [8]. Owing to these properties, ILs are ideal solvents for making conductive gels, which could be used at open atmosphere over a wide range of temperatures [9–12]. The stimuli-responsive properties can be readily tailored by changing the chemical structures of the ions or by mixing with different solvents, without modification of polymer structures [13,14]. Such convenience of manipulating gels for a target purpose generates interest in developing polymeric materials in ILs, especially the thermoreversible ionogels [15–20].

The ionic liquid, 1-alkyl-3-methylimidazolium bis(trifluoromethanesulfonyl)imide ([XMIM][TFSI], X = E for ethyl, and X = B for butyl), has proven to be a solvent for PNIPAMs [21]. The PNIPAM/IL mixture exhibits an upper critical solution temperature (UCST) phase behavior [21–24], opposite to aqueous solutions [3,25]. Based on this property, PNIPAM and its copolymers have been employed to make thermoresponsive ionogels in ILs by self-assembly. He et al. [26] synthesized poly (*N*-isopropyl acrylamide-*b*-ethylene oxide-*b*-*N*-isopropyl acrylamide) (PNIPAM-*b*-PEO-*b*-PNIPAM) triblock copolymers and blended them with [EMIM][TFSI]. Above the UCST-type transition temperature, PNIPAM blocks are expanded coils, while below the transition temperature, PNIPAM blocks experience coil-to-globule transition and become aggregates, which are connected by PEO coils, finally forming a gel. Ueki et al. [16] produced poly (benzyl methacrylate-*b*-*N*-isopropyl acrylamide) (PBnMA-*b*-PNIPAM) diblock copolymers, with PBnMA and PNIPAM blocks exhibiting LCST and UCST phase separations in ILs, respectively. As a result, a doubly thermosensitive self-assembly was observed in [EMIM][TFSI], where a homogeneous solution only exists in a window between UCST and LCST phase boundary. Likewise, Lee et al. [27] reported a doubly thermoresponsive diblock copolymer, poly (ethylene oxide-*b*-*N*-isopropylacrylamide) (PEO-*b*-PNIPAM), that exhibits both an upper critical micellization temperature (UCMT) and a lower critical micellization temperature (LCMT) in 1-ethyl-3-methylimidazolium tetrafluoroborate ([EMIM][BF$_4$]), 1-butyl-3-methylimidazolium tetrafluoroborate ([BMIM][BF$_4$]), and their blends. By varying the mixing ratio of two ILs, both UCMT and LCMT can be readily tuned in a mixed solvent. Ueki et al. [28] further developed a triblock ABA copolymer, denoted as P(AzoMA-*r*-NIPAM)-*b*-PEO-*b*-P(AzoMA-r-NIPAM), with AzoMA being 4-phenylazophenyl methacrylate, which is sensitive to photo-stimulus. By combining the photosensitive (AzoMA) and thermosensitive (NIPAm) blocks, a doubly reversible gel in response to either photo-and thermo-stimuli, was obtained in the IL 1-butyl-3-methylimidazolium hexafluorophosphate ([BMIM][PF$_6$]).

To clarify the mechanism of thermoresponse in PNIPAM/IL mixtures and guide the design of ionogels, fundamental researches have been performed to investigate the effect of molecular weight, concentration, and structure on their phase behavior. For example, Asai et al. [22] reported the cloud point of PNIPAM/[EMIM][TFSI] increases with molecular weight (M_n) and polymer concentration by dynamics light scattering. Using small-angle neutron scattering, they found the chain size at dilute regime decreases, while the correlation length at semidilute regime increases as approaching UCST. The Flory–Huggins interaction parameter, χ, becomes larger with cooling, and exceeds 0.5 around 45 °C. Besides, molecularly dispersed PNIPAM chains are found to still remain in ILs even after macroscopic phase separation, distinct from a complete phase separation in a molecular level in aqueous solutions. De Santis et al. [23] confirmed the abovementioned M_n dependence and firstly reported the increase in phase transition temperature with isotacticity, although the meso-diad content (denoted as *m*) in their PNIPAMs only ranges a narrow variation from 55% to 66% and the concentration is fixed at 1%. Wang et al. [29] experimentally investigated the phase separation mechanism in PNIPAM/[EMIM][TFSI] solutions by infrared spectroscopy and attributed it to the desolvation of PNIPAMs. Ueki et al. [30] reported the isomeric effect on the UCST-type separation of random copolymers comprising NIPAM and AzoMA in [EMIM][TFSI] and confirmed that trans-isomer leads to an increase in the phase separation temperature, whereas the cis-isomer contributes to a decrease in the phase separation temperature. So et al. [31] showed how the incorporation of comonomers with different hydrogen-bonding capacities into PNIPAM affects the UCST behavior and the corresponding sequential and reversible self-folding. Very recently, we have studied the phase diagram of two series (depending on molecular weight) of PNIPAMs with varying isotacticity in [BMIM][TFSI] by turbidity measurement using a concentration range of 1–12.5% (w/v) [24]. It was observed that the UCST-type phase separation temperature increases with isotacticity and molecular weight. The phase diagram was firstly determined for a highly isotactic PNIPAM ($m \approx 78\%$), where a clear UCST peak was observed at low concentrations, similar with the conventional polymer solution, implying an entropy-driven dissolution mechanism.

In spite of the abovementioned research on the coil-globule transition, however, the sol-gel transition in PNIPAM/IL mixtures has not been systematically studied to the best of our knowledge. Heretofore, the study on sol-gel transition of PNIPAMs mainly exists in aqueous and organic solutions. Nakano et al. [32] studied the thermoreversible gelation of isotactic-rich poly (N-isopropylacrylamide) in water and observed the gelation happens below the LCST cloud-point temperatures. Upon heating, solutions undergo transparent gels first before they become turbid. Tanaka et al. [33] attributed this phase diagram to the preferential dehydration of meso-diad segments on the chain, whose partial phase separation creates physical crosslinkers for a transparent gel. Upon further heating, other segments are also dehydrated, so that the whole chain transfers to globules and the system becomes turbid. Such partial phase separation was also reported in isotactic-rich PNIPAM/benzyl alcohol blends [34], where a transparent gel was formed below the UCST-type transition temperatures, and the transition temperature increases with the meso-diad content. As molten salts, ILs have different interactions with PNIPAM in comparison to water and organic solvents. The structure-property relationship in aqueous and organic gel does not necessarily apply for the IL gels. Therefore, the gelation behavior in PNIPAM/IL is highly desired to be investigated for both scientific interest and application-based importance in guiding the design of smart ionogels. To this end, the effect from several key factors, including tacticity, molecular weight, and concentration, on the sol-gel transition need to be clarified. The phase diagram from gelation should be established and compared with that from cloud-points.

In this work, we for the first time studied the reversible gelation in isotactic-rich PNIPAM/IL blends using rheological technique. PNIPAMs with a wide range of tacticity ($m = 47$–79%) and different molecular weights were employed to mix with [BMIM][TFSI]. The dependence of sol-gel transition on tacticity, concentration, and molecular weight was examined and compared with the coil-globule transition determined from cloud points. A complete phase diagram was established accordingly. This research will provide a guideline for the design of thermoresponsive ionogels based on PNIPAMs.

2. Experiments

2.1. Sample Information

The synthesis and characterization of stereo-controlled PNIPAMs was described in Reference [34]. Table 1 lists the number-average molecular weight (M_n), polydispersity (PDI), and tacticity of the investigated PNIPAMs. The tacticity is defined as the percentage of the meso-diad content, which was determined by ^{1}H NMR [34]. Based on molecular weight, PNIPAMs are denoted as m and Hm series. For m-series, M_n is from 34,900 to 42,600 g/mol with PDI in between 1.22 and 1.26. Their tacticities are 47%, 58%, 66%, and 79%, respectively. For Hm-series, M_n is doubled and ranging from 60,200 to 85,700 g/mol with PDI in between 1.26 and 1.49. Their tacticities are 48%, 57%, 67%, and 78%, respectively.

Table 1. Characteristics of poly (*N*-isopropylacrylamides) (PNIPAM) Samples.

Sample	M_n (g/mol)	PDI	Meso-Diad Content (*m*) (%)
m47	35,400	1.25	47
m58	39,500	1.24	58
m66	34,900	1.22	66
m79	40,300	1.26	79
Hm48	62,900	1.49	48
Hm57	60,200	1.26	57
Hm67	62,800	1.32	67
Hm78	85,700	1.38	78

The ionic liquid was 1-butyl-3-methylimidazolium bis(trifluoromethanesulfonyl)imide ([BMIM][TFSI]), from the Center for Green Chemistry and Catalysis of Lanzhou Institute of Chemical Physics, Chinese Academy of Sciences. [BMIM][TFSI] was dried at 80 °C under vacuum to remove residue water before use. The polymer/IL solution with a particular concentration (w/v %) was prepared with the help of a cosolvent (tetrahydrofuran). After sufficient evaporation in a fume-hood, mixtures were dried in a vacuum to remove residue cosolvent and bubbles. Details for mixture preparation was reported in Reference [24]. The structures of PNIPAM and [BMIM][TFSI] are presented in Scheme 1.

PNIPAM   [BMIM][TFSI]

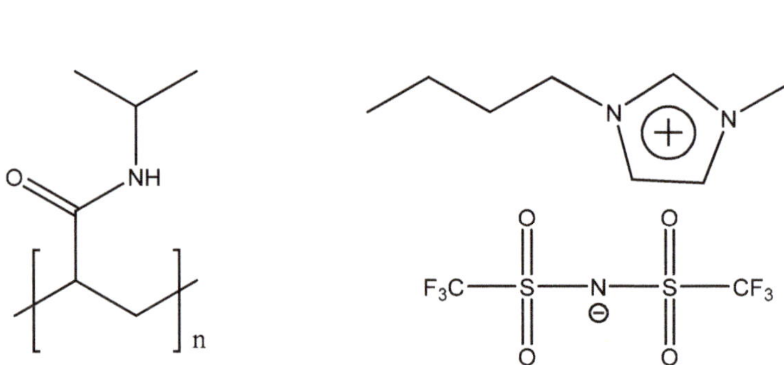

Scheme 1. Structure of PNIPAM and [BMIM][TFSI].

2.2. Rheological Measurements

The solutions were characterized rheologically using a cone-plate geometry with cone angle of 2° and diameter of 25 mm on an Anton Paar 302 rheometer. Temperatures were controlled by Peltier plate. Samples were subjected to temperature ramps between 10 and 60 °C at a heating and cooling rate of 1 °C/min with an angular frequency of 10 rad/s. Heating was performed first followed by a cooling procedure. The strain amplitude is kept at 0.1%, within the linear viscoelastic regime at all temperatures. Dynamic frequency sweep was also performed on each sample from 10 to 40 °C with proper strain to ensure the linear viscoelasticity and enough waiting time to allow the material to equilibrate.

3. Result and Discussion

3.1. Tacticity Dependence

The temperature dependence of storage (*G′*) and loss (*G″*) moduli for 5% m series in ILs is shown in Figure 1. In heating procedure, moduli first exhibit a moderate decrease at the regime where *G′* > *G″*. Then, an abrupt decrease in moduli occurs with *G″* gradually becoming larger than *G′*, reflecting a gel-to-sol transition. At higher temperatures (in solution state), data are scattered because the phase angle in oscillatory measurement is close to 90°, i.e., close to Newtonian, which makes G' difficult to be precisely determined. Furthermore, the magnitude of the complex modulus $|G^*|$ is below 1 Pa, which corresponds to a torque of 1 nN·m for the setup and parameters used, close to the resolution limit of the rheometer. In the cooling procedure, the sample state changes from solution to gel, proving the reversibility of the transition. An increase in moduli is observed followed by the recovery that G' becomes larger than G''. The drop of moduli once the gel state is obtained is due to the slippage of sample, considering the sample becomes heterogeneous and solvophobic, which leads to a phase separation, partially at the interface between geometry [35]. Despite this artefact, the crossover of G' and G'', which corresponds to the gelation transition, appears before this drop, so the final result is not affected. The gel-to-sol transition is located at a higher temperature than the sol-to-gel transition for all samples. This hysteresis is consistent with that in transmittance measurement [24], where the globule-to-coil transition is believed to be retarded by the extra hydrogen bonding among polymers in the aggregates with respect to the coil-to-globule transition [36,37]. Such hydrogen bonding mechanism may also account for the hysteresis in the sol-gel transition here, since the extra hydrogen bonding can kinetically delay the breaking of clusters that are the crosslinker of the gel network.

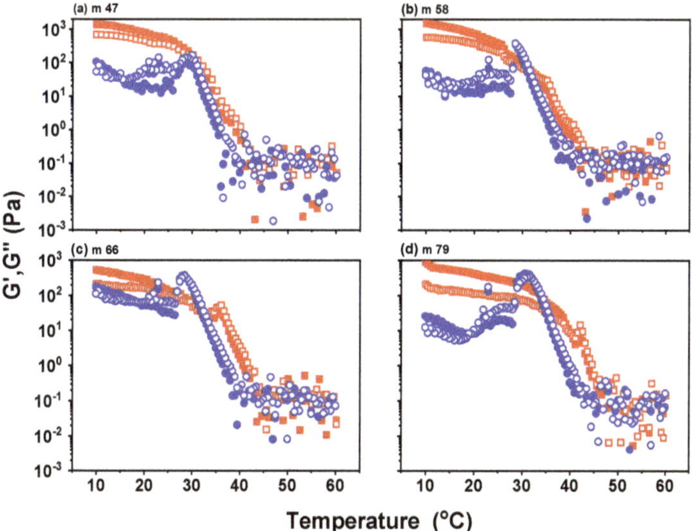

Figure 1. Variations in storage (G', filled symbol) and loss (G'', open symbol) moduli of 5% (**a**) m47, (**b**) m58, (**c**) m66, and (**d**) m79 solutions as a function of temperature at a frequency of $\omega = 10$ rad/s and a strain amplitude of $\gamma = 0.1\%$. Symbols (■, □) and (●, ○) are for heating and cooling sweeps, respectively.

For Hm series, similar heating/cooling evolutions in rheological behavior are observed in Figure 2. Compared with m-series, the sol-gel transition in Hm appears at higher temperature, as the case in transmittance measurement [22–24,38]. This is attributed to a strong entropic contribution to the phase separation for high-M_n polymers in the thermodynamic aspect [24,39], which favors the forming of clusters as crosslinkers. The hysteresis in Hm also increases with isotacticity, and the amplitude is similar with that in m series, implying its insensitivity to molecular weight. In previous reports [23,24], the hysteresis in transmittance is not M_n dependent above 44,000 g/mol, because chain associations within aggregates are similar for different M_n. This could be used to account for the gelation hysteresis and implies the forming/breaking of aggregates (crosslinkers of gels) governs the gelation/melting procedure.

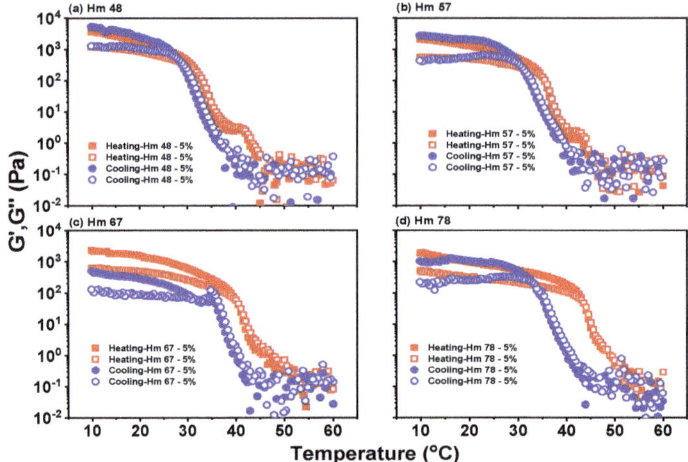

Figure 2. Variations in storage (G', filled symbol) and loss (G'', open symbol) moduli of 5% (**a**) Hm48, (**b**) Hm58, (**c**) Hm67, and (**d**) Hm78 solutions as a function of temperature at a frequency of $\omega = 10$ rad/s and a strain amplitude of $\gamma = 0.1\%$. Symbols ($\blacksquare$, $\square$) and ($\bullet$, $\circ$) are for heating and cooling sweeps, respectively.

To assess the gel network of different tacticity, we compared their storage moduli at 10 °C on heating curves in Figure 3, which are proportional to the density of network. For both m and Hm series, G' slightly decreases with isotacticity, showing a scaling power law slope −2.2 and −1.2, respectively. This dependence is opposite to the case of PNIPAMs in benzyl alcohol [34], where the plateau G' increases with isotacticity. However, we note that PNIPAM/benzyl alcohol gel is transparent, while the PNIPAM/IL gel is opaque. In a transparent gel, partial phase separation takes place only in solvophobic isotactic segments. These desolvated segments act as junctions to connect other segments that are still dissolved. With increasing content of meso-diad on the chain, the number of junctions increases, and hence the gel network is better linked. In an opaque gel, however, macroscopic phase separation takes places to a much larger extent. Segments are mainly aggregated into globules rather than left in ILs to act as a bridging chain. This situation is intensified with increasing isotacticity, reducing the number of connections among globules, and thus thinning out the network. The storage moduli of Hm are higher than m series, reflecting a better-connected gel because long chains have a higher probability to bridge different globules and increase the number of interglobular association [34,40,41].

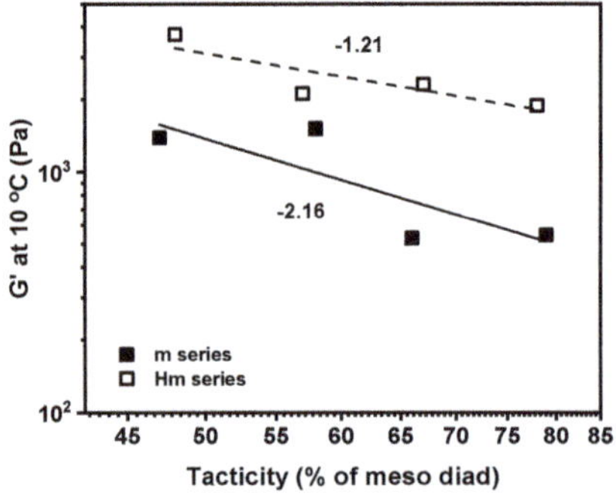

Figure 3. Comparison between G' of m and Hm series at 10°C (heating scan) as function of tacticity.

Two methods are employed to extract the phase transition temperatures from rheological data. First, we adopt the crossover of G' and G'' at dynamic temperature sweep, which has been widely used to define transition temperature in hydro- and iono-gels formed by polymers' phase separation [34,42–52]. Such a method can reflect the scanning rate dependence, making its results comparable with cloud points, which were also measured at 1 °C/min [24]. However, we note that this method suffers the problem of frequency dependence, making it not a rigorous determination of transition point. Therefore, we only refer to this point as "crossover point" in the following discussion. Another method is based on Winter and Chambon criterion [53–56], where the transition point is determined rheologically by finding the temperature at which the storage and loss moduli share the same power-law dependence on frequency. This method is independent of test frequency and can reflect the equilibrium transition temperature, which is insensitive to scanning rate. Lodge and coworkers [44] compared these two methods by investigating the gelation of PNIPAM-*b*-PEO-*b*-PNIPAM triblock copolymers in [EMIM][TFSI], where polymers have a molecular weight of 29,000 g/mol and concentration of 10%, close to our systems. They found both methods generate similar transition temperatures, with the crossover method being 17 °C and the Winter-Chambon criterion being 20 °C. They further observed the consistency of two methods in poly (ethylene-alt-propylene-*b*-ethylene oxide-*b*-*N*-isopropylacrylamide) aqueous solutions [46] and poly (phenylethyl methacrylate-*b*-methyl mechacrylate-*b*-phenylethyl methacrylate) IL solutions [45]. In the next sections, we use both methods to extract transition temperatures and compare their results.

The transition points corresponding to both sol-to-gel and gel-to-sol procedures are first determined by the crossover of G' and G'' and plotted in Figure 4 as a function of meso-diad content (tacticity). Both temperatures monotonically increase with isotacticity, consistent with the tendency of coil-globule transition [23,24], implying the structural factors for phase separation also apply to the gelation procedure. Ray et al. [25] proposed that highly isotactic PNIPAMs have more side groups aligned on the same side of backbones, which improves hydrogen bonding among amide groups and solvophobic association among isopropyl groups in the aggregates. Since aggregates act as the crosslinker within gels, the formation and melting of the gel is affected by tacticity. Likewise, the crossover point for melting PNIPAM/benzyl alcohol gels also increases with isotacticity [34]. In spite of the same tendency in both gels, however, the underlying mechanisms are different. The increase in crossover temperatures for melting PNIPAM/IL gels reflects the increasing difficulty to disaggregate globules, which are formed by the whole polymer chain, whereas the same dependence for PNIPAM/benzyl alcohol is attributed to the increasing number of small-size junctions in the network of transparent gels, which are induced by the partial phase separation of isotactic-rich segments on the chain. The crossover point for melting procedure is higher than that for gelation. With increasing isotacticity, their difference, ΔT_c, is enlarged, which is inconsistent with the transmittance results where hysteresis is insensitive to tacticity [24]. The value of ΔT_c is in the same magnitude as in transmittance transition only when m is lower than 60%. For highly isotactic PNIPAMs ($m > 65\%$), however, ΔT_c in sol-gel transition is significantly larger. This result implies the chain that associates clusters may introduce long-range interactions, which can improve the difficulty in both forming and breaking of clusters. Such a long-range effect is negligible at low isotacticity since the inter-and intra-chain interactions are relatively weak, but becomes important at high isotacticity where polymer interactions are strengthened. The transition temperatures of Hm series are higher than those of m series, which can be attributed to a stronger entropic contribution to the phase transition from high molecular weight [24]. The difference in transition temperatures between Hm and m series also increases with isotacticity, because the intrachain interactions in high-M_n polymers are relatively important and hence prone to be strengthened by the extra interaction from isotactic structures, with respect to low-M_n polymers.

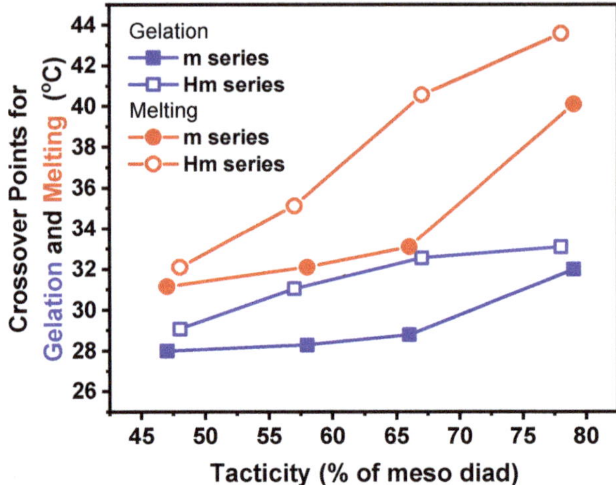

Figure 4. The crossover point for melting (●,○) and gelation (■, □) procedures vs the tacticity for m (solid symbol) and Hm (open symbol) series.

Dynamic frequency sweep was performed at different temperatures by a heating route (Figures S1 and S2 for m and Hm series, respectively, shown in Supporting Information). The melting temperature was determined through Winter-Chambon criterion [53–56]. Since the sol-gel transition point is characterized by the identical power-law dependence on frequency for G' and G'', the frequency-independent loss tangent tanδ (= G''/G') has been extensively employed to determine the critical transition temperature, time, and concentration [57–59]. In this spirit, tanδ is plotted as a function of temperature at different frequencies (Figures S3 and S4 for m and Hm series, respectively, shown in Supporting Information), and the melting temperature was determined via an intersecting or closest point of tanδ curves [60]. Figure 5a illustrates the melting points from Winter-Chambon criterion, G'-G'' crossover points for melting, as well as the clear point from transmittance measurement for m series [24]. The crossover point shares the same tacticity dependence as the clear point, considering both were measured at a 1 °C/min scanning rate. The transition point determined by Winter-Chambon criterion, however, is higher than the G'-G'' crossover point at m = 47~66%, because the percolation of small-size globules from partial phase separation of segments (frequency-independent tanδ) takes place at higher temperature than the further development of the elastic network ($G' > G''$). At m = 79%, the Winter-Chambon transition temperature is close to the G'-G'' crossover value, implying the percolation of globules and the formation of the elastic gel happen simultaneously. Such synchronization could be attributed to the fact that high-m PNIPAMs favor the phase separation of the global chain rather than local segments upon percolation, which bypasses the grow-up of globule junctions and instantly forms an elastic gel. Figure 5b compared the abovementioned three kinds of temperatures for Hm series. Likewise, the crossover points monotonically increase with isotacticity as clear points. The Winter-Chambon melting points are higher than crossover values except for m = 78%, consistent with the m series and can be explained in the same spirit. The average value of all Winter-Chambon melting points in both m and Hm series in Figure 5 is 40 ± 2.2 °C, indicating the percolation temperature is lack of obvious tacticity and molecular weight dependence, different from the monotonic increase in the G'-G'' crossover point, which reflects the formation of elasticity.

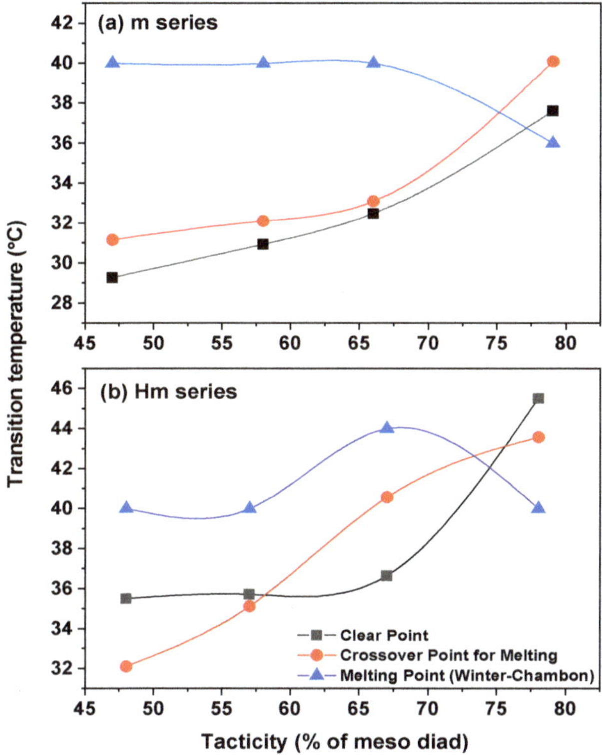

Figure 5. The *G′-G″* crossover point for melting procedure (●), the melting point determined by Winter-Chambon criterion (▲), and the clear point determined by transmittance measurement (■), as a function of the tacticity for (**a**) m and (**b**) Hm series with concentration of 5%.

3.2. Concentration Dependence

The temperature dependence of moduli for m79 at different concentrations is shown in Figure 6. The reversible gelation procedure is observed at all concentrations. At 7.5~12.5%, the recovered moduli are even higher than those before melting. This is probably because the recovered gel bears a better-established network than the original gel, which experienced both long-term storage and structural disturbance during sample loading. The heating transition happens at higher temperatures than cooling transition, with their difference increasing first and then decreasing, reaching the maximum at the intermediate concentration (7.5%). Such concentration dependence of hysteresis is different from the concentration-independent result in the transmittance measurements [24]. The larger hysteresis at intermediate concentrations can be attributed to the strengthened long-range interactions due to the higher possibility of inter-cluster association. However, the increase in the inter-cluster contact is at the cost of intra-cluster association, so at the highest concentrations, the hysteresis is reduced since the inside of clusters is not as strongly bonded as at low concentrations and easier to be disaggregated during heating.

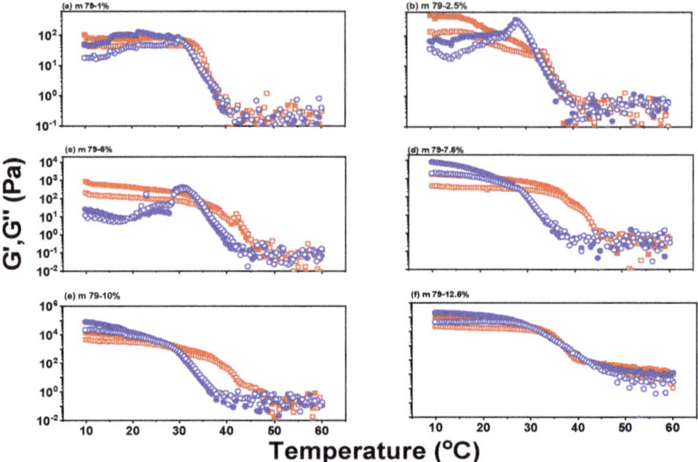

Figure 6. Variations in storage (G', filled symbol) and loss (G'', open symbol) moduli for m79 solutions at concentrations of (**a**) 1%, (**b**) 2.5%, (**c**) 5%, (**d**) 7.5%, (**e**) 10%, and (**f**) 12.5% as a function of temperature at a frequency of $\omega = 10$ rad/s and a strain amplitude of $\gamma = 0.1\%$. Symbols (■, □) and (●, ○) are for heating and cooling sweeps, respectively.

The temperature dependence of moduli for Hm78 at different concentrations is shown in Figure 7. Since Hm series have the higher molecular weight, their moduli at gel regime are higher and the transition regime locates at higher temperature in comparison to m series for entropic reason. Like m series, the gel-state moduli of Hm monotonically increase with concentration, and the hysteresis between heating and cooling exhibits a maximum at the intermediate concentration. Rheological data of Hm are less scattered owing to the high stiffness of the gel. Besides, the recovered moduli are higher than the original ones at high concentrations, consistent with the results in m series, confirming a denser mesh in reestablished networks.

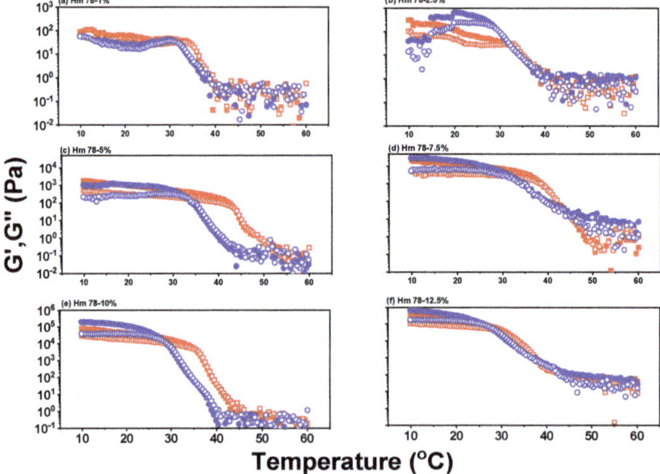

Figure 7. Variations in storage (G', filled symbol) and loss (G'', open symbol) moduli for Hm78 solutions at concentrations of (**a**) 1%, (**b**) 2.5%, (**c**) 5%, (**d**) 7.5%, (**e**) 10%, and (**f**) 12.5% as a function of temperature at a frequency of $\omega = 10$ rad/s and a strain amplitude of $\gamma = 0.1\%$. Symbols (■, □) and (●, ○) are for heating and cooling sweeps, respectively.

The concentration dependence of G' at 10°C on heating curves is shown in Figure 8. With increasing polymer concentrations, the moduli at gel regime gradually increase because the network mesh becomes denser. Unlike entangled polymer solutions [39], there is not a scaling-law type relationship between moduli and concentration. Instead, moduli increase moderately at low concentrations and rise rapidly at higher concentrations. Interestingly, moduli of m79 and Hm78 are similar at low concentration, suggesting that isolated globules, rather than associating chains, are preferentially formed when polymers are scarce, which makes the gel network too loose to be affected by molecular weight. Until concentration is high enough, the association among globules could be established and the gel becomes increasingly strong. This phenomenon is different from that of PNIPAMs in benzyl alcohol, where moduli increase rapidly at low concentrations and then tend to saturate at high concentrations [34]. This is because the partial phase separation mainly generates small-size junctions rather than large globules. The number of junctions significantly increases at low concentrations but approaches saturation at high concentrations. The upturn of data occurs at a lower concentration for Hm78 than m79, because the high molecular weight of Hm78 allows for a lower network percolation concentration. Similar results were also observed in PNIPAM/benzyl alcohol [34], where the plateau G' moduli for Hm78 appears at lower concentrations than that for m79, indicating a lower percolation concentration for Hm.

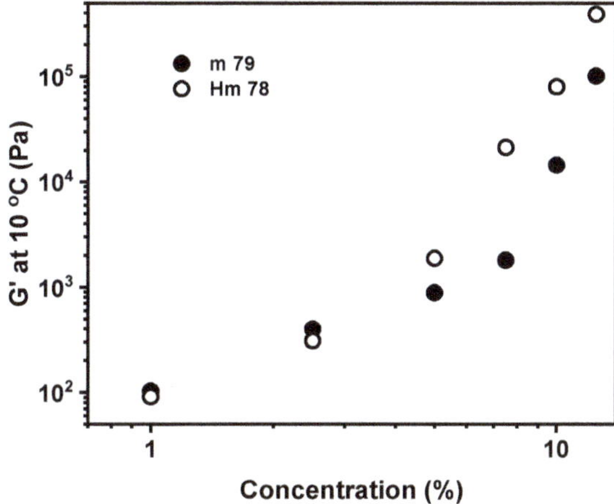

Figure 8. Comparison between G' of m79 and Hm78 at 10 °C as a function of concentration.

The crossover points for melting and gelation procedure are plotted as function of concentration in Figure 9. The two kinds of temperatures exhibit a UCST-type phase diagram, consistent with our previous observation in the transmittance method [24]. This asymmetric UCST phase transition is similar with that of the conventional polymer solutions [39], so an entropy driven dissolution mechanism was expected to be responsible for this behavior. Additionally, the similarity between sol-gel and coil-globule transitions implies the aggregation and disaggregation of clusters govern the gelation and melting procedures, respectively. The crossover points for melting are higher than those for gelation because the disaggregation of clusters is relatively delayed due to extra interactions inside clusters [23,36,37]. Furthermore, the difficulties of both bridging and breaking the inter-cluster associations further enlarge the hysteresis. As for the molecular weight effect, the crossover points for gelation of m79 and Hm78 are approximately overlapped at low and high concentration limits, while at intermediate concentrations, crossover points of Hm are higher than m because of the entropically induced poor solubility for higher-M_n samples. On the other hand, the crossover points for melting of m79 and Hm78 are close at low concentration limit, but at higher concentrations, the peak of m79 locates at the right side of Hm78 samples, with the data at the highest three concentrations being even larger than those of Hm78. This result is in conflict with the cloud-point phase diagram where phase separation temperatures of Hm78 are always higher than those of m79 at a fixed concentration [24]. This anomalous phenomenon could be tentatively attributed to the significant hysteresis for m79 at high concentrations. At a given concentration, the number density of chains in solutions is higher for low-M_n PNIPAMs than high-M_n PNIPAMs. As a result, low-M_n PNIPAMs favors the establishment of the stable interchain-associated aggregates, while high-M_n PNIPAMs are relatively isolated so that they prefer to form the less stable intrachain-associated aggregates [23]. This molecular weight effect seems to be amplified in the sol-gel transition where the stability of interchain associations plays an important role in connecting clusters rather than just agglomerating chains into globules. At last, we note that in PNIPAM/benzyl alcohol [34], the crossover point for gelation monotonically increases with concentration and lacks an UCST peak. Such difference lies in the distinct constitution within two kinds of gels. While the gelation of PNIPAM/IL is governed by the coil-to-globule procedure, the gelation of PNIPAM/benzyl alcohol mainly depends on the number of junctions formed by partial phase separation of isotactic-rich segments, which monotonically increases with polymer concentration and results in a higher transition temperature at high concentrations.

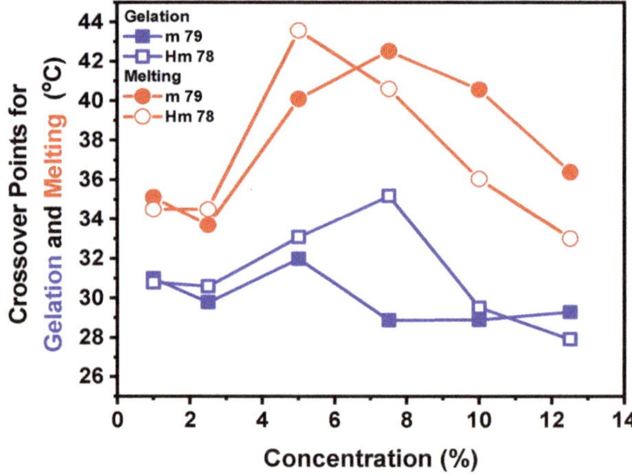

Figure 9. The crossover point for melting (●,○) and gelation (■, □) procedures vs the concentration for m79 (solid symbol) and Hm78 (open symbol) solutions.

In Figure 10, we compared the crossover points for gelation measured by rheology and the cloud points measured by transmittance test. For m79, the crossover-point line intersects the cloud-point line and phase diagram is divided into four kinds of status. At high enough temperatures that are above both crossover-point and cloud-point boundaries, samples are homogenous solutions. With temperatures decreasing, two different transition routes will happen (illustrated in Scheme 2). At two ends of the phase diagram, samples experience gelation first and then become turbid. The status between gelation and phase separation boundaries is called clear gel, while the status after complete phase separation is called opaque gel. Around the UCST peak, cloud points are higher than the crossover temperature. Therefore, samples experience an opaque sol first before they become an opaque gel. Such interference phenomenon between sol-to-gel and coil-to-globule lines was also reported in isotactic-rich PNIPAM in water [32]. In this system, the gelation phase boundary intersects the cloud-point phase boundary at a concentration of 2% and then drops below at higher concentrations. Their phase diagram was divided into four phases, the same as ours. Tanaka et al. [33] proposed a mechanism to predict such a diagram. In their model, the isotactic-rich PNIPAMs contain two kinds of hydrogen-bonding sites, i.e., isotactic diad and syndiotactic diad, with the latter showing a greater bonding strength with solvents than the former. Therefore, when the temperature is approaching the phase separation boundary, isotactic segments on polymers are desolvated preferentially in comparison to stereo-irregular segments. The desolvated isotactic segments are driven into intermolecular junctions even though other segments are still solvated. By the infrared spectroscopic analysis of the v (C–H) region of imidazole ring, Wang et al. [29] experimentally confirmed the desolvation mechanism applies in PNIPAM/IL system, where the interaction experiences a changing process from the dissociation of polymer-ion interaction to the formation of chain-chain bonding, followed by the final chain shrink to aggregation. This mechanism successfully rationalizes the formation of clear gel, which takes place when junctions are formed faster than globules. At concentrations around the CST peak, the formation of globules is faster than the junction, so an opaque sol was formed first across the phase boundary. As gelation further approaches the boundary, the junctions among globules are established so the sol is transformed to an opaque gel.

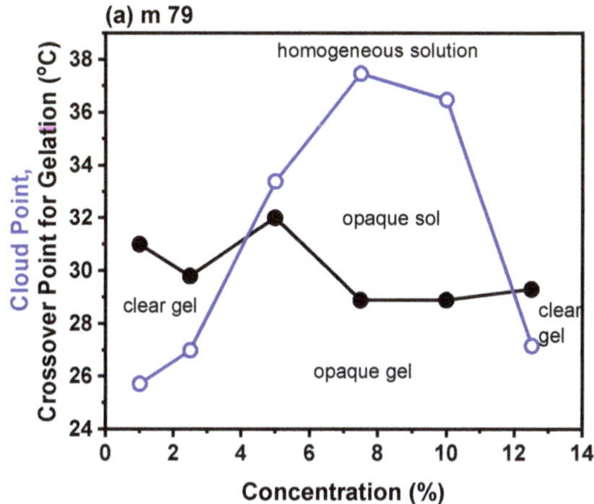

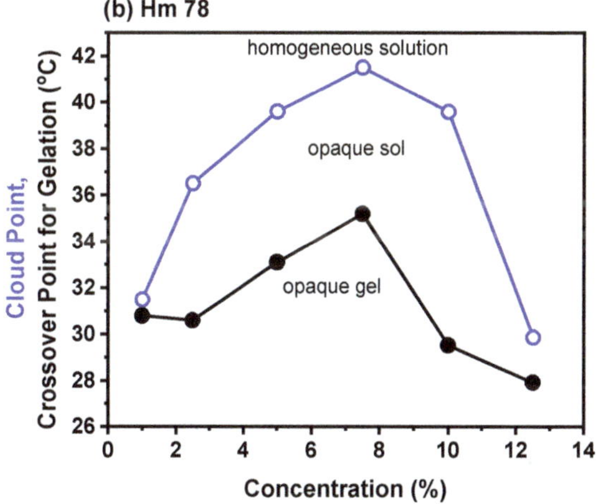

Figure 10. The comparison of the crossover point for gelation (●) and cloud point (○, Reference [24]) for (**a**) m79 and (**b**) Hm78 solutions.

Route I. m series (low and high concentrations)

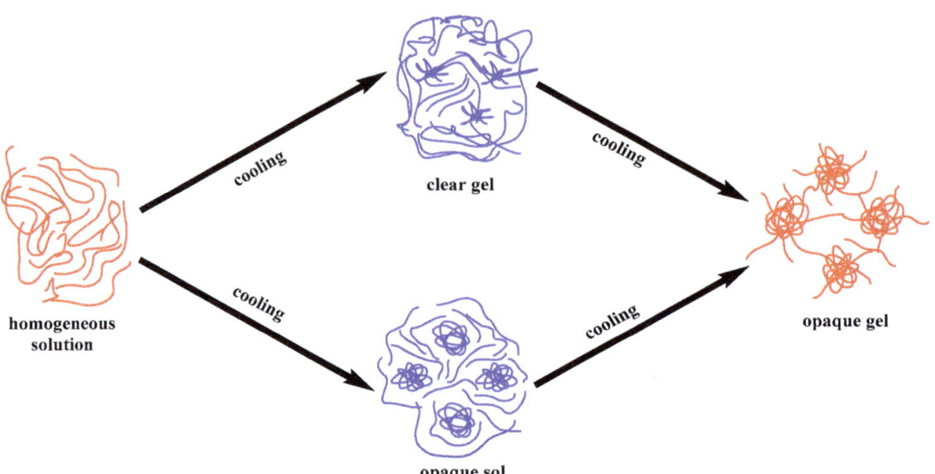

Route II. m series (intermediate concentrations) and Hm series

Scheme 2. The gelation procedure of m79 and Hm78 in [BMIM][TFSI].

For Hm78, all crossover points for gelation are located below the cloud points. As a result, only one transition procedure exists, i.e., samples transfer from solutions to opaque sols and then to opaque gels with temperature decreasing. The absence of the clear gel phase could be attributed to the poorer solubility of high-M_n polymers for entropic reasons, which makes formation of globules as fast as junctions. The crossover-point phase diagram has a similar shape with the cloud-point one. That implies the formation of globules is a dominating procedure for the following gelation.

The dynamic frequency sweep at different temperatures was performed on m79 and Hm78 solutions of different concentrations (Figure S5 for m79 and Figure S6 for Hm78 solutions, shown in Supporting Information). The melting temperature based on Winter-Chambon criterion was extracted from the intersecting or closest point of tanδ curves as a function of temperature at different frequencies (Figure S7 for m79 and Figure S8 for Hm78 solutions, shown in Supporting Information). In Figure 11a, the phase diagram of m79 solutions based on Winter-Chambon melting points is established, along with the G'-G'' crossover points for melting and clear points from transmittance measurement. At all investigated concentrations, the difference of transition temperatures between Winter-Chambon and G'-G'' crossover definitions is as small as 4°C, which is within experimental uncertainty. Similar with the case in Figure 10a, both the melting-point line (Winter-Chambon) and the crossover-point line intersect the clear-point line, so the phase diagram is divided into four parts, i.e., homogeneous solution, clear gel, opaque sol, and opaque gel. Such consistency in gelation and melting phase diagrams confirms the validity of the phase transition procedure in Scheme 2. For Hm78 solutions, the phase diagram based on the aforementioned three kinds of transition temperatures is shown in Figure 11b. Again, the difference between Winter-Chambon melting and G'-G'' crossover points is within 4 °C. All melting points are lower than clear points except the Winter-Chambon value in 12.5% solution, which, however, could be attributed to the experimental uncertainty. Therefore, the validity of gelation phase diagram in Figure 10b is basically confirmed by Figure 11b, in which the phase mainly transfers from homogeneous solutions, to opaque sols, and then to opaque gels, or conversely.

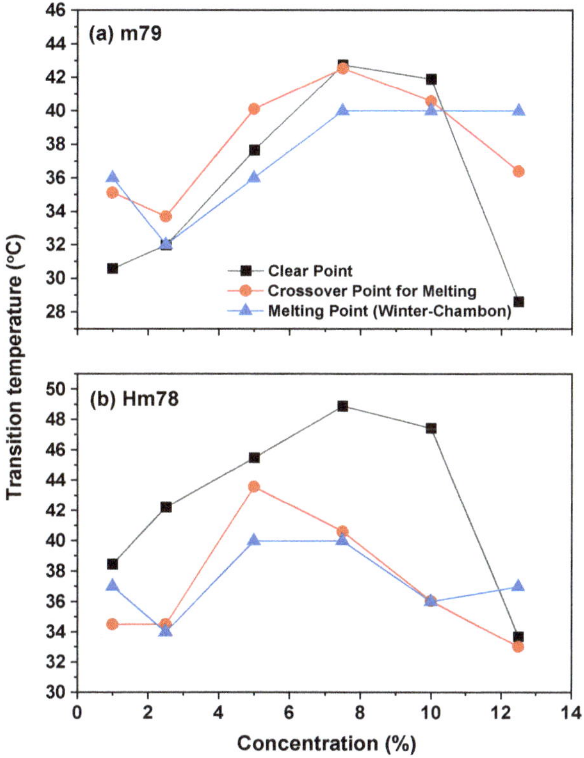

Figure 11. The G'-G'' crossover point for melting (●), the melting point determined by Winter-Chambon criterion (▲), and the clear point determined by transmittance measurement (■), as a function of the concentration for (**a**) m79 and (**b**) Hm78 solutions.

4. Conclusions

The sol-gel transition of isotactic-rich PNIPAMs in ionic liquid was investigated by rheological technique. The G'-G'' crossover points for both gelation and melting procedures increase with the meso-diad content, consistent with the coil-globule transition determined by cloud points at the same scanning rate, while the melting point based on Winter-Chambon criterion is of a higher average value (40 ± 2.2 °C) and approximately tacticity and M_n independent because it corresponds to the percolation point rather than the establishment of elasticity for gels ($G' > G''$). The UCST-type phase diagrams composed of both crossover points for gelation and cloud points are determined. For low-M_n PNIPAMs (m79), the crossover-point line intersects the cloud-point line, so the phase diagram is divided into four parts: solution, clear gel, opaque sol, and opaque gel. The clear gel is formed because the isotactic segments on the chain experience desolvation preferentially and then act as junctions to connect the solvated segments. When the junction formation is faster than the globule formation, a clear gel appears. Otherwise, an opaque sol will be formed instead of the clear gel prior to the formation of the opaque gel. For high-M_n PNIPAMs (Hm78), crossover points are below cloud points at all investigated concentrations, so solutions transfer to opaque sol first and then opaque gel with decreasing temperatures. This difference in phase diagram could be attributed to the poorer solubility of high-M_n polymers for entropic reason. Phase diagrams composed of Winter-Chambon melting points, G'-G'' crossover points for melting, and clear points follow the similar tendency as those based on gelation points, confirming the above transition procedure.

Polymers **2019**, *11*, 783

Supplementary Materials: The following are available online at http://www.mdpi.com/2073-4360/11/5/783/s1, Figure S1: The storage (closed symbol) and loss (open symbol) moduli of m series with concentration of 5% and different tacticity, Figure S2: The storage (closed symbol) and loss (open symbol) moduli of Hm series with concentration of 5% and different tacticity, Figure S3: The loss factor as a function of temperature for m series with concentration of 5% and different tacticity, Figure S4: The loss factor as a function of temperature for Hm series with concentration of 5% and different tacticity. Symbols are the same as in Figure S3, Figure S5: The storage (closed symbol) and loss (open symbol) moduli of m79 solutions with different concentration, Figure S6: The storage (closed symbol) and loss (open symbol) moduli of Hm78 solutions with different concentration, Figure S7: The loss factor as a function of temperature for m79 solutions with different concentration. Symbols are the same as in Figure S3, Figure S8: The loss factor as a function of temperature for Hm78 solutions with different concentration. Symbols are the same as in Figure S3.

Author Contributions: Conceptualization, Z.C.Y.; Synthesis, characterization, and rheology, C.S.B.; Data preparation, C.S.B. and Y.Z.C.; Writing—original draft, Z.C.Y.; Writing—review & editing, Z.C.Y., C.S.B., and F.J.S.; Funding acquisition, Z.C.Y. and F.J.S.

Funding: This research was funded by the National Natural Science Foundation of China (Grant No. 21803039), the Natural Science Foundation of SZU (Grant No. 2017002), and the Nanshan District Key Lab for Biopolymers and Safety Evaluation (No. KC2014ZDZJ0001A).

Acknowledgments: This work was supported by the National Natural Science Foundation of China (Grant No. 21803039), the Natural Science Foundation of SZU (Grant No. 2017002), and the Nanshan District Key Lab for Biopolymers and Safety Evaluation (No. KC2014ZDZJ0001A).

Conflicts of Interest: The authors declare no conflict of interest.

References and Note

1. Stuart, M.A.C.; Huck, W.T.S.; Genzer, J.; Müller, M.; Ober, C.; Stamm, M.; Sukhorukov, G.B.; Szleifer, I.; Tsukruk, V.V.; Urban, M.; et al. Emerging applications of stimuli-responsive polymer materials. *Nat. Mater.* **2010**, *9*, 101. [CrossRef]

2. Liu, F.; Urban, M.W. Recent advances and challenges in designing stimuli-responsive polymers. *Prog. Polym. Sci.* **2010**, *35*, 3–23. [CrossRef]

3. Hirokawa, Y.; Tanaka, T. Volume phase transition in a nonionic gel. *J. Chem. Phys.* **1984**, *81*, 6379–6380. [CrossRef]

4. Gil, E.S.; Hudson, S.M. Stimuli-reponsive polymers and their bioconjugates. *Prog. Polym. Sci.* **2004**, *29*, 1173–1222. [CrossRef]

5. Keerl, M.; Smirnovas, V.; Winter, R.; Richtering, W. Interplay between hydrogen bonding and macromolecular architecture leading to unusual phase behavior in thermosensitive microgels. *Angew. Chem. Int. Ed.* **2008**, *47*, 338–341. [CrossRef] [PubMed]

6. Schild, H.G. Poly(*N*-isopropylacrylamide): Experiment, theory and application. *Prog. Polym. Sci.* **1992**, *17*, 163–249. [CrossRef]

7. Lutz, J.-F.; Akdemir, Ö.; Hoth, A. Point by point comparison of two thermosensitive polymers exhibiting a similar lcst: is the age of poly(NIPAM) over? *J. Am. Chem. Soc.* **2006**, *128*, 13046–13047. [CrossRef]

8. Rogers, R.D.; Seddon, K.R. Ionic liquids -solvents of the future? *Science* **2003**, *302*, 792–793. [CrossRef] [PubMed]

9. Lu, J.; Yan, F.; Texter, J. Advanced applications of ionic liquids in polymer science. *Prog. Polym. Sci.* **2009**, *34*, 431–448. [CrossRef]

10. Ueki, T.; Watanabe, M. Macromolecules in ionic liquids: Progress, challenges, and opportunities. *Macromolecules* **2008**, *41*, 3739–3749. [CrossRef]

11. Wang, H.; Wang, Z.; Yang, J.; Xu, C.; Zhang, Q.; Peng, Z. Ionic gels and their applications in stretchable electronics. *Macromol. Rapid Commun.* **2018**, *39*, 1800246. [CrossRef] [PubMed]

12. Lodge, T.P.; Ueki, T. Mechanically tunable, readily processable ion gels by self-assembly of block copolymers in ionic liquids. *Acc. Chem. Res.* **2016**, *49*, 2107–2114. [CrossRef]

13. Lee, H.-N.; Lodge, T.P. Lower critical solution temperature (LCST) phase behavior of poly(ethylene oxide) in ionic liquids. *J. Phys. Chem. Lett.* **2010**, *1*, 1962–1966. [CrossRef]

14. Kodama, K.; Nanashima, H.; Ueki, T.; Kokubo, H.; Watanabe, M. Lower critical solution temperature phase behavior of linear polymers in imidazolium-based ionic liquids: Effects of structural modifications. *Langmuir* **2009**, *25*, 3820–3824. [CrossRef] [PubMed]

15. He, Y.Y.; Boswell, P.G.; Buhlmann, P.; Lodge, T.P. Ion gels by self-assembly of a triblock copolymer in an ionic liquid. *J. Phys. Chem. B* **2007**, *111*, 4645–4652. [CrossRef]

16. Ueki, T.; Watanabe, M.; Lodge, T.P. Doubly thermosensitive self-assembly of diblock copolymers in ionic liquids. *Macromolecules* **2009**, *42*, 1315–1320. [CrossRef]

17. Noro, A.; Matsushima, S.; He, X.D.; Hayashi, M.; Matsushita, Y. Thermoreversible supramolecular polymer gels via metal-ligand coordination in an ionic liquid. *Macromolecules* **2013**, *46*, 8304–8310. [CrossRef]

18. Ueki, T.; Usui, R.; Kitazawa, Y.; Lodge, T.P.; Watanabe, M. Thermally reversible ion gels with photohealing properties based on triblock copolymer self-assembly. *Macromolecules* **2015**, *48*, 5928–5933. [CrossRef]

19. Zhang, Y.D.; Fan, X.H.; Shen, Z.H.; Zhou, Q.F. Thermoreversible ion gel with tunable modulus self-assembled by a liquid crystalline triblock copolymer in ionic liquid. *Macromolecules* **2015**, *48*, 4927–4935. [CrossRef]

20. Tamate, R.; Hashimoto, K.; Ueki, T.; Watanabe, M. Block copolymer self-assembly in ionic liquids. *Phys. Chem. Chem. Phys.* **2018**, *20*, 21803–21808. [CrossRef]

21. Ueki, T.; Watanabe, M. Upper critical solution temperature behavior of poly(*N*-isopropylacrylamide) in an ionic liquid and preparation of thermo-sensitive nonvolatile gels. *Chem. Lett.* **2006**, *35*, 964–965. [CrossRef]

22. Asai, H.; Fujii, K.; Ueki, T.; Sawamura, S.; Nakamura, Y.; Kitazawa, Y.; Watanabe, M.; Han, Y.-S.; Kim, T.-H.; Shibayama, M. Structural study on the UCST-type phase separation of poly(*N*-isopropylacrylamide) in ionic liquid. *Macromolecules* **2013**, *46*, 1101–1106. [CrossRef]

23. De Santis, S.; La Mesa, C.; Masci, G. On the upper critical solution temperature of PNIPAAM in an ionic liquid: Effect of molecular weight, tacticity and water. *Polymer* **2017**, *120*, 52–58. [CrossRef]

24. Biswas, C.S.; Stadler, F.J.; Yan, Z.-C. Tacticity effect on the upper critical solution temperature behavior of poly(*N*-isopropylacrylamide) in an imidazolium ionic liquid. *Polymer* **2018**, *155*, 101–108. [CrossRef]

25. Ray, B.; Okamoto, Y.; Kamigaito, M.; Sawamoto, M.; Seno, K.; Kanaoka, S.; Aoshima, S. Effect of tacticity of poly(*N*-isopropylacrylamide) on the phase separation temperature of its aqueous solutions. *Polym. J.* **2005**, *37*, 234–237. [CrossRef]

26. He, Y.; Lodge, T.P. Thermoreversible ion gels with tunable melting temperatures from triblock and pentablock copolymers. *Macromolecules* **2008**, *41*, 167–174. [CrossRef]

27. Lee, H.N.; Bai, Z.F.; Newell, N.; Lodge, T.P. Micelle/inverse micelle self-assembly of a PEO-PNIPAM block copolymer in ionic liquids with double thermoresponsivity. *Macromolecules* **2010**, *43*, 9522–9528. [CrossRef]

28. Ueki, T.; Nakamura, Y.; Usui, R.; Kitazawa, Y.; So, S.; Lodge, T.P.; Watanabe, M. Photoreversible gelation of a triblock copolymer in an ionic liquid. *Angew. Chem. Int. Ed.* **2015**, *54*, 3018–3022. [CrossRef]

29. Wang, Z.; Wu, P. Spectral insights into gelation microdynamics of PNIPAM in an ionic liquid. *J. Phys. Chem. B* **2011**, *115*, 10604–10614. [CrossRef]

30. Ueki, T.; Nakamura, Y.; Yamaguchi, A.; Niitsuma, K.; Lodge, T.P.; Watanabe, M. UCST phase transition of azobenzene-containing random copolymer in an ionic liquid. *Macromolecules* **2011**, *44*, 6908–6914. [CrossRef]

31. So, S.; Hayward, R.C. Tunable upper critical solution temperature of poly(*N*-isopropylacrylamide) in ionic liquids for sequential and reversible self-folding. *ACS Appl. Mater. Interfaces* **2017**, *9*, 15785–15790. [CrossRef]

32. Nakano, S.; Ogiso, T.; Kita, R.; Shinyashiki, N.; Yagihara, S.; Yoneyama, M.; Katsumoto, Y. Thermoreversible gelation of isotactic-rich poly(*N*-isopropylacrylamide) in water. *J. Chem. Phys.* **2011**, *135*, 114903. [CrossRef] [PubMed]

33. Tanaka, F.; Katsumoto, Y.; Nakano, S.; Kita, R. LCST phase separation and thermoreversible gelation in aqueous solutions of stereo-controlled poly(*N*-isopropylacrylamide)s. *React. Funct. Polym.* **2013**, *73*, 894–897. [CrossRef]

34. Biswas, C.S.; Wu, Y.; Wang, Q.; Du, L.; Mitra, K.; Ray, B.; Yan, Z.-C.; Du, B.; Stadler, F.J. Effect of tacticity and molecular weight on the rheological properties of poly(*N*-isopropylacrylamide) gels in benzyl alcohol. *J. Rheol.* **2017**, *61*, 1345–1357. [CrossRef]

35. Hashmi, S.; Vatankhah-Varnoosfaderani, M.; GhavamiNejad, A.; Obiweluozor, F.O.; Du, B.; Stadler, F.J. Self-associations and temperature dependence of aqueous solutions of zwitterionically modified *N*-isopropylacrylamide copolymers. *Rheol. Acta* **2015**, *54*, 501–516. [CrossRef]

36. Cheng, H.; Shen, L.; Wu, C. LLS and FTIR studies on the hysteresis in association and dissociation of poly(*N*-isopropylacrylamide) chains in water. *Macromolecules* **2006**, *39*, 2325–2329. [CrossRef]

37. Wu, C. A comparison between the "coil-to-globule" transition of linear chains and the "volume phase transition" of spherical microgels. *Polymer* **1998**, *39*, 4609–4619. [CrossRef]

38. Tong, Z.; Zeng, F.; Zheng, X.; Sato, T. Inverse molecular weight dependence of cloud points for aqueous poly(*N*-isopropylacrylamide) solutions. *Macromolecules* **1999**, *32*, 4488–4490. [CrossRef]

39. Rubinstein, M.; Colby, R.H. *Polymer Physics*; Oxford University Press: New York, NY, USA, 2003.

40. Stadler, F.J. Quantifying primary loops in polymer gels by linear viscoelasticity. *Proc. Natl. Acad. Sci. USA* **2013**, *110*, E1972. [CrossRef]

41. Zhou, H.; Woo, J.; Cok, A.M.; Wang, M.; Olsen, B.D.; Johnson, J.A. Counting primary loops in polymer gels. *Proc. Natl. Acad. Sci. USA* **2012**, *109*, 19119. [CrossRef]

42. Noro, A.; Matsushita, Y.; Lodge, T.P. Thermoreversible supramacromolecular ion gels via hydrogen bonding. *Macromolecules* **2008**, *41*, 5839–5844. [CrossRef]

43. Noro, A.; Matsushita, Y.; Lodge, T.P. Gelation mechanism of thermoreversible supramacromolecular ion gels via hydrogen bonding. *Macromolecules* **2009**, *42*, 5802–5810. [CrossRef]

44. He, Y.Y.; Lodge, T.P. A thermoreversible ion gel by triblock copolymer self-assembly in an ionic liquid. *Chem. Commun.* **2007**, *26*, 2732–2734. [CrossRef]

45. Kitazawa, Y.; Ueki, T.; Imaizumi, S.; Lodge, T.P.; Watanabe, M. Tuning of sol gel transition temperatures for thermoreversible ion gels. *Chem. Lett.* **2014**, *43*, 204–206. [CrossRef]

46. Zhou, C.; Hillmyer, M.A.; Lodge, T.P. Efficient formation of multicompartment hydrogels by stepwise self-assembly of thermoresponsive abc triblock terpolymers. *J. Am. Chem. Soc.* **2012**, *134*, 10365–10368. [CrossRef]

47. Verber, R.; Blanazs, A.; Armes, S.P. Rheological studies of thermo-responsive diblock copolymer worm gels. *Soft Matter* **2012**, *8*, 9915–9922. [CrossRef]

48. Hamley, I.W. *Block Copolymers in Solution: Fundamentals and Applications*; John Wiley and Sons: Chichester, UK, 2005.

49. Fu, W.; Bai, W.; Jiang, S.; Seymour, B.T.; Zhao, B. UCST-type thermoresponsive polymers in synthetic lubricating oil polyalphaolefin (PAO). *Macromolecules* **2018**, *51*, 1674–1680. [CrossRef]

50. Tamate, R.; Usui, R.; Hashimoto, K.; Kitazawa, Y.; Kokubo, H.; Watanabe, M. Photo/thermoresponsive ABC triblock copolymer-based ion gels: Photoinduced structural transitions. *Soft Matter* **2018**, *14*, 9088–9095. [CrossRef]

51. Xuan, S.; Lee, C.-U.; Chen, C.; Doyle, A.B.; Zhang, Y.; Guo, L.; John, V.T.; Hayes, D.; Zhang, D. Thermoreversible and injectable ABC polypeptoid hydrogels: Controlling the hydrogel properties through molecular design. *Chem. Mat.* **2016**, *28*, 727–737. [CrossRef]

52. Drzal, P.L.; Shull, K.R. Origins of mechanical strength and elasticity in thermally reversible, acrylic triblock copolymer gels. *Macromolecules* **2003**, *36*, 2000–2008. [CrossRef]

53. Winter, H.H. Can the gel point of a cross-linking polymer be detected by the G'–G'' crossover? *Polym. Eng. Sci.* **1987**, *27*, 1698–1702. [CrossRef]

54. Chambon, F.; Winter, H.H. Linear viscoelasticity at the gel point of a crosslinking PDMS with imbalanced stoichiometry. *J. Rheol.* **1987**, *31*, 683–697. [CrossRef]

55. Winter, H.H.; Chambon, F. Analysis of linear viscoelasticity of a crosslinking polymer at the gel point. *J. Rheol.* **1986**, *30*, 367–382. [CrossRef]

56. Chambon, F.; Petrovic, Z.S.; MacKnight, W.J.; Winter, H.H. Rheology of model polyurethanes at the gel point. *Macromolecules* **1986**, *19*, 2146–2149. [CrossRef]

57. Liu, C.; Zhang, J.; He, J.; Hu, G. Gelation in carbon nanotube/polymer composites. *Polymer* **2003**, *44*, 7529–7532. [CrossRef]

58. Sun, F.; Huang, Q.; Wu, J. Rheological behaviors of an exopolysaccharide from fermentation medium of a Cordyceps sinensis fungus (Cs-HK1). *Carbohydr. Polym.* **2014**, *114*, 506–513. [CrossRef]

59. Mo, G.; Zhang, R.; Wang, Y.; Yan, Q. Rheological and optical investigation of the gelation with and without phase separation in PAN/DMSO/H_2O ternary blends. *Polymer* **2016**, *84*, 243–253. [CrossRef]

60. In some cases, the intersection cannot be reached because the phase angle δ above 40 °C approaches 90°, which is beyond the resolution of rheometer and thus generates unreasonably high tanδ value. Therefore, we select 40 °C, at which tanδ points are closest, as their transition points.

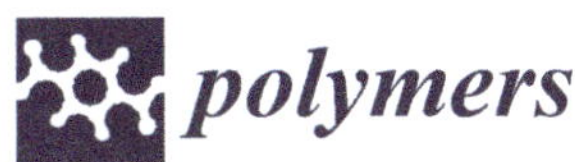 

Article

Effect of Hydrophobic Interactions on Lower Critical Solution Temperature for Poly(*N*-isopropylacrylamide-co-dopamine Methacrylamide) Copolymers

Alberto García-Peñas [1,2], Chandra Sekhar Biswas [1], Weijun Liang [1], Yu Wang [1], Pianpian Yang [3,*] and Florian J. Stadler [1,*]

[1] College of Materials Science and Engineering, Shenzhen Key Laboratory of Polymer Science and Technology, Guangdong Research Center for Interfacial Engineering of Functional Materials, Nanshan District Key Laboratory for Biopolymers and Safety Evaluation, Shenzhen University, Shenzhen 518055, China; alberto@szu.edu.cn (A.G.-P.); chandra123@szu.edu.cn (C.S.B.); weijunliang1116@hotmail.com (W.L.); one.wangyucoak@outlook.com (Y.W.)

[2] Key Laboratory of Optoelectronic Devices and Systems of Ministry of Education and Guangdong Province, College of Optoelectronic Engineering, Shenzhen University, Shenzhen 518060, China

[3] Department of Management, Shenzhen University, Shenzhen 518060, China

* Correspondence: yangpianpian2008@hotmail.com (P.Y.); fjstadler@szu.edu.cn (F.J.S.); Tel.: +86-0755-2653-5179 (P.Y.); +86-0755-8671-3986 (F.J.S.)

Received: 16 April 2019; Accepted: 23 May 2019; Published: 4 June 2019

Abstract: For the preparation of thermoresponsive copolymers, for e.g., tissue engineering scaffolds or drug carriers, a precise control of the synthesis parameters to set the lower critical solution temperature (LCST) is required. However, the correlations between molecular parameters and LCST are partially unknown and, furthermore, LCST is defined as an exact temperature, which oversimplifies the real situation. Here, random N-isopropylacrylamide (NIPAM)/dopamine methacrylamide (DMA) copolymers were prepared under a systematical variation of molecular weight and comonomer amount and their LCST in water studied by calorimetry, turbidimetry, and rheology. Structural information was deduced from observed transitions clarifying the contributions of molecular weight, comonomer content, end-group effect or polymerization degree on LCST, which were then statistically modeled. This proved that the LCST can be predicted through molecular structure and conditions of the solutions. While the hydrophobic DMA lowers the LCST especially the onset, polymerization degree has an important but smaller influence over all the whole LCST range.

Keywords: *N*-isopropylacrylamide; lower critical solution temperature; thermoresponsive polymers; hydrophobic interactions; statistical modeling

---

1. Introduction

The research on smart polymers is growing, owing to new advances of the scientific community as well as current and future applications. Some thermoresponsive polymers present an extraordinary sensitivity to external stimuli related to temperature, pH, light, or solvents, among others. These properties opened new interesting applications regarding to scaffolds or drug carriers [1–4].

The use of poly(*N*-isopropylacrylamide) (PNIPAM) as a water-soluble polymer is extended along numerous publications because its lower critical solution temperature (LCST) is near human physiological temperature (around 33 °C) [5,6]. Generally, a homogenous and transparent solution is observed below LCST, but higher temperatures lead to insolubility of the polymer in water. This transition can be explained based on hydrogen bonding and hydrophobic interactions between polymer and water, whose balance is temperature dependent [7–9].

Various ways for varying LCST behavior are used in practice—such as adding comonomer, modification of the end group or introduction of additives [10]. Block copolymers (vs. random copolymers) can provide a second phase transition promoting new features to the final material and according to the monomers involved, but synthesis can require more steps than making other structures.

Generally, the preparation of random copolymers is used as a valuable method for adding new functionalities or increasing the range of properties. In this way, bioinspired materials grow a great interest owing to simulate the living tissues very useful for developing specific properties and getting non-invasive therapies. For instance, the use of catechol groups was inspired by marine mussels, which produce proteinaceous adhesive materials for attaching themselves to rocks in the intertidal zone [11]. The catechol groups show a great versatility in terms of applications such as biosensors or biomedicine [12–17]. Moreover, the use of dopamine methacrylamide (DMA) as a comonomer is suitable due to its main chain structure being similar to N-isopropylacrylamide (NIPAM) [18].

The introduction of reversible addition-fragmentation chain transfer (RAFT) polymerization allowed getting custom polymers due to a higher control on the final molecular features, namely polydispersity and molecular weight [19,20]. Additionally, this potential method is convenient for preparing copolymers, whose properties, such as the LCST, can be tailored to the specific necessities. At present, the influence of some parameters on LCST is clear but there are still many inconsistences related to the influence of molecular weights and end-group effects, among others. Important information associated with LCST transitions is missed, which could partially explain those influences. In general, all research works are focused on a single temperature value for defining the LCST, but information about the range of temperature or time of this process is obviated. Thus, a complete evaluation of LCST transition could be very useful for this purpose.

There are different procedures for detecting LCST as turbidimetry, calorimetry, proton nuclear magnetic resonance, rheology, or dynamic light scattering, which provide rich information. Specifically, calorimetric analysis can provide an estimate of the number of hydrogen bonds involved on this process [21]. In terms of applications, slow LCST transitions could be desired for a specific slow drug release, while fast LCST transitions could be useful for biosensors.

This work is focused on the preparation of several random copolymers, where molecular weight and comonomer composition were accurately varied in line to investigate the influence of these parameters on LCST transition. The use of different characterization methods (calorimetry analysis, UV-visible spectroscopy, and rheology) allowed for detecting the sudden change from hydrophilic to hydrophobic behavior, i.e., the LCST. Those changes will be modeled with a statistical regression analysis to check for the diverse contributions of parameters involved in LCST.

2. Experimental

2.1. Materials

Sodium borate (99%, Macklin Reagent Company, Shanghai, China, sodium bicarbonate (99.8%, Macklin Reagent Company, Shanghai, China), 3,4-dihydroxyphenethylamine hydrochloride (98%, Sigma-Aldrich, Hamburg, Germany), azobisisobutyronitrile (99%, Aladdin, Shanghai, China) were used without pretreatment. Diverse solvents, as tetrahydrofuran (99.9%, Aladdin, Shanghai, China) and N,N-dimethylformamide (99%, Aladdin, Shanghai, China), were distilled under sodium and calcium chloride with nitrogen bubbling. N-Isopropylacrylamide (98%, Aladdin, Shanghai, China) was recrystallized from a mixture of hexane and benzene (65:35). The RAFT-agent I-phenylethyl phenyldithioacetate was prepared according to literature [22].

2.2. Synthesis of Dopamine Methacrylamide (DMA)

The preparation of dopamine methacrylamide (DMA) was carried out according to the procedure of Glass et al. [23]. The resulting powder was purified under a solution of methyl acetate (40 mL), and subsequently, the obtained monomer was precipitated in 600 mL hexane.

2.3. Synthesis of Random Poly(NIPAM-co-DMA) Copolymers

The different RAFT copolymerizations were carried out under inert conditions in dry Schlenk tubes where *N*-isopropylacrylamide (NIPAM) and DMA were placed. Different amounts of azobisisobutyronitrile (AIBN) and I-phenylethyl phenyldithioacetate (PEPD) were used as initiator and RAFT agent, respectively. Then, *N,N*-dimethylformamide was added as a solvent and the mixture was kept in a nitrogen environment using a Schlenk system. Subsequently, Schlenk tubes were placed in a thermostatted bath at 70 °C for 48 h, and reactions were stopped by freezing in liquid nitrogen.

The random copolymers (SIScheme 1) were purified three times by precipitation in diethyl ether, and finally, samples were dried under vacuum for 48 h. The samples were stored at room temperature.

Samples were denominated according to the DMA content and the molecular weight. For example, C5M4000 is associated with a DMA content of 5.5 mol % and a molecular weight of 4300 g/mol. The NIPAM homopolymers were prepared under similar conditions and used as reference. These were exclusively denominated according to the molecular weight.

Aqueous polymer solutions with 4, 8, 10, and 15 wt % concentrations were prepared and stored in a refrigerator for 12 h to ensure the complete dissolution before LCST analysis.

2.4. Analytical Methods

2.4.1. Nuclear Magnetic Resonance

The DMA content was estimated by proton nuclear magnetic resonance for the diverse copolymers (Table 1). Proton nuclear magnetic resonance spectra were recorded with an AVANCE III 600 MHz spectrometer (Bruker, Switzerland) at 25 °C using deuterated DMSO as a solvent [24].

Table 1. Data of reversible addition-fragmentation chain transfer (RAFT)-polymerization runs and molecular features.

Samples	Monomer Feed Ratio	DMA	M_n *	PDI (M_w/M_n) *	T_g
		mol %	g/mol	-	°C
M7000	-	-	7000	1.12	131.4
C5M4000	0.06	5.5	4300	1.16	129.0
C5M24000	0.06	5.3	23,900	1.45	137.6
C1M18000	0.0075	0.6	17,900	1.49	134.8
C5M18000	0.06	5.9	17,500	1.44	132.1
M35000	-	-	35,400	1.25	137.7

* The molecular weight (Mn) and polydispersity (PDI) were determined from gel permeation chromatography calibrated with narrow molar mass distribution polystyrenes.

2.4.2. Gel Permeation Chromatography (GPC)

The molecular weights and polydispersities (PDI) were estimated by size exclusion chromatography in a Beijing Wenfen LC98IIRI (Beijing, China; Table 1). Two polystyrene gel columns (Shodex, KD-803 and KD-806, Detector: RI-201H) were used, and tetrahydrofuran was selected as a solvent. The measurements were carried out at 40 °C and at a flow rate of 1 mL/min. Narrow molecular mass distribution polystyrene standards were used to calibrate the experiments. In our previous paper [11], we found that a polystyrene-calibrated GPC produces ca. 35% lower values for copolymers with 2.5 mol% and 5 mol% DMA in comparison to vapor pressure osmometry—an absolute method for the determination of molar mass M_n. Hence, while the PS-calibrated values are not absolute, they can be considered to be good approximations of the absolute values.

2.4.3. Thermogravimetric Analysis

Thermo-gravimetric (TGA—TA Instruments Q50, New Castle, PA, USA.) analysis was used for the confirmation of the real sample concentration in water. The polymer content was evaluated in a broad range of temperature defined between 25 and 250 °C at a heating rate of 20 °C/min in nitrogen atmosphere.

2.4.4. Differential Scanning Calorimetry (DSC)

The calorimetric analysis was performed in a Q200 Differential Scanning Calorimeter (TA Instruments, New Castle, PA, USA) with a cooling system. The sample weight was around 5 mg and the machine was calibrated with different standards.

The temperature of the phase transition was analyzed at a heating rate of 5 °C/min in the range of 0–60 °C for the diverse polymer solutions in water in sealed crucibles to avoid evaporations. Three scans were performed for each sample to evaluate the reproducibility of results. The resulting data were normalized, and the baseline was corrected with another experiment performed in the same conditions with the same amount of pure water.

Moreover, the glass transition temperature was estimated by calorimetric analysis. The experiments were carried out at 20 °C/min with a flow of 40 mL/min nitrogen from 20 to 250 °C.

2.4.5. UV-Visible Spectroscopy

The thermal transitions were studied using a UV-vis spectrophotometer, PerkinElmer UV/VIS Lambda 365 (Seoul, Korea), with temperature control. The transmittance of the diverse polymer solutions in water was tested in a wide range of temperatures from 15 to 40 °C with a heating rate of 1 °C/min at 500 nm. Finally, the cloud points were calculated at 50% transmittance for each polymer solution.

2.4.6. Rheological Measurements

LCSTs were also evaluated by rheological measurements with a cone plate geometry (15 mm/2°) where temperature dependence was evaluated with $q = 1$ K min^{-1}, $\omega = 0.16$ s^{-1}, and $\gamma_0 = 2$–5% under linear conditions in an Anton Paar MCR 302 rheometer (Graz, Austria) in a humidity saturated atmosphere.

2.5. Statistical Modeling

To analyze the relationship between diverse values associated with the LCST (onset temperatures T_{onset}, peak temperatures T_{peak}, cloud temperatures T_{cloud}, offset temperatures T_{onset}, and the LCST ranges LSCT$_{range}$), obtained from calorimetry and turbidimetry, and the characteristics of the polymeric solutions (comonomer content, polymerization degree, and the polymer concentration in water) a regression analysis was done by the SPSS software.

3. Results

3.1. Synthesis and Molecular Characterization

The DMA content was estimated by ^{1}H-NMR for the copolymers in d$_6$-DMSO. The proton signals were elucidated in relation to the data reported in the literature where a good equivalence was observed (Figure S1) [11]. The integration of NH-CH-(CH$_3$)$_2$ ($\delta = 3.85$ ppm) and benzene signals ($\delta = 6.52$–6.70 ppm) allowed to estimate the content of DMA (Figure S2). The DMA contents are displayed in Table 1 where C5M4000, C5M18000, and C5M24000 samples show similar values.

The GPC results show a narrow molar mass distribution (PDI) for all materials as can be deduced from PDI values in Table 1, together the different molecular weights (M_n). The polydispersity values can be explained through the benefits of RAFT polymerization. Figure 1 shows unimodal curves of the molecular weight distribution profiles for the copolymers of this research work. Similarities between C5M18000 and C1M18000 curves allow for getting a great uniformity between size and molecular weight of polymeric chains. Therefore, these features can be essential for excluding the influence of polydispersity on the LCST behavior of C5M18000 and C1M18000 samples (Figure 1).

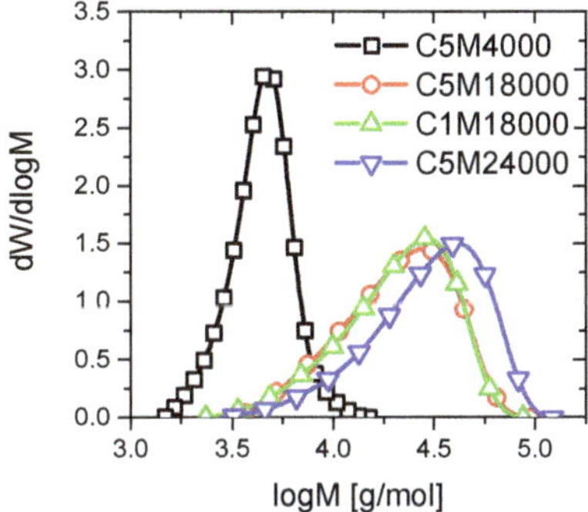

Figure 1. Molecular weight distribution profiles obtained from GPC curves for the copolymers with different comonomer content and molecular weights.

A low amount of RAFT agent during polymerization could justify a slight increment into the polydispersity of C5M24000, C1M18000, and C5M18000 samples (Table 1) because control agents provide narrow polydispersities [25]. Nevertheless, the polydispersities are relatively narrow for all materials owing to RAFT polymerization vs. traditional free radical polymerization.

3.2. LCST Behavior

The preparation of diverse aqueous solutions (4, 8, 10, and 15 wt %) was carried out in relation to studying the LCST behavior for the copolymers by conventional calorimetry, turbidimetry (UV-visible spectroscopy), and rheology. The real polymer concentration in water was confirmed by TGA analysis, where as expected a clear equivalence was observed between theoretical concentrations of solutions and the data estimated from TGA-curves (Figure S2).

The coil to globule transition can be easily observed visually (Figure 2), where a sample below LCST is a transparent aqueous solution due to hydrogen bonds between polymer–water (Figure 2b), whereas the polymer–polymer interactions increase over LCST and the hydrogen bonds are broken, consequently polymer globules are formed and the sample becomes solid and opaque (Figure 2a) [26]. Moreover, the polymers are shown to have statistical DMA distributions—as expected—because block copolymers tend to form hairy micellar solutions in selective solvents, which likely are often turbid (Figure 2a) [27–31]. Furthermore, it is clear from Figure 2 that the DMA is not oxidized, as this would lead to lead to catechol tanning, which turns the sample dark red after an initial pink hue at very low oxidation levels.

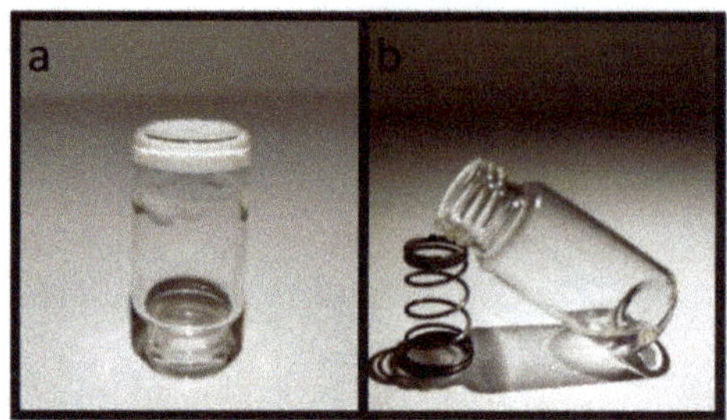

Figure 2. Polymer solution in water (C5M24000, c = 10 wt %) above LCST (**a**) and below LCST (**b**).

3.3. Influence of DMA on LCST

The different endothermic processes of the diverse polymer solutions (4, 8, 10, and 15 wt %) were carried out by calorimetric analysis for C1M18000 and C518000 (Figure 3). The differences between both polymers, whose comonomer amount was varied, are evident from the thermograms.

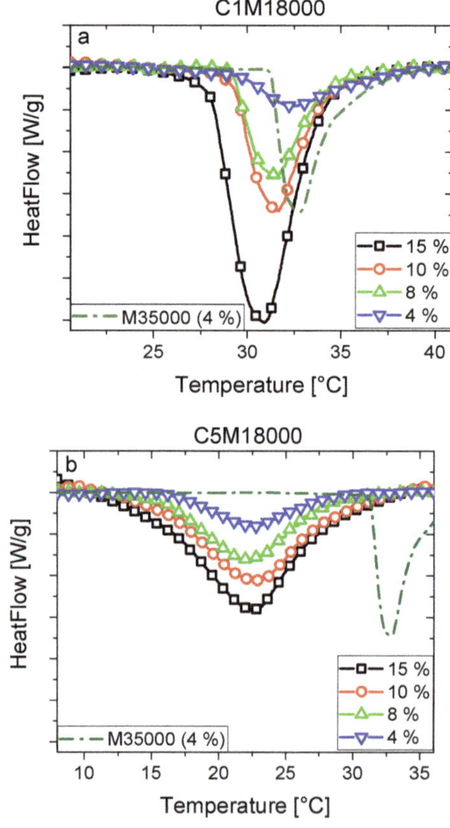

Figure 3. DSC endotherms of aqueous solutions (4, 8, 10, and 15 wt %) for C1M18000 (**a**) and C5M18000 (**b**) samples. The results were compared with M35000 homopolymer (c = 4 wt %, dash-dot green line).

Firstly, the peak temperatures of C1M18000 are over 30 °C, whilst LCSTs are found below 25 °C for C5M18000. A higher content of DMA in C5M18000 could explain the decrease of LCST regarding to C1M18000. Other influences such as polydispersity or end-group effect were dismissed according to the GPC results. In general, the incorporation of the hydrophobic comonomers reduces the LCST whereas hydrophilic comonomers increase the LCST values [32–36]. As the hydrophobicity of DMA is higher than NIPAM, the LCST tends to decrease when comonomer content increases [11]. The homopolymer showed a slightly higher LCST than C1M18000 due to low DMA content and high polymerization degree. A possible end-group effect caused by the differences in molecular weight between C1M18000 and M35000 was rather small, as will be discussed later. C5M18000 exhibited lower LCST due to the higher comonomer amount, increasing its hydrophobicity.

Secondly, C1M18000 has sharp and narrow transitions for all solutions with regard to the endotherms associated with C5M18000. Furthermore, C1M18000 seems to show a better thermal sensitivity [10] than C5M18000 as endotherms in Figure 3 show. Higher hydrophobic interactions could promote longer transitions when the DMA content is increased, i.e., lower thermal sensitivity such as C5M18000 endotherms displayed (Figure 3b). Generally, the LCST transition defines the temperature where the hydrogen bonds are breaking [37] as C1M18000 endotherms clearly show (Figure 3a). Nevertheless, C5M18000 displays another effect during the LCST transition, i.e., hydrophobic DMA could partially slow down the LCST leading to wider transitions where LCST occurs.

Thirdly, DSC shows clearly transitions for C1M18000 in comparison to C5M18000, but less clear than for M35000. Generally, the calorimetric analysis provides information about the LCST, i.e., these endotherms allow for determining the number of hydrogen bonds broken during the LCST transition [21]. The presence of hydrophobic DMA interferes with the LCST transition, and consequently, the transition width increases (Figure 3b). This can be understood through the fact that the DMA monomer itself in the chain is hydrophobic enough that it would render the DMA homopolymer water-insoluble. Thus, in the vicinity of the DMA monomer, the hydrophobicity was higher ((Figure 4, green circles), the blue line stands for NIPAM monomers, red circles are the RAFT-end groups and the green circles are the hydrophobic dopamine rings of DMA. The orange shading stands for the local hydrophobicity). Obviously, this effect becomes more pronounced, when 2 DMA monomers are adjacent to each other, at 1–6 mol % DMA content is rather unlikely. As a consequence, the chain has to be considered to have hydrophobicity variation along the chain, depending on the local comonomer distribution. Furthermore, also the end groups play a role, as their hydrophobicity is even greater than DMA (Figure 4, red circles), which leads to the same hydrophobicity fluctuations.

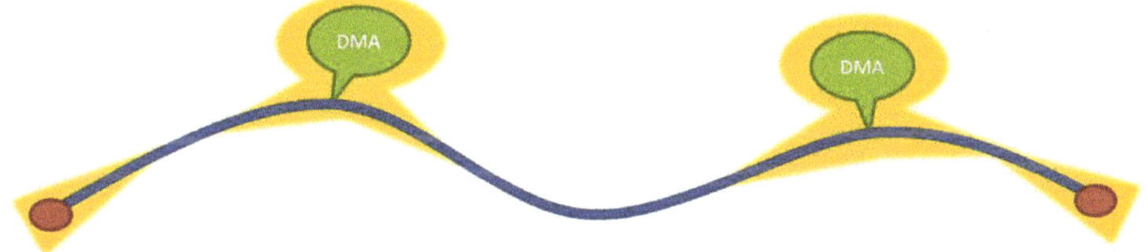

Figure 4. Scheme of the hydrophobicity fluctuations along the chain.

On the other hand, C1M18000 endotherms show a clear LCST dependence with polymer concentration in water, i.e., decreasing peak temperatures as polymer concentration increases probably because a dilute medium reduces the polymer–polymer interactions, and thus hydrogen bonds could be stronger.

Figure 5 shows the transmittance curves vs. temperature of the diverse polymer solutions in water (4, 10, and 15 wt %) of C1M18000 (Figure 5a) and C5M18000 (Figure 5b) samples. LCST transitions, studied by transmittance (50%), are in good agreement with the endotherm peaks (Figure 3).

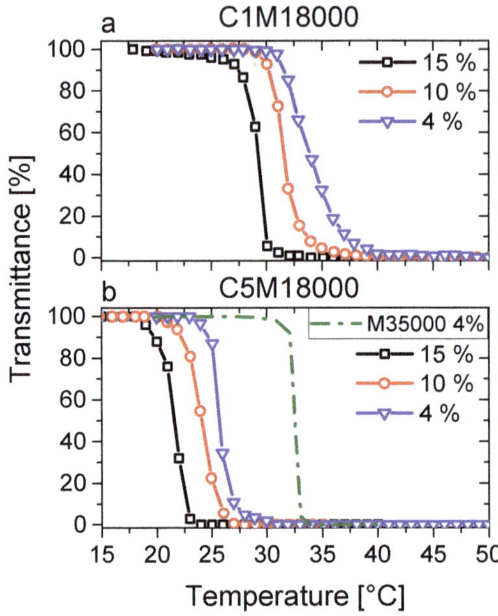

Figure 5. Transmittance as a function of temperature of aqueous solutions (4 wt %, 10 wt %, and 15 wt %) for C1M18000 (**a**) and C5M18000 (**b**) samples.

However, here, LCST values describe a decrease with increasing polymer concentration in water, which is observed to a much lower degree in DSC data. The reason for the stronger temperature dependence of the UV-visible spectroscopic data was concluded to be a lower percentage of the dissolved polymer will lead to a 50% transmission loss at higher polymer concentration, which means that the UV-vis data have to be regarded to be less reliable in giving the true LCST temperatures than DSC, due to their different response characteristics.

The LCST transitions, analyzed from calorimetric curves (Figure 3) and transmittance data (Figure 5), were evaluated not only with respect to the peak point ($T_{peakDSC}$) and the cloud point ($T_{cloudUV}$), defined as the half-height of the transmittance transition, as classically done, but also from the onset and offset temperatures ($T_{onsetDSC}$ and $T_{onsetUV}$/$T_{offsetDSC}$ and $T_{offsetUV}$). The results were collected and displayed in Figure 6, where an interesting equivalence can be deduced from peak temperatures ($T_{peakDSC}$) and cloud points ($T_{cloudUV}$).

Both peak temperatures ($T_{peakDSC}$) and cloud points ($T_{cloudUV}$) exhibit similar values for C1M18000 (Figure 6a) and C5M18000 (Figure 6b), indicating that phase transitions happen in analogous ranges as the endotherms and transmittance data show. Hence, the different heating rates used (5 vs. 1 °C/min) do not play a significant role for the polymeric aqueous solutions from a physicochemical standpoint. Therefore, results can be discussed according to the molecular features and the sample conditions.

The cloud points of C1M18000 ranged between 29 and 33 °C and seem to describe a trend in relation to polymer concentration in water, closely resembling the DSC data ($T_{peakDSC}$). In contrast to this, while the cloud points of 5M18000 are located between 21.5 and 26 °C, again following a concentration-dependent trend, the peak temperatures ($T_{peakDSC}$) display constant values for all concentrations in water. The DSC sensitivity to polymer concentrations in water could depend on the hydrophobic interactions. Nevertheless, calorimetry provides important information of these interactions such as endotherms exhibited for C5M18000.

It is obvious that a difference between $T_{onsetDSC}$ and $T_{offsetDSC}$ exists, which is higher for C5M18000 vs. C1M18000. This effect could be explained according to hydrophobic interactions [32–35], produced by the presence of DMA in relation to its higher hydrophobicity in comparison with NIPAM [11] promoting long LCST transitions because these interfere along LCST transition, i.e., the aforementioned hydrophobicity fluctuations induced by the statistical distribution of DMA induce a distribution of local hydrophobicity, which will lead to local LCST fluctuations, leading to a broadening of the global LCST process (Figure 4). This is similar to the work on ethylene–hexene copolymers, where a precise placement of the comonomer, e.g., every 15th main chain carbon leads to a significantly more narrow and higher melting point than found for the standard random copolymers [38,39]. Secondly, the $T_{onsetDSC}$ decreases, while $T_{offsetDSC}$ increases when the polymer concentration rises for both samples. The difference between $T_{onsetDSC}$ and $T_{offsetDSC}$ achieves a maximum for the highest concentration of 15 wt %, especially for C5M18000 where a great number of hydrophobic interactions occur.

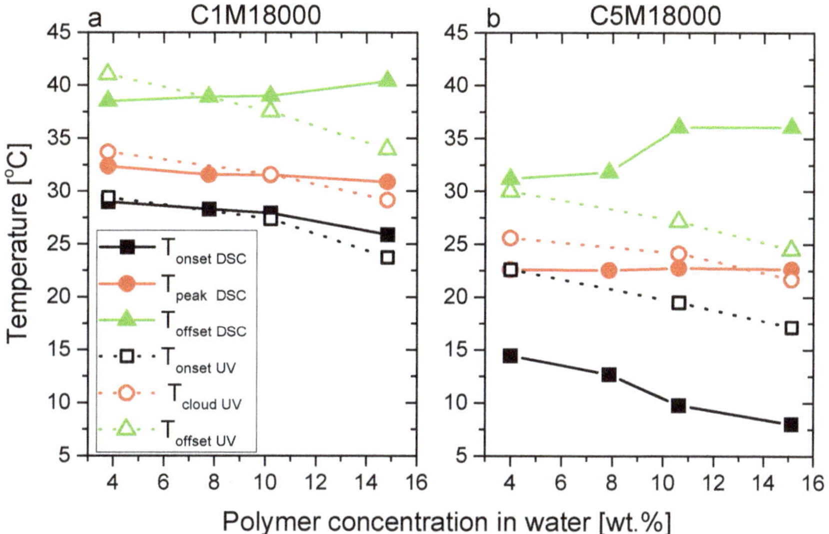

Figure 6. Peak temperatures ($T_{peakDSC}$), cloud points ($T_{cloudUV}$), onset temperatures ($T_{onsetDSC}$ and $T_{onsetUV}$), and offset temperatures ($T_{offsetDSC}$ and $T_{offsetUV}$) derived from calorimetric analysis and transmittance curves, respectively.

Temperature ramps (Figure 7) were tested by rheology for both samples at the same polymer concentration in water of 15 wt %, with both solutions showing a clear increase in storage and loss moduli around the LCST. The LCST transitions are located at almost identical temperatures as those from conventional calorimetry and UV-measurements. Thus, consistent data is achieved from the diverse detection methods used along this research work. The LCST transitions are wider than transitions related to UV-measurements and calorimetric analysis. This fact could be related to rheology, which can detect the primary traces to the coil-to-globule transition earlier than the other techniques and, furthermore, rheology can also detect the offsets while UV-vis cannot find any more increment due to complex opacity of the sample. The reasons for the higher sensitivity of rheology are multifarious. They mainly come from the fact that in general rheology reacts highly sensitively to the slightest changes in sample structure, such as the first traces of globule formation. Those globule onsets are too small to be detected by UV-vis due to their size being below the wavelength of visible light. Furthermore, supposedly the number of H-bonds being formed is too low to produce a clear calorimetric signal.

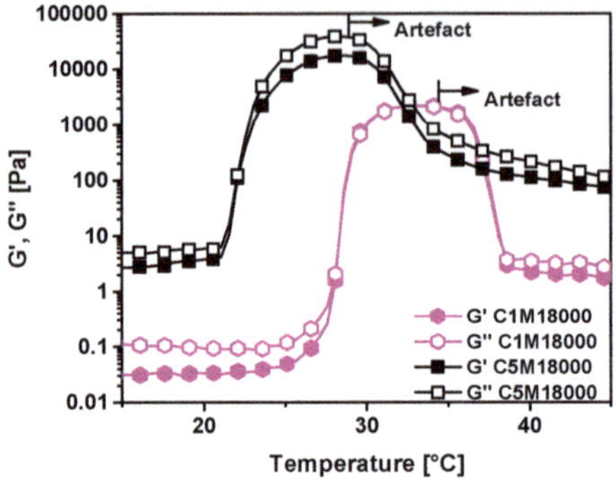

Figure 7. Temperature ramps of homopolymers solutions in water at 15 wt.% for C1M18000 and C5M18000.

It should be mentioned that ca. 8 K above the LCST, the data go through a maximum, which is a well-known artefact related to the sample loosing contact with both plates due to its LCST induced contraction [11,40,41]. Consequently, all data at temperatures above this peak should be ignored.

The moduli for C5M18000 are significantly higher than for C1M18000 below the LCST, which can be understood as the consequence of the sample C5M18000 being able to complex with each other through radical crosslinking of the dopamine groups as well as through capturing of ions, which would eventually lead to crosslinking. These ions diffuse away from the surface of the stainless steel plates and are, thus, supposedly mostly iron and nickel ions in nature, which are known to have strong interactions with dopamine groups [42]. As Fe^{3+} ions are known to form di-dopamine complexes with a dirty greenish color, it is not surprising that the sample C5M18000 had a slightly greenish hue at the end of the experiment, which of course does not mean that the sample's dopamine groups were saturated with ions, but only that a rather small amount of ions diffused in. This can also be seen from the fact that the samples are not gels.

3.4. LCST for Copolymers with Different Molecular Weight

C5M24000 and C5M4000 aqueous solutions were studied by conventional calorimetric analysis (Figure 8), and results were also compared with C5M18000. Differences between endotherms were evident due to LCST transitions of C5M4000 were located around to 5 °C below to C5M18000 and C5M24000. Nevertheless, LCSTs were somewhat lower than it could be expected by the results previously reported [11]. In that case, the molecular weight was lower (2800 g/mol), as well as the comonomer content was also somewhat lower (≈4.9 mol %). Nevertheless, some considerations should take account in order to understand the different results.

On the one hand, the previous work reported a copolymer prepared by free radical polymerization, whose polydispersity was 2.1 and whose chain ends were far less hydrophobic. Thus, the presence of the RAFT agent leads to an additional influence on the LCST behavior for C5M24000, C5M18000, and C5M4000. Furthermore, it needs to be taken into consideration that the data were measured by a rheometer, whose temperature calibration might be somewhat different from the one used here. In addition, the polymer concentration was 20 wt % [11] and not max. 15 wt % as studied in this paper.

C5M24000 and C5M18000 endotherms (Figures 3b and 8a) showed higher LCSTs than C5M4000 (Figure 8b) due to a molecular weight increase, leading to a lower hydrophobicity difference as previously discussed by us for PNIPAM [11], because end-groups are reduced and consequently hydrophobic interactions. The M7000 homopolymer exhibited the highest LCST showing the strong effect of the DMA content in C5M24000, C5M18000, and C5M4000. This effect was slightly perceived for the differences between LCSTs of C5M24000 and C5M18000, but the changes between DMA content made difficult extracting conclusions. However, it should also be mentioned that, while the comonomer contents were similar, they were slightly different. In the next section, we will discuss these effects in detail.

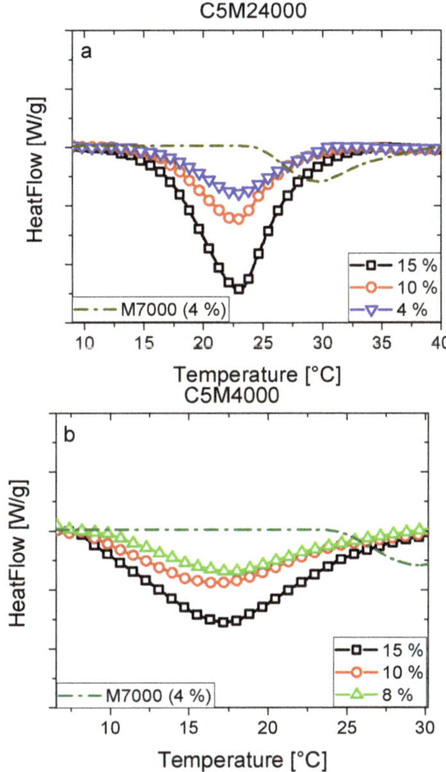

Figure 8. DSC endotherms of aqueous solutions for C5M24000 (**a**) and C5M4000 (**b**) samples. The results were compared with M7000 homopolymer (c = 4 wt %).

Nevertheless, the influence of molecular weight on LCST was discussed along the literature where some inconsistences and contradictions can be observed between trends and behaviors [32]. Some parameters as polymerization degree or end-group effect could significantly contribute on LCST over the molecular weight, and consequently, could explain the different results reported. Thus, it is necessary to carry out a deep study, where the influence of each parameter can be explained separately before reaching conclusions focused only around to molecular weight.

The endotherms of C5M4000 appear more smeared out and wider than transitions from C5M24000 or C5M18000. As said above (Figure 4), the hydrophobic interactions can lead to this kind of LCST transitions, because they are caused by hydrogen bonds cleavage. Nevertheless, very similar endotherms should be expected due to very similar DMA contents. Thus, the presence of the RAFT-agent at the end of the polymer chain must be the responsible of this effect, as the RAFT-agent content increases as polymerization degree decreases promoting the end-group effect.

The LCST transitions were studied in detail through the estimation of onset, peak, and offset temperatures ($T_{peakDSC}$, $T_{onsetDSC}$, and $T_{offsetDSC}$, analogously to Figure 5) of calorimetric curves for C5M4000 and C5M24000 solutions with different concentrations. Furthermore, some onset, cloud, and offset points ($T_{cloudUV}$, $T_{onsetUV}$, and $T_{offsetUV}$) obtained from transmittance data were included and an excellent correspondence between LCSTs can be observed. In this way, both peak temperatures

(T$_{peakDSC}$) and cloud points (T$_{cloudUV}$) exhibit similar values for C5M4000 (Figure 9a) and C5M24000 (Figure 9b).

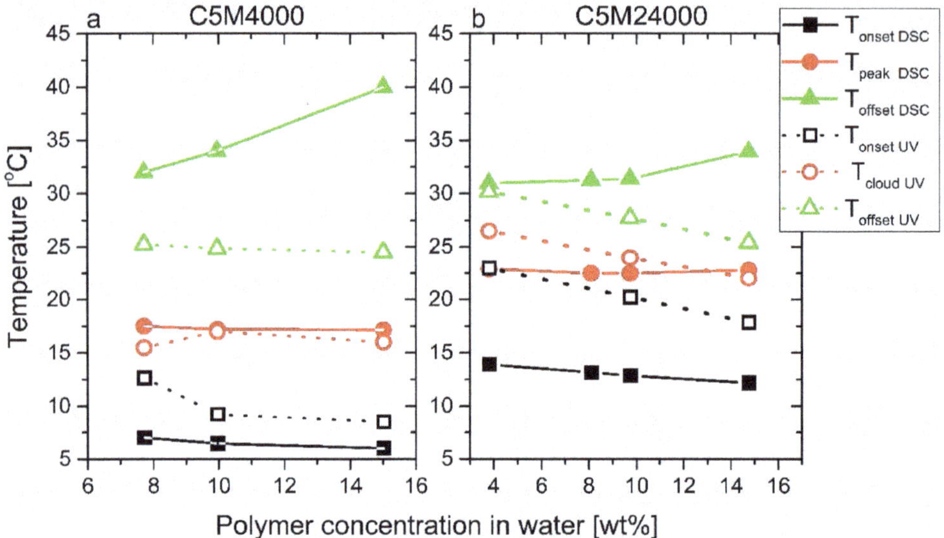

Figure 9. Peak temperatures (T$_{peakDSC}$), cloud points (T$_{cloudUV}$), onset temperatures (T$_{onsetDSC}$ and T$_{onsetUV}$), and offset temperatures (T$_{offsetDSC}$ and T$_{offsetUV}$) derived from calorimetric analysis and transmittance curves for C5M4000 and C5M24000.

Again, the C5M24000 aqueous solutions (Figure 9b) show different cloud points between 22.5 °C and 27 °C, whose trend corresponds to the polymer concentration in water. Nevertheless, this tendency is not observed from peak temperatures of calorimetric analysis. Likewise, this trend is not shown for C5M4000, whose T$_{cloudUV}$ and T$_{peakDSC}$ values are around 17 °C, probably due to strong hydrophobic interactions explained before for the copolymers with different DMA amount.

It is obvious from Figure 10a that the differences between onset and offset temperatures increased significantly with decreasing molar mass, which needs to be explained. For this purpose it was determined how many DMA groups are on the different chains, assuming a binomial distribution probability, i.e., that the monomers are incorporated 100% randomly. The result was that for C5M4000 (DP$_n$ = 31, end groups' molar mass 272.42 g/mol) 20.3% of the chains did not have any DMA in them, 33.3% had 1, 26.2% had 2, 13.3% had 3, and 6.7% had more than 3 (the numbers are not entirely correct, as it is assumed that the average molar mass per monomer unit was constant, but that was clearly not the case, as DMA was about twice as heavy as NIPAM and, therefore, chains with fewer DMA groups but the same DP$_n$ as those with more DMA groups had a lower molar mass. However, this influence was considered to be minor (estimated to be <±2%) and, therefore, ignored for further discussions). Consequently, there are several species of chains in C5M4000, which differed in their hydrophilicity—most prominently, the ca. 20% of pure PNIPAM in C5M4000 should have a slightly lower LCST than M7000 (ca. 25 °C), while the observed LCST is at 17 °C, ca. 8 K lower. Considering this difference alone, it is logical that such a material has a much broader LCST than a homopolymer material or a copolymer with significantly higher comonomer content. For C5M24000, the six times higher molar mass leads to a negligible homopolymer fraction and 80% of the chains contain 6–13 DMA functionalities, suggesting the inhomogeneity in comonomer distribution is much lower. Furthermore, as pointed out previously, the rather hydrophobic RAFT-agent residues at the chain ends lead to a further gradient in hydrophobicity, which also explains the same kind of effect for the PNIPAM-homopolymers M7000 and M35000. For low molar masses (e.g., C5M4000 has a DP$_n$

of only ca. 31), this end group effect plays a significant role, while for C5M24000 (DP$_n$ ca. 200) this effect is much less pronounced. These two effects lead to a higher LCST of C5M24000 taking place in a narrower temperature range than C5M4000. Hence, the LCST of C5M24000 has to be considered to be the normal LCST value of copolymer with ca. 5 mol% DMA and the RAFT end groups of C5M4000 lower the LCST due to the end group effect.

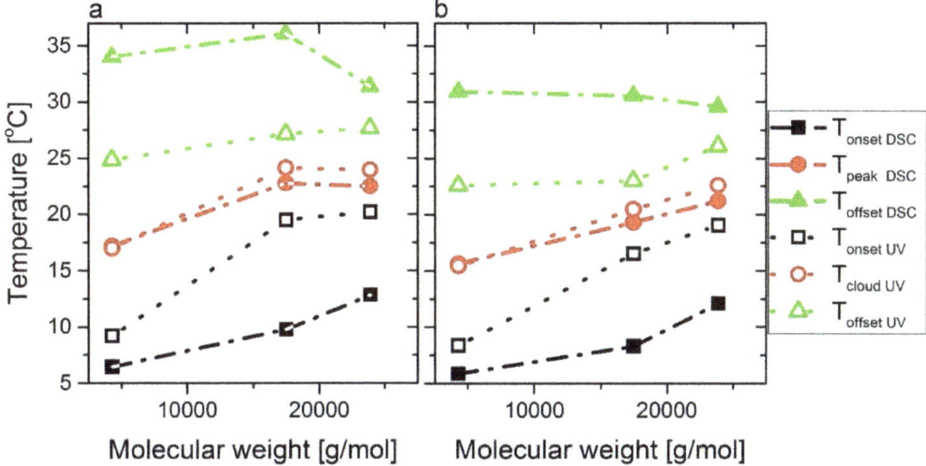

Figure 10. (**a**) Peak temperatures (T$_{peakDSC}$), cloud points (T$_{cloudUV}$), onset temperatures (T$_{onsetDSC}$ and T$_{onsetUV}$), and offset temperatures (T$_{offsetDSC}$ and T$_{offsetUV}$) derived from calorimetric analysis and transmittance curves vs. molecular weights for C5M4000, C5M18000, and C5M24000 at the same polymer concentration in water (10 wt %). (**b**) Same data but normalized to 5 mol % DMA content using a statistical model.

This trend seems slightly disturbed by C5M18000 but this fact can be perfectly understood from its higher amount of DMA. Nevertheless, if the data is normalized according to comonomer amount, clear trends in relation to molecular weights will be observed for the LCST derived from calorimetry (T$_{peakDSC}$) and UV-vis spectroscopy (T$_{cloudUV}$) as Figure 10b shows, where the data is normalized for a DMA content of 5 mol %. Moreover, tendencies are defined according to hydrophobic interactions for T$_{onsetDSC}$, and T$_{offsetDSC}$ regarding to T$_{onsetUV}$ and T$_{offsetUV}$. Again, calorimetry seems more sensitive than UV-vis spectroscopy to these hydrophobic transitions, which could partially explain the contributions of the end-group effects on LCST.

The rheological properties were tested for both C5M4000 and C5M24000 at the polymer concentration in water of 10 wt % (Figure 11). Again, the LCST transitions show an excellent correspondence with the data collected from conventional calorimetric analysis and turbidimetry.

It should be mentioned that just like in our previous paper, a small shoulder is obvious between 15 and 20 °C for C5M24000, which might be the consequence of high local DMA concentrations along the chain [11]. Otherwise, the curves follow the same pattern and exhibit very similar artefacts as previously discussed.

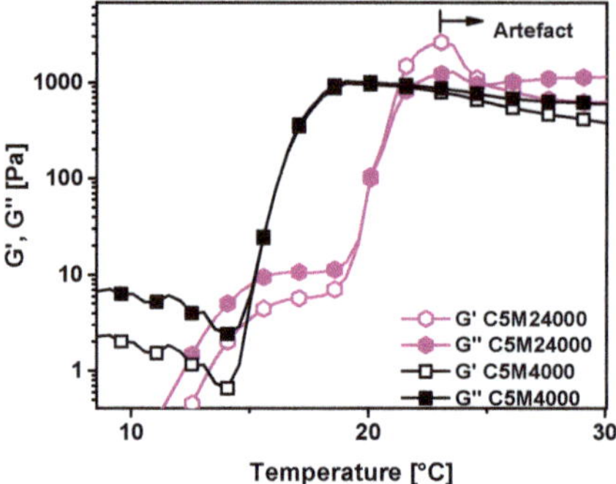

Figure 11. Temperature ramps of homopolymers solutions in water at 10 wt % for C5M4000 and C5M24000.

3.5. LCST Ranges

Commonly, LCST is defined as an exact temperature where the transition to coil to globule takes place. Nevertheless, LCST is a process where the hydrogen bonds are breaking according to the influence of various factors discussed before [1,37]. Consequently, the LCST is not occurring at a single temperature, but in a certain interval. Thus, important information can be lost and, consequently, the study of the transition as a whole is for understanding the real influence of some parameters on the LCST.

This work defined the LCST ranges as the difference between offset and onset temperatures ($T_{onsetDSC}$ and $T_{offsetDSC}$) derived from DSC endotherms. While other LCST ranges could be estimated from turbidimetry or rheology, calorimetric analysis could provide an important information because the endotherm is more precise than the other methods and, furthermore, directly related to the hydrogen bonds involved on LCST process [21] and, thus, the data are more related to the molecular processes. As mentioned above, the endotherms could reflect other mechanisms as hydrophobic interactions shown from C5M4000 and C5M24000.

Figure 12 shows the diverse LCST ranges estimated from the different endotherms (Figures 3 and 8) for all the samples and were displayed as a function of polymer concentration in water. The LCST ranges changed in respect of the polymer concentration in water. This effect could be ascribed to the coil-to-globule transition, as less concentrated solutions tend to aggregate and settle easier. On the other hand, a lower percentage of the dissolved polymer could lead to increase the transition temperature at higher polymer concentration. Moreover, the LCST range dependence on concentration of the sample seemed to be connected to the LCST range at high concentrations. When extrapolating the data to zero concentration, the LCST range for all samples appeared to be around 10 °C, suggesting that this was the minimum width observable for the LCST. As C5M4000 had the broadest LCST, it also showed the highest slope regarding to other samples in relation to great hydrophobic interactions and the polymer concentration in water.

Hence, these transitions open a new perspective on LCST behavior because the influence of end-group effect, presence of comonomer and/or molecular weight could be explained through the hydrophobic interactions.

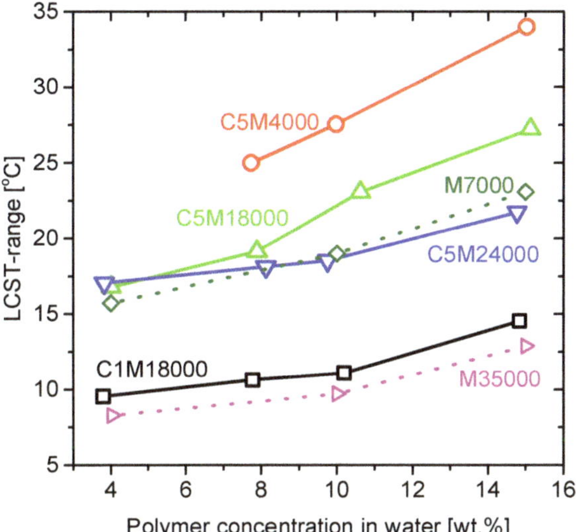

Figure 12. LCST transition ranges versus polymer concentration in water for all samples.

4. Statistical Modeling

In order to understand these complex correlations better, a series of regression analyses were conducted to study the influence of the comonomer content (DMA), polymerization degree (DPn) and the polymer concentration in water (wt %; independent variables) on the dependent variables. Those were the parameters involved in the LCST, such as $T_{onsetDSC}$, $T_{peakDSC}$, $T_{offsetDSC}$, and $LSCT_{rangeDSC}$, DSC integral derived from calorimetry, and $T_{onsetUV}$, $T_{cloudUV}$, $T_{offsetUV}$, and $LSCT_{rangeUV}$, obtained by turbidimetry.

In general, the regression analysis model can be explained through equation [1], where the comonomer content (DMA), polymerization degree (DPn), and the polymer concentration in water (wt %) were independent variables, and the parameters involved in the LCST, such as $T_{onsetDSC}$, $T_{peakDSC}$, $T_{offsetDSC}$, and $LSCT_{rangeDSC}$, derived from calorimetry, and $T_{onsetUV}$, $T_{cloudUV}$, $T_{offsetUV}$, and $LSCT_{rangeUV}$, were dependent variables.

$$\text{Dependent variable (one of the LCST characteristics)} = B_0 + B_1\,(DMA) + B_2\,(DPn) + B_3\,(wt\,\%) \quad (1)$$

All parameters were displayed in Table 2, were supplemented by the standard error (std. error), t (t should be higher than 1.96 when independent variable has a significant effect on dependent variable), the beta distribution (Beta), and Sig (refers to significance level, which indicates whether independent variable has a significant effect on dependent variable). The regression analysis routine deletes independent variables that are deemed not relevant, which is denoted in Table 2.

The LCST parameters showed interesting results where the influence of comonomer content, polymerization degree and polymeric solution in water was clarified. Firstly, LCST value is general studied by the T_{peak} obtained from the endotherms of calorimetry, and/or the T_{cloud} estimated from turbidimetry, as the next equations show:

$$T_{peakDSC} = 27.586\,°C - 1.617\,°C/\text{mol}\,\%\,(DMA) + 0.020\,°C/\text{monomer}\,(DPn) \quad (2)$$

$$T_{cloudUV} = 20.776\,°C - 0.806\,°C/\text{mol}\,\%\,(DMA) + 0.042\,°C/\text{monomer}\,(DPn) \quad (3)$$

The results of regression analysis of calorimetry showed that DMA had a negative correlation on T_{peak} (b = −1.617, *p* = 0.000), while DPn had a positive correlation on T_{peak} (b = 0.02, *p* = 0.000), and polymer concentration in water (wt %) had no significant effect on T_{peak}. Similarly, the results of regression analysis of turbidimetry showed that DMA was negatively correlated to T_{cloud} (b = −0.806, p = 0.000), while DPn had a positive effect on T_{cloud} (b = 0.042, p = 0.000), and polymer concentration in water (wt %) had no significant effect on T_{peak}. Specifically, the regression analysis of these T_{peak} and T_{cloud} values, as dependent variables, showed that the comonomer content plays an important role in the LCST behavior due to its negative correlation with the LCST value. Consequently, while Figures 6 and 10 seem to suggest a correlation between $T_{peakDSC}$ and $T_{cloudUV}$, the regression analysis clearly shows that DPn and DMA content were significantly more important than the concentration. The reason can be easily understood from the fact that DPn and DMA content change the molecules' hydrophobicity, while the concentration only brings them somewhat closer or farther apart. The effect of the former was logically much stronger than that of the latter, which does not mean that the concentration does not have an effect—it was just so small that its influence was within the experimental error.

Table 2. Statistic data obtained for the analysis of the influence of comonomer content (DMA), polymerization degree (DPn), and the polymer concentration in water (wt %).

Input \ Output		Calorimetry				UV-Spectroscopy			
		$T_{onsetDSC}$	$T_{peakDSC}$	$T_{offsetDSC}$	$LSCT_{rangeDSC}$	$T_{onsetUV}$	$T_{cloudUV}$	$T_{offsetUV}$	$LSCT_{rangeUV}$
Constant	B_0	24.374	27.586	36.907	13.608	13.607	20.776	24.508	13.783
	Std. Error	1.712	0.943	1.263	2.318	1.975	1.791	2.008	0.924
	Beta	-	-	-	-	-	-	-	-
	t	14.241	29.245	29.226	5.871	6.89	11.601	12.208	14.916
	Sig	0	0	0	0	0	0	0	0
DMA	B_1	−2.618	−1.617	−1.157	1.417	−0.696	−0.806	n.s.	n.s.
	Std. Error	0.182	0.139	0.169	0.246	0.305	0.276	n.s.	n.s.
	Beta	−0.813	−0.821	−0.807	0.547	−0.277	−0.384	n.s.	n.s.
	t	−14.401	−11.64	−6.86	5.756	−2.285	−2.917	n.s.	n.s.
	Sig	0	0	0	0	0.037	0.011	n.s.	n.s.
DPn	B_2	0.034	0.02	n.s.	−0.039	0.058	0.042	0.03	−0.033
	Std. Error	0.006	0.004	n.s.	0.008	0.009	0.008	0.011	0.005
	Beta	0.333	0.33	n.s.	−0.483	0.774	0.682	0.551	−0.845
	t	5.889	4.685	n.s.	−5.077	6.391	5.176	2.643	−6.321
	Sig	0	0	n.s.	0	0	0	0.018	0
Polymer concen-tration in water	B_3	−0.367	-	0.319	0.678	n.s.	n.s.	n.s.	n.s.
	Std. Error	0.117	-	0.112	0.159	n.s.	n.s.	n.s.	n.s.
	Beta	−0.172	-	0.336	0.395	n.s.	n.s.	n.s.	n.s.
	t	−3.135	-	2.856	4.275	n.s.	n.s.	n.s.	n.s.
	Sig	0.006	-	0.01	0.001	n.s.	n.s.	n.s.	n.s.

n.s. = not significant.

Obviously, the presence of the comonomer decreases the LCST as was reported before and we described it along this work (Figure 6). Nevertheless, the content of DMA seems to influence the LCST detected calorimetry more than turbidimetry, such as could be observed from a higher b_1 value associated with the calorimetry whose value (−1.617) was double the value related to turbidimetry (−0.806). This fact can be easily understood because turbidimetry measures optical signals and some changes can be produced below the detection range, whilst calorimetry can still detect them. On the other hand, the polymerization degree showed a minor positive correlation with LCST, i.e., the LCST increased with rising polymerization degree. The data shows that the polymerization degree influence on LCST was lower than other parameters, which were responsible of the diverse trends reported and the discussion in the literature.

In order to visualize the quality of the fits, Figure 13 shows a comparison between input data and the modeled prediction. The thick dashed line represents an ideal correlation, while the thin dashed lines indicate a deviation by ±3, which encompasses almost all results. The r^2 value of all data in Figure 13 was found to be 0.95598, suggesting all data can be modeled well. Furthermore, it becomes obvious that the correlations for the DSC derived quantities were more precise, especially for the T_{peak}. This can be explained from firstly, the cleaner calorimetric signal, especially for the peak, and secondly from the much higher data density of calorimetry with ca. 100 data per K, while only ca. 1 point could be measured per K for UV-vis spectroscopy.

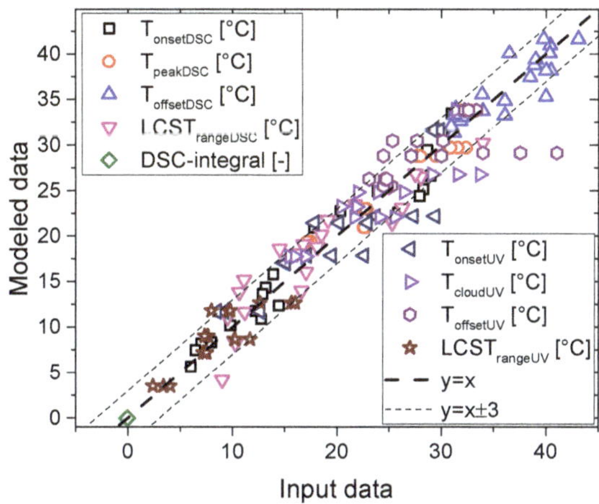

Figure 13. Comparison of input data and modeled data.

The LCST range was also analyzed using the statistical method for calorimetry and turbidimetry data. The results associated with calorimetry displayed a positive influence of the comonomer content on LCST range, a lower effect of the polymer solution in water, and finally, a small negative effect of polymerization degree was observed from the data obtained by calorimetry. Nevertheless, the data acquired by turbidimetry exclusively showed the negative influence of the polymerization degree on the range LCST range as shown in Table 2. Thus, calorimetry seems more sensitive to the LCST transition than turbidimetry for these materials, i.e., endotherms contained a lot of information of all transition due to the main parameters associated with the molecular features of the samples and the conditions of the samples involved. This fact is really interesting, because the LCST ranges were not deeply studied previously, which we concluded to be the consequence of the majority of the LCST studies being carried out by turbidimetry, where LCST ranges did not seem to be affected by the diverse parameters involved.

Onset and offset temperatures, derived from calorimetry and turbidimetry, were also examined by this statistical method. The calorimetry data showed an interesting relation with onset and offset temperatures through the comonomer content, the polymerization degree and the polymer solution in water. Both temperatures were negatively correlated with DMA content and polymer content in water, but positively influenced by polymerization degree. Nevertheless, less information was obtained by turbidimetry, especially for the offset temperature where polymerization degree effect was exclusively significant.

Another statistical study was also performed using the areas obtained from the endotherms of DSC studies (Figure 14), which were previously normalized to the polymer concentration in water and averaged by concentration. A clear linear regression can be observed from these results, where the error bars were estimated through the comparison between experimental data and regression model. The results clearly exhibited a dependence between polymerization degree and the area of the endotherms, as those could be associated with the number of hydrogens involved in the LCST [21]. Obviously, the comonomer content will not affect the final trend so much, because the number of comonomer units along the polymeric branches was rather low, as all the samples present DMA contents below 6%.

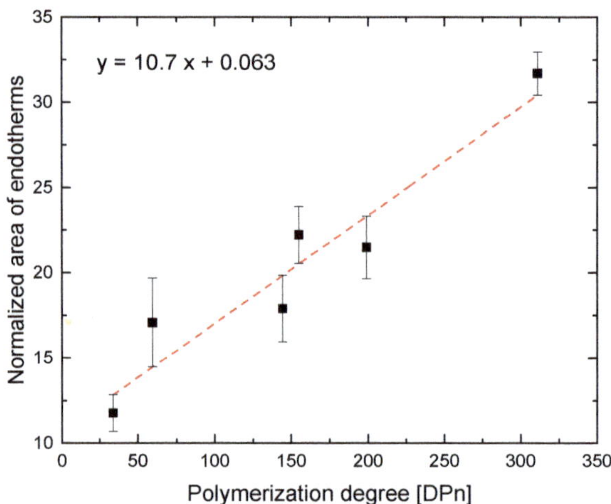

Figure 14. Comparison between polymerization degree vs. normalized are of the endotherms obtained from calorimetry data (dots) and from the simulation (line).

In general, statistical modeling can predict the LCST behavior of DSC results according to the molecular features (DMA content and DPn) and the polymer concentration in water. Onset temperatures could exhibit the strongest dependence on those, as these parameters play an important role during the first stages of the LCST transition, especially the comonomer content, which could disrupt the starting point of LCST transition. Further, peak temperature has a similar influence, but the offset temperatures do not seem intensively affected by these factors. Furthermore, the LCST range can exhibit a good resolution, as the onset temperature plays an important role in that value.

5. Conclusions

RAFT polymerization allowed for obtaining interesting copolymers to study the effect of the comonomer content and the molecular weight on LCST. A good agreement was observed for the results obtained from calorimetry, turbidimetry, and rheology.

The content of more hydrophobic DMA decreased the LCST for the copolymers, where the comonomer amount was varied. Influence of other parameters, such as polydispersity and the end-group effect were found to be negligible in line with molecular features owing to the high molar mass of those samples.

LCST transitions derived from calorimetric analysis showed a great sensitivity to the change in hydrophobic interactions promoting to wider and more smeared out LCST endotherms, which was quantified in terms of LCST range. Hence, LCST transitions can clearly show the contributions from various molecular parameters in relation to hydrophobic interactions improving the reported analysis just focused on a simple value.

The LCSTs clearly showed the end-group effect for low molar masses, related to the presence of the hydrophobic RAFT-agent, which can be understood as the weight fraction of the RAFT-agent fragments residing at the chain ends decreases with increasing polymerization degree. The LCST ranges increase for low polymerization degrees and increasing concentration. While the influence of polymerization degree and comonomer content are clearly related to the inhomogeneity of the polymer chain (RAFT-residues and DMA leads to locally increased hydrophobicity), the influence of the concentration is probably due to an increasing tendency to phase separate into more or less hydrophobic chain segments with increasing concentration. The driving force behind this kind of process is that at higher concentrations, the statistical probability of chain segments of like hydrophobicity meets each other with a higher probability. That allows for local LCSTs in a wider temperature range.

The statistical model showed that the LCST behavior could be directly related to the comonomer content, polymerization degree and polymer solution in water with an accuracy of less than ±3 K. The statistical modeling allowed for discerning the different contributions of these factors on LCST.

The study of the complete transition associated with LCST could elucidate the diverse and partially contradictory results reported in literature in respect to the influence of parameters involved on LCST. Furthermore, LCST ranges could be very useful for new applications as biosensors or drug carriers where the control of the response will be essential.

Lastly, the application of a statistical model, frequently used in humanities and economics, was able to discern the influences on the different LCST parameters, which could prove a very useful strategy for many cases, where mixed influences prevail or where the trends are not as clear and particularly, where the synthesis or production of exactly defined model systems is not feasible.

Supplementary Materials: The supplementary materials are available online at http://www.mdpi.com/2073-4360/11/6/991/s1.

Author Contributions: A.G.-P. has synthetized, characterized, and explained the results of the resulting materials. C.S.B., W.L., and Y.W. helped in the preparation of the materials and the characterization. P.Y. performed the statistical analysis and the interpretation of the results. A.G.-P. and F.J.S. devised the original idea of the present research work together. A.G.-P. and P.Y. wrote the sections corresponding to their contributions and F.J.S. revised and checked the manuscript.

Funding: This research was funded by Shenzhen Fundamental Research Funds number KC2014ZDZJ0001A, the Shenzhen Sci & Tech research grant number ZDSYS201507141105130, and the China Postdoctoral Science Foundation Grant number 2018M633119.

Conflicts of Interest: The authors declare no conflict of interest.

References

1. Kotsuchibashi, Y.; Ebara, M.; Aoyagi, T.; Narain, R. Recent Advances in Dual Temperature Responsive Block Copolymers and Their Potential as Biomedical Applications. *Polymers* **2016**, *8*, 380. [CrossRef] [PubMed]

2. Stuart, M.A.C.; Huck, W.T.; Genzer, J.; Müller, M.; Ober, C.; Stamm, M.; Sukhorukov, G.B.; Szleifer, I.; Tsukruk, V.V.; Urban, M.; et al. Emerging applications of stimuli-responsive polymer materials. *Nat. Mater.* **2010**, *9*, 101–113. [CrossRef] [PubMed]

3. Hoffman, A.S. Stimuli-responsive polymers: Biomedical applications and challenges for clinical translation. *Adv. Drug Deliv. Rev.* **2013**, *65*, 10–16. [CrossRef] [PubMed]

4. Elsabahy, M.; Wooley, K.L. Design of polymeric nanoparticles for biomedical delivery applications. *Chem. Soc. Rev.* **2012**, *41*, 2545–2561. [CrossRef] [PubMed]

5. Schild, H.G. Poly(N-isopropylacrylamide): Experiment, theory and application. *Prog. Polym. Sci.* **1992**, *17*, 163–249. [CrossRef]

6. Saeed, A.; Georget, D.M.R.; Mayes, A.G. Synthesis, characterisation and solution thermal behaviour of a family of poly (N-isopropyl acrylamide-co-N-hydroxymethyl acrylamide) copolymers. *React. Funct. Polym.* **2010**, *70*, 230–237. [CrossRef]

7. Obeso-Vera, C.; Cornejo-Bravo, J.M.; Serrano-Medina, A.; Licea-Claverie, A. Effect of crosslinkers on size and temperature sensitivity of poly(N-isopropylacrylamide) microgels. *Polym. Bull.* **2012**, *70*, 653–664. [CrossRef]

8. Oh, J.K.; Drumright, R.; Siegwart, D.J.; Matyjaszewski, K. The development of microgels/nanogels for drug delivery applications. *Prog. Polym. Sci.* **2008**, *33*, 448–477. [CrossRef]

9. Ballauff, M.; Lu, Y. "Smart" nanoparticles: Preparation, characterization and applications. *Polymer* **2007**, *48*, 1815–1823. [CrossRef]

10. Chen, S.; Jiang, X.; Sun, L. Effects of end groups on the thermal response of poly(N-isopropylacrylamide) microgels. *J. Appl. Polym. Sci.* **2013**, *130*, 1164–1171. [CrossRef]

11. Vatankhah-Varnoosfaderani, M.; Hashmi, S.; GhavamiNejad, A.; Stadler, F.J. Rapid self-healing and triple stimuli responsiveness of a supramolecular polymer gel based on boron–catechol interactions in a novel water-soluble mussel-inspired copolymer. *Polym. Chem.* **2014**, *5*, 512–523. [CrossRef]

12. Brassinne, J.; Fustin, C.A.; Gohy, J.F. Control over the assembly and rheology of supramolecular networks via multi-responsive double hydrophilic copolymers. *Polym. Chem.* **2017**, *8*, 1527–1539. [CrossRef]

13. Tugulu, S.; Arnold, A.; Sielaff, I.; Johnsson, K.; Klok, H.A. Protein-functionalized polymer brushes. *Biomacromolecules* **2005**, *6*, 1602–1607. [CrossRef] [PubMed]

14. Zhou, Y.; Jiang, K.; Song, Q.; Liu, S. Thermo-induced formation of unimolecular and multimolecular micelles from novel double hydrophilic multiblock copolymers of N,N-dimethylacrylamide and N-isopropylacrylamide. *Langmuir* **2007**, *23*, 13076–13084. [CrossRef] [PubMed]

15. Sivakumaran, D.; Maitland, D.; Hoare, T. Injectable microgel-hydrogel composites for prolonged small-molecule drug delivery. *Biomacromolecules* **2011**, *12*, 4112–4120. [CrossRef]

16. Pan, Y.; Bao, H.; Sahoo, N.G.; Wu, T.; Li, L. Water-Soluble Poly(N-isopropylacrylamide)-Graphene Sheets Synthesized via Click Chemistry for Drug Delivery. *Adv. Funct. Mater.* **2011**, *21*, 2754–2763. [CrossRef]

17. Bigot, J.; Charleux, B.; Cooke, G.; Delattre, F.; Fournier, D.; Lyskawa, J.; Sambe, L.; Stoffelbach, F.; Woisel, P. Tetrathiafulvalene end-functionalized poly(N-isopropylacrylamide): A new class of amphiphilic polymer for the creation of multistimuli responsive micelles. *J. Am. Chem. Soc.* **2010**, *132*, 10796–10801. [CrossRef]

18. Oyeneye, O.O.; Xu, W.Z.; Charpentier, P.A. Designing catechol-end functionalized poly(DMAm-co-NIPAM) by RAFT with tunable LCSTs. *J. Polym. Sci. Part A Polym. Chem.* **2017**, *55*, 4062–4070. [CrossRef]

19. Schilli, C.M.; Müller, A.H.E.; Rizzardo, E.; Thang, S.H.; Chong, Y.K. RAFT Polymers: Novel Precursors for Polymer—Protein Conjugates. In *Advances in Controlled/Living Radical Polymerization*; American Chemical Society: Washington, DC, USA, 2003; Volume 854, pp. 603–618.

20. Kujawa, P.; Segui, F.; Shaban, S.; Diab, C.; Okada, Y.; Tanaka, F.; Winnik, F.M. Impact of End-Group Association and Main-Chain Hydration on the Thermosensitive Properties of Hydrophobically Modified Telechelic Poly(N-isopropylacrylamides) in Water. *Macromolecules* **2006**, *39*, 341–348. [CrossRef]

21. Zhang, Q.; Weber, C.; Schubert, U.S.; Hoogenboom, R. Thermoresponsive polymers with lower critical solution temperature: From fundamental aspects and measuring techniques to recommended turbidimetry conditions. *Mater. Horiz.* **2017**, *4*, 109–116. [CrossRef]

22. Quinn, J.F.; Davis, T.P.; Rizzardo, E. Ambient temperature reversible addition–fragmentation chain transfer polymerisation. *Chem. Commun.* **2001**, *11*, 1044–1045. [CrossRef]

23. Glass, P.; Chung, H.; Washburn, N.R.; Sitti, M. Enhanced reversible adhesion of dopamine methacrylamide-coated elastomer microfibrillar structures under wet conditions. *Langmuir* **2009**, *25*, 6607–6612. [CrossRef] [PubMed]

24. Hou, L.; Wu, P. Comparison of LCST-transitions of homopolymer mixture, diblock and statistical copolymers of NIPAM and VCL in water. *Soft matter* **2015**, *11*, 2771–2781. [CrossRef] [PubMed]

25. Moad, G.; Rizzardo, E.; Thang, S.H. Toward living radical polymerization. *Acc. Chem. Res.* **2008**, *41*, 1133–1142. [CrossRef] [PubMed]

26. Sun, X.; Tyagi, P.; Agate, S.; Lucia, L.; McCord, M.; Pal, L. Unique thermo-responsivity and tunable optical performance of poly(N-isopropylacrylamide)-cellulose nanocrystal hydrogel films. *Carbohydr. Polym.* **2019**, *208*, 495–503. [CrossRef] [PubMed]

27. Guillet, P.; Mugemana, C.; Stadler, F.J.; Schubert, U.S.; Fustin, C.A.; Bailly, C.; Gohy, J.F. Connecting micelles by metallo-supramolecular interactions: Towards stimuli responsive hierarchical materials. *Soft Matter* **2009**, *5*, 3409. [CrossRef]

28. Abetz, V.; Simon, P.F.W. Phase Behaviour and Morphologies of Block Copolymers. In *Block Copolymers I*; Springer: Berlin/Heidelberg, Germany, 2005.

29. Alexandridis, P.; Alan Hatton, T. Poly(ethylene oxide) poly(propylene oxide) poly(ethylene oxide) block copolymer surfactants in aqueous solutions and at interfaces: Thermodynamics, structure, dynamics, and modeling. *Colloids Surf. A* **1995**, *96*, 1–46. [CrossRef]

30. Castelletto, V.; Caillet, C.; Fundin, J.; Hamley, I.W.; Yang, Z.; Kelarakis, A. The liquid–solid transition in a micellar solution of a diblock copolymer in water. *J. Chem. Phys.* **2002**, *116*, 10947–10958. [CrossRef]

31. Chiper, M.; Hoeppener, S.; Schubert, U.S.; Fustin, C.-A.; Gohy, J.-F. Self-Assembly Behavior of Bis(terpyridine) and Metallo-bis(terpyridine) Pluronics in Dilute Aqueous Solutions. *Macromol. Chem. Phys.* **2010**, *211*, 2323–2330. [CrossRef]

32. Luan, B.; Muir, B.W.; Zhu, J.; Hao, X. A RAFT copolymerization of NIPAM and HPMA and evaluation of thermo-responsive properties of poly(NIPAM-co-HPMA). *RSC Adv.* **2016**, *6*, 89925–89933. [CrossRef]

33. Yoshida, R.; Sakai, K.; Okano, T.; Sakurai, Y. Modulating the phase transition temperature and thermosensitivity in N-isopropylacrylamide copolymer gels. *J. Biomater. Sci. Polym. Ed.* **2012**, *6*, 585–598. [CrossRef]

34. Neradovic, D.; Hinrichs, W.L.J.; Kettenes-van den Bosch, J.J.; Hennink, W.E. Poly(N-isopropylacrylamide) with hydrolyzable lactic acid ester side groups: A new type of thermosensitive polymer. *Macromol. Rapid Commun.* **1999**, *20*, 577–581. [CrossRef]

35. Feil, H.; Bae, Y.H.; Feijen, J.; Kim, S.W. Effect of comonomer hydrophilicity and ionization on the lower critical solution temperature of N-isopropylacrylamide copolymers. *Macromolecules* **1993**, *26*, 2496–2500. [CrossRef]

36. Deng, Y.; Pelton, R. Synthesis and Solution Properties of Poly(N-isopropylacrylamide-co-diallyldimethylammonium chloride). *Macromolecules* **1995**, *28*, 4617–4621. [CrossRef]

37. Tanaka, F.; Koga, T.; Winnik, F.M. Competitive Hydrogen Bonds and Cononsolvency of Poly(N-isopropylacrylamide)s in Mixed Solvents of Water/Methanol. In *Gels: Structures, Properties, and Functions*; Springer: Berlin, Germany, 2009; Volume 136, pp. 1–7.

38. Rojas, G.; Berda, E.B.; Wagener, K.B. Precision polyolefin structure: Modeling polyethylene containing alkyl branches. *Polymer* **2008**, *49*, 2985–2995. [CrossRef]

39. Rojas, G.; Wagener, K.B. Precisely and Irregularly Sequenced Ethylene/1-Hexene Copolymers: A Synthesis and Thermal Study. *Macromolecules* **2009**, *42*, 1934–1947. [CrossRef]

40. Hashmi, S.; Vatankhah-Varnoosfaderani, M.; GhavamiNejad, A.; Obiweluozor, F.O.; Du, B.; Stadler, F.J. Self-associations and temperature dependence of aqueous solutions of zwitterionically modified N-isopropylacrylamide copolymers. *Rheologica Acta* **2015**, *54*, 501–516. [CrossRef]

41. Du, L.; GhavamiNejad, A.; Yan, Z.-C.; Biswas, C.S.; Stadler, F.J. Effect of a functional polymer on the rheology and microstructure of sodium alginate. *Carbohydr. Polym.* **2018**, *199*, 58–67. [CrossRef]

42. Stadler, F.J.; Hashmi, S.; GhavamiNejad, A.; Vatankhah Varnoosfaderani, M. Rheology of dopamine containing polymers. *Annu. Trans. Nord. Rheol. Soc.* **2017**, *25*, 115–120.

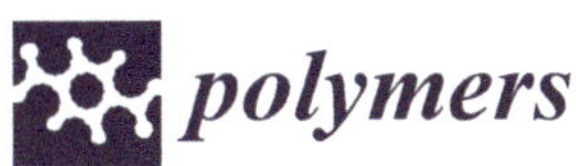 

Article

Reynolds Stress Model for Viscoelastic Drag-Reducing Flow Induced by Polymer Solution

Yi Wang

National Engineering Laboratory for Pipeline Safety/MOE Key Laboratory of Petroleum Engineering/Beijing Key Laboratory of Urban Oil and Gas Distribution Technology, China University of Petroleum, Beijing 102249, China; wangyi1031@cup.edu.cn; Tel.: +86-1881-090-5760

Received: 10 July 2019; Accepted: 9 October 2019; Published: 11 October 2019

Abstract: Viscoelasticity drag-reducing flow by polymer solution can reduce pumping energy of pipe flow significantly. One of the simulation manners is direct numerical simulation (DNS). However, the computational time is too long to accept in engineering. Turbulent model is a powerful tool to solve engineering problems because of its fast computational ability. However, its precision is usually low. To solve this problem, we introduce DNS to provide accurate data to construct a high-precision turbulent model. A Reynolds stress model for viscoelastic polymer drag-reducing flow is established. The rheological behavior of the drag-reducing flow is described by the Giesekus constitutive Equation. Compared with the DNS data, mean velocity, mean conformation tensor, drag reduction, and stresses are predicted accurately in low Reynolds numbers and Weissenberg numbers but worsen as the two numbers increase. The computational time of the Reynolds stress model (RSM) is only 1/120,960 of DNS, showing the advantage of computational speed.

Keywords: Reynolds stress model; polymer; turbulent model; drag reduction; DNS

---

1. Introduction

Drag reduction (DR) phenomenon was first discovered by Toms [1]. He observed in his experiment that the addition of a long-chain polymer (polymethyl methacrylate) in monochlorobenzene dramatically reduced the turbulent skin friction by as high as 80%. The flow rate could be increased by the addition of the polymer at constant pressure gradient. Then, he reported these results at the First International Rheological Congress, so it is usually referred to as the "Toms Effect". The polymers that can reduce skin friction were later called drag-reducing agents (DRAs). The energy-saving effect of DRAs attracts many applications. The first famous application for polymer drag reduction was its use in the 48 inch diameter 800 mile length Alaska pipeline, carrying crude oil from the North slope in Alaska to Valdez in the south of Alaska [2]. After injecting a concentrated solution of a high-molecular-weight polymer downstream of pumping stations at homogeneous concentrations as low as 1 ppm [3], crude throughput was increased by up to 30%. Polymer DRAs were also successfully applied in other crude oil pipelines such as Iraq-Turkey, Bass Strait in Australia, Mumbai Offshore [4], and North Sea Offshore [5], and in finished hydrocarbon product lines [6].

The drag-reducing flow induced by polymer solution usually appears viscoelastic. Direct numerical simulation (DNS) can simulate this kind of viscoelastic turbulent flow in high precision [7–11]. More recent progresses are as follow. Dubief et al. [12] investigated the energetics of turbulence by correlating the work done by polymers on the flow with turbulent structures. Polymers are found to store and to release energy to the flow in a well-organized manner. Graham [13] proposed a tentative unified description of rheological drag reduction based on his numerical observations. Thais et al. [14] found that the spectra of its cross-flow component in viscoelastic flows exhibit a significantly higher energy level at a large scale. Pereira et al. [15] studied the polymer–turbulence interactions from an energetic standpoint for a range of Weissenberg numbers and found a cyclic mechanism of energy exchange between the polymers and turbulence that drives the flow through an oscillatory behavior. They also addressed the numerical simulation of thermo-fluid characteristics of triangular jets [16]. However, DNS needs numerous storages of the computer because very dense mesh is required to resolve small eddies in turbulent drag-reducing flow, such that computational time is too long to be accepted in engineering. Turbulent model computes the turbulent flow very quickly. Compared with DNS, turbulent model for viscoelastic drag-reducing flows develops slowly. There were zero-Equation models established by Edwards et al. [17] and Azouz et al. [18]. One-Equation models and two-Equation models were established by Durst et al. [19], and Hassid and Poreh [20–22]. Cruz et al. [23] and Pinho et al. [24] considered elongation thickening of drag-reducing fluid based on Newtonian fluid turbulent flow and derived a new low Reynolds number k-ε model for polymer drag-reducing flow. Elongation thickening is a very important factor of drag reduction. Thus, Pinho's work promoted the turbulent model greatly, but the precision is still not high enough due to the complexity of viscoelastic turbulent flow. Reynolds stress model (RSM) can simulate Newtonian turbulent flow in high precision, so that it has potential advantages to deal with the complex viscoelastic turbulent flow with polymer additives. If the RSM is established based on DNS data, the precision may be better.

In this paper, an RSM for viscoelastic drag-reducing flow is established based on DNS data. The goal is to find a new modeling way to solve drag-reducing flows in engineering with fast computation and good accuracy.

2. Governing Equations

2.1. Instantaneous Equations

Viscoelastic drag-reducing flow can be described by the following governing Equations [25].
(1) Continuity Equation:

$$\frac{\partial u_i}{\partial x_i} = 0 \tag{1}$$

(2) Momentum Equation:

$$\frac{\partial u_i}{\partial t} + u_k \frac{\partial u_i}{\partial x_k} = -\frac{1}{\rho}\frac{\partial p}{\partial x_i} + \frac{\mu}{\rho}\frac{\partial^2 u_i}{\partial x_k^2} + \frac{1}{\lambda \rho}\frac{\partial c_{ik}}{\partial x_k} \tag{2}$$

(3) Giesekus constitutive Equation:

$$\frac{\partial c_{ij}}{\partial t} + \frac{\partial u_m c_{ij}}{\partial x_m} - c_{mj}\frac{\partial u_i}{\partial x_m} - c_{im}\frac{\partial u_j}{\partial x_m} + \frac{1}{\lambda}\left[-\eta\delta_{ij} + c_{ij} + \frac{\alpha}{\eta}\left(c_{im} - \eta\delta_{im}\right)\left(c_{mj} - \eta\delta_{mj}\right)\right] = 0 \tag{3}$$

where ρ and λ are the density and the relaxation time of drag-reducing fluid respectively. α is the mobility factor, which determines the extensional viscosity. p is pressure. u_i ($i = x, y, z$) are the velocity components in the x, y, z directions. c_{ij} is the conformation tensor. μ is the zero-shear-rate viscosity of solvent. η is the zero-shear-rate viscosity of drag-reducing polymer. We choose the following dimensionless transformation: $x_i^* = x_i/h$, $t^* = t/(h/u_\tau)$, $u_i^+ = u_i/u_\tau$, $p^+ = p/\left(\rho u_\tau^2\right)$, $c_{ij}^+ = c_{ij}/\eta$ and use the definition of Weissenberg number $We_\tau = \rho\lambda u_\tau^2/(\mu + \eta)$, the definition of frictional Reynolds number $Re_\tau = \rho u_\tau h/(\mu + \eta)$, and the ratio $\beta = \mu/(\mu + \eta)$. u_τ is the frictional velocity ($u_\tau = \sqrt{\tau_w/\rho}$, τ_w is the wall shear stress). h is the half height of the channel, as shown in Figure 1. Using these definitions, Equations (1)–(3) can be transformed to be the following dimensionless Equations:

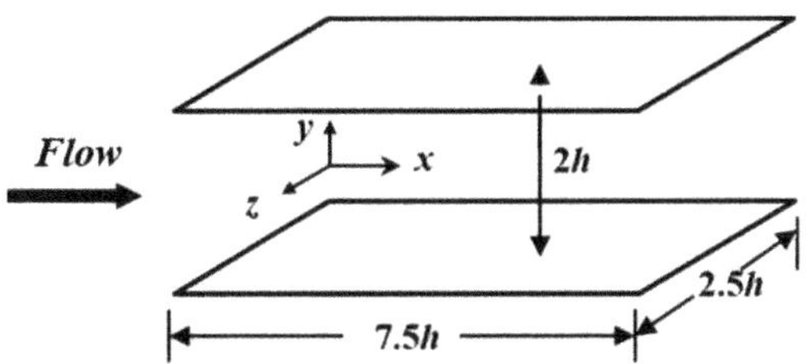

Figure 1. Computational domain.

(4) Dimensionless continuity Equation:

$$\frac{\partial u_i^+}{\partial x_i^*} = 0 \tag{4}$$

(5) Dimensionless momentum Equation:

$$\frac{\partial u_i^+}{\partial t^*} + u_k^+ \frac{\partial u_i^+}{\partial x_k^*} = -\frac{\partial p^+}{\partial x_i^*} + \frac{\beta}{Re_\tau}\frac{\partial^2 u_i^+}{\partial x_k^{*2}} + \frac{1-\beta}{We_\tau}\frac{\partial c_{ik}^+}{\partial x_k^*} \tag{5}$$

(6) Dimensionless Giesekus constitutive Equation:

$$\frac{\partial c_{ij}^+}{\partial t^*} + \frac{\partial u_m^+ c_{ij}^+}{\partial x_m^*} = \frac{Re_\tau}{We_\tau}\left[\delta_{ij} - c_{ij}^+ - \alpha\left(c_{im}^+ - \delta_{im}\right)\left(c_{mj}^+ - \delta_{mj}\right)\right] + \frac{\partial u_i^+}{\partial x_m^*}c_{mj}^+ + \frac{\partial u_j^+}{\partial x_m^*}c_{mi}^+ \tag{6}$$

2.2. Time-Average Equations

Reynolds stress model focuses on the time-average effect of turbulent flow. Thus, instantaneous variables in the above Equations can be considered as the summation of time-average variables and fluctuation variables, that is, $\varphi^+ = \overline{\varphi^+} + \varphi^{+\prime}$ (φ^+ represents instantaneous variables (u_i^+, p^+, c_{ij}^+ etc.); superscripts "—" and "′" represent time-average variables and fluctuation variables respectively). After this decomposition, time-average operations can be made for the instantaneous Equations to obtain the following time-average Equations:

(1) Time-average momentum Equations:

$$\frac{\partial \overline{u_i^+}}{\partial t^*} + \overline{u_k^+}\frac{\partial \overline{u_i^+}}{\partial x_k^*} + \frac{\partial \overline{u_i^{+\prime} u_k^{+\prime}}}{\partial x_k^*} = -\frac{\partial \overline{p^+}}{\partial x_i^*} + \frac{\beta}{Re_\tau}\frac{\partial^2 \overline{u_i^+}}{\partial x_k^{*2}} + \frac{1-\beta}{We_\tau}\frac{\partial \overline{c_{ik}^+}}{\partial x_k^*} \tag{7}$$

(2) Time-average constitutive Equations:

$$\frac{\partial \overline{c_{ij}^+}}{\partial t^*} + \frac{\partial \overline{u_m^+ c_{ij}^+}}{\partial x_m^*} + \underbrace{\frac{\partial \overline{u_m^{+\prime} c_{ij}^{+\prime}}}{\partial x_m^*}}_{A_{ij}} = \frac{Re_\tau}{We_\tau}\left[\delta_{ij} - \overline{c_{ij}^+} - \alpha\left(\overline{c_{im}^+ c_{mj}^+} - 2\overline{c_{ij}^+} + \delta_{ij}\right)\right] - \underbrace{\alpha\frac{Re_\tau}{We_\tau}\overline{c_{im}^{+\prime} c_{mj}^{+\prime}}}_{B_{ij}}$$

$$+ \underbrace{\frac{\partial \overline{u_i^+}}{\partial x_m^*}\overline{c_{mj}^+} + \frac{\partial \overline{u_j^+}}{\partial x_m^*}\overline{c_{mi}^+} + \overline{\frac{\partial u_i^{+\prime}}{\partial x_m^*}c_{mj}^{+\prime}} + \overline{\frac{\partial u_j^{+\prime}}{\partial x_m^*}c_{mi}^{+\prime}}}_{C_{ij}} \tag{8}$$

Equations (2)–(7) so that we can obtain the following fluctuation Equations.

(3) Fluctuation Equations:

$$\frac{\partial u_i^{+\prime}}{\partial t^*} + u_k^{+\prime}\frac{\partial \overline{u_i^+}}{\partial x_k^*} + \overline{u_k^+}\frac{\partial u_i^{+\prime}}{\partial x_k^*} = -\frac{\partial p^{+\prime}}{\partial x_i^*} + \frac{\beta}{Re_\tau}\frac{\partial^2 u_i^{+\prime}}{\partial x_k^{*2}} + \frac{1-\beta}{We_\tau}\frac{\partial c_{ik}^{+\prime}}{\partial x_k^*} - \frac{\partial}{\partial x_k^*}\left(u_i^{+\prime}u_k^{+\prime} - \overline{u_i^{+\prime}u_k^{+\prime}}\right) \tag{9}$$

Equation (9) can also be rewrote as:

$$\frac{\partial u_j^{+\prime}}{\partial t^*} + u_k^{+\prime}\frac{\partial \overline{u_j^+}}{\partial x_k^*} + \overline{u_k^+}\frac{\partial u_j^{+\prime}}{\partial x_k^*} = -\frac{\partial p^{+\prime}}{\partial x_j^*} + \frac{\beta}{Re_\tau}\frac{\partial^2 u_j^{+\prime}}{\partial x_k^{*2}} + \frac{1-\beta}{We_\tau}\frac{\partial c_{jk}^{+\prime}}{\partial x_k^*} - \frac{\partial}{\partial x_k^*}\left(u_j^{+\prime}u_k^{+\prime} - \overline{u_j^{+\prime}u_k^{+\prime}}\right) \tag{10}$$

Equation (9) $\times u_j^{+\prime}$ + Equation (10) $\times u_i^{+\prime}$ and do the time average, we can obtain the following Reynolds stress transport Equations.

(4) Reynolds stress transport Equations:

$$\frac{\partial \overline{u_i^{+\prime}u_j^{+\prime}}}{\partial t^*} + \overline{u_k^+}\frac{\partial \overline{u_i^{+\prime}u_j^{+\prime}}}{\partial x_k^*} = \underbrace{-\overline{u_i^{+\prime}u_k^{+\prime}}\frac{\partial \overline{u_j^+}}{\partial x_k^*} - \overline{u_j^{+\prime}u_k^{+\prime}}\frac{\partial \overline{u_i^+}}{\partial x_k^*}}_{P_{ij}} + \underbrace{\overline{p^{+\prime}\left(\frac{\partial u_i^{+\prime}}{\partial x_j^*} + \frac{\partial u_j^{+\prime}}{\partial x_i^*}\right)}}_{\phi_{ij}}$$

$$\underbrace{-\frac{\partial}{\partial x_k^*}\left(\overline{p^{+\prime}u_i^{+\prime}}\delta_{jk} + \overline{p^{+\prime}u_j^{+\prime}}\delta_{ik} + \overline{u_i^{+\prime}u_j^{+\prime}u_k^{+\prime}}\right) + \frac{\beta}{Re_\tau}\frac{\partial^2 \overline{u_i^{+\prime}u_j^{+\prime}}}{\partial x_k^{*2}}}_{T_{ij}} \underbrace{- 2\frac{\beta}{Re_\tau}\overline{\frac{\partial u_i^{+\prime}}{\partial x_k^*}\frac{\partial u_j^{+\prime}}{\partial x_k^*}}}_{\varepsilon_{ij}} \tag{11}$$

$$\underbrace{+\frac{1-\beta}{We_\tau}\frac{\partial}{\partial x_k^*}\left(\overline{u_i^{+\prime}c_{jk}^{+\prime}} + \overline{u_j^{+\prime}c_{ik}^{+\prime}}\right)}_{E_{ij}} \underbrace{-\frac{1-\beta}{We_\tau}\left(\overline{c_{ik}^{+\prime}\frac{\partial u_j^{+\prime}}{\partial x_k^*}} + \overline{c_{jk}^{+\prime}\frac{\partial u_i^{+\prime}}{\partial x_k^*}}\right)}_{F_{ij}}$$

All these Equations contain high-order moments A_{ij}, B_{ij}, C_{ij}, ϕ_{ij}, T_{ij}, ε_{ij}, E_{ij}, F_{ij} and $\overline{u_i^{+\prime}u_j^{+\prime}}$. They are new unknowns. Apparently, the number of unknowns exceeds the number of Equations, so that all the high-order unknowns need to be modeled as functions of time-average variables.

3. Modeling of High-Order Moments of Fluctuations

Different from Newtonian turbulent flow, the above high-order moments of drag-reducing flow are all related to viscoelasticity. The terms directly related to viscoelasticity (A_{ij}, B_{ij}, C_{ij}, E_{ij}, F_{ij}) do not appear in the turbulent models of Newtonian flow, such that they are modeled for the first time. The other terms (ϕ_{ij}, T_{ij}, ε_{ij}) have implicit and complex relations with viscoelasticity, such that the modeling manner of these terms uses the same way of Newtonian flow. The modeling process is as follows.

3.1. High-Order Moments Directly Related to Viscoelasticity

The additional turbulent diffusion terms introduced by viscoelasticity (A_{ij}, E_{ij}) and the correlation of conformation tensor B_{ij} are much smaller than turbulent diffusion. Thus, they can be neglected. To conform the physical process and consider the easy use of the model, C_{ij} is assumed to have relations with Reynolds stress, mean strain rate, and so on. According to the positive definiteness of Reynolds stress and nonpositive definiteness of mean strain rate, the dimensional form of C_{ij} can be expressed as:

$$C_{ij}^{\ \dim} = \varphi_1 f_w \lambda \frac{\rho}{T_t}\overline{u_i'u_j'} + \varphi_2 \frac{1}{T_t}\left(\overline{c_{ij}} - \eta\delta_{ij}\right) \tag{12}$$

where $C_{ij}^{\text{dim}} = C_{ij}\eta u_\tau/h$, $f_w = 1 - \exp(-y^+/100)$, y^+ is the dimensionless distance to the lower wall of the channel, $T_t (= (v/\varepsilon)^{1/2})$ is Kolmogorov time scale calculated from Kolmogorov length scale $((v^3/\varepsilon)^{1/4})$ and Kolmogorov velocity scale $((\varepsilon v)^{1/4})$. $v (= \mu/\rho)$ is kinetic viscosity of fluid.

Equation (8) is nondimensionalized using $\varepsilon^+ = \varepsilon/(u_\tau^3/h)$, fluid density, viscosity, and frictional velocity and combined with the definitions of Re_τ, We_τ, and β. The model of C_{ij} is as follows:

$$C_{ij} = \varphi_1 f_w \frac{We_\tau}{(1-\beta)\sqrt{\beta/(Re_\tau \varepsilon^+)}} \overline{u_i^{+\prime} u_j^{+\prime}} + \varphi_2 \frac{1}{\sqrt{\beta/(Re_\tau \varepsilon^+)}} \left(\overline{c_{ij}^+} - \delta_{ij}\right) \tag{13}$$

where $\varphi_1 = 0.05$, $\varphi_2 = -0.01$. From observations, F_{ij} and C_{ij} have the following simple relation:

$$F_{ij} = -\frac{1-\beta}{We_\tau} C_{ij} \tag{14}$$

3.2. High-Order Moments Indirectly Related to Viscoelasticity

For turbulent diffusion terms T_{ij}, Daly and Harlow's model [26] is applied:

$$T_{ij} = \frac{\partial}{\partial x_k^*}\left(C_s \frac{k^+}{\varepsilon^+} \overline{u_k^{+\prime} u_l^{+\prime}} \frac{\partial \overline{u_i^{+\prime} u_j^{+\prime}}}{\partial x_l^*}\right) \tag{15}$$

where k^+ and ε^+ are dimensionless turbulent kinetic energy and energy dissipation rate. The model parameter C_s is 0.22.

Turbulent dissipation term ε_{ij} and redistribution term ϕ_{ij} are modeled together:

$$-\varepsilon_{ij} + \phi_{ij} = -\frac{2}{3}\delta_{ij}\varepsilon^+ + \phi_{(1)ij} + \phi_{(2)ij} \tag{16}$$

where $\phi_{(1)ij} = -C_1 \frac{\varepsilon^+}{k^+}\left(\overline{u_i^{+\prime} u_j^{+\prime}} - \frac{2}{3}k^+\delta_{ij}\right)$, $\phi_{(2)ij} = -C_2\left(P_{ij} - \frac{2}{3}\delta_{ij}P\right) - C_3\left(D_{ij} - \frac{2}{3}\delta_{ij}P\right) - C_4 k^+\left(\frac{\partial \overline{u_i^+}}{\partial x_j^*} + \frac{\partial \overline{u_j^+}}{\partial x_i^*}\right)$,
$P = P_{kk}/2$, $D_{ij} = -\left(\overline{u_j^{+\prime} u_k^{+\prime}}\frac{\partial \overline{u_k^+}}{\partial x_i^*} + \overline{u_i^{+\prime} u_k^{+\prime}}\frac{\partial \overline{u_k^+}}{\partial x_j^*}\right)$. Shima [27] gave the expressions of coefficients:

$$C_1 = 1 + 2.45A_2^{1/4}A^{3/4}\left\{1 - \exp\left[-(7A)^2\right]\right\}\left\{1 - \exp\left[-(R_T/60)^2\right]\right\} \tag{17}$$

$$C_2 = 0.7A \tag{18}$$

$$C_3 = 0.3A^{1/2} \tag{19}$$

$$C_4 = 0.65A(0.23C_1 + C_2 - 1) + 1.3A_2^{1/4}C_3 \tag{20}$$

where $A = 1 - 9A_2/8 + 9A_3/8$, $A_2 = a_{ij}a_{ji}$, $A_3 = a_{ij}a_{jk}a_{ki}$, $a_{ij} = \overline{u_i^{+\prime} u_j^{+\prime}}/k^+ - 2\delta_{ij}/3$, $R_T = k^{+2}/(\varepsilon^+\beta/Re_\tau)$.

The above modeling introduces two new unknowns, k^+ and ε^+, which need be modeled. k^+ can be solved from the definition of turbulent kinetic energy ($k^+ = \overline{u_i^{+\prime} u_i^{+\prime}}/2$), where $\overline{u_i^{+\prime} u_i^{+\prime}}$ can be solved from the Reynolds stress transport Equation (Equation (11)). ε^+ can be solved from Shima's model [27]:

$$\frac{\partial \varepsilon^+}{\partial t^*} + \overline{u_k^+}\frac{\partial \varepsilon^+}{\partial x_k^*} = C_{\varepsilon 1}\frac{\varepsilon^+}{k^+}P - C_{\varepsilon 2}\frac{\varepsilon^+\widetilde{\varepsilon}^+}{k^+} + \frac{\partial}{\partial x_k^*}\left(C_\varepsilon \frac{k^+}{\varepsilon^+}\overline{u_k^{+\prime} u_l^{+\prime}}\frac{\partial \varepsilon^+}{\partial x_l^*} + \frac{\beta}{Re_\tau}\frac{\partial \varepsilon^+}{\partial x_k^*}\right) \tag{21}$$

where $\widetilde{\varepsilon}^+ = \varepsilon^+ - 2\frac{\beta}{Re_\tau}\left[\frac{\partial (k^+)^{1/2}}{\partial x_i^*}\right]^2$, $C_{\varepsilon 1} = 1.44 + \beta_1 + \beta_2$, $C_\varepsilon = 0.15$, $C_{\varepsilon 2} = 1.92$,

$$\beta_1 = 0.25A\min(\lambda'/2.5 - 1, 0) - 1.4A\min(P/\varepsilon^+ - 1, 0), \beta_2 = 1.0A\lambda'^2\max(\lambda'/2.5 - 1, 0),$$

$$\lambda' = \min(\lambda^*, 4), \lambda^* = \left[\frac{\partial}{\partial x_i}\left(\frac{k^{+3/2}}{\varepsilon^+}\right)\frac{\partial}{\partial x_i}\left(\frac{k^{+3/2}}{\varepsilon^+}\right)\right]^{1/2}.$$

All the high-order moments are modeled to be the functions of time-average variables. Equations (8)–(17) combined with Equations (4), (5), and (7) compose a Reynolds stress model.

4. Results and Discussion

The above Reynolds stress model was used to simulate the fully developed viscoelastic drag-reducing channel flow. The computational domain is shown in Figure 1. Periodic boundary conditions were imposed in both the streamwise (x-) and spanwise (z-) directions, while nonslip boundary conditions were adopted for the top and bottom walls. Computational parameters were: $Re_\tau = 150$, $We_\tau = 10$, $\alpha = 0.001$, $\beta = 0.8$.

The numerical method of DNS is a fractional step method. Adams–Bashforth scheme was used to ensure the second-order accuracy of velocity. An implicit scheme was used for the pressure term. Staggered grid was applied to avoid unphysical oscillations of pressure. A second-order finite difference scheme was used for spatial discretization. Uniform mesh was used in the x and z directions due to the periodic boundary condition. To capture small eddies near the walls, nonuniform mesh was used in the y direction. Grid number was $64 \times 64 \times 64$. Drag reduction is defined as:

$$DR\% = \frac{C_{fDean} - C_f}{C_{fDean}} \times 100\% \tag{22}$$

where C_f is the calculated friction factor, C_{fDean} is evaluated by Dean's correlation [28], $DR\%$ is the drag reduction.

Mean streamwise velocity, drag reduction, Reynolds stress, and fluctuation intensity were obtained and compared with the results of DNS to validate the model. Bulk mean variables are listed in Table 1. The results obtained by the turbulent model (RSM) agree well with the DNS results. The relative deviations of mean streamwise velocity, Reynolds number, frictional factor, and drag reduction were 0.57%, 0.57%, 1.27%, and 9.3%. Thus, the prediction of the bulk mean variables by RSM is accurate.

Table 1. Comparison of bulk mean variables.

	u_m^+	Re_m	C_f	$DR\%$
Direct numerical simulation (DNS)	15.87	4762	0.0079	9.7%
Reynolds stress model (RSM)	15.96	4789	0.0078	10.6%

Figure 2 is the comparison of time-average streamwise velocity profiles between RSM and DNS. The DNS results for Newtonian turbulent flow agree well with the typical distribution of viscous sublayer ($u^+ = y^+$), buffer layer ($u^+ = 5\ln y^+ - 3.05$), and turbulent core region ($u^+ = 2.5\ln y^+ + 5.5$) [29], showing that the fluid has achieved fully developed turbulent flow. The time-average velocity of RSM coincides well with DNS in the viscous sublayer, is lower than DNS in the buffer layer, and is higher than DNS in the turbulent core region. The mean deviation is only 9.1%, which is much lower than previous turbulent models (as high as 30–50% or more). This indicates the established model can predict the time-average streamwise velocity profile accurately. Fluctuation intensities are compared in Figure 3. Predicted v_{rms}^+ and w_{rms}^+ are larger than DNS, while u_{rms}^+ is smaller than DNS. The peak value positions of u_{rms}^+ v_{rms}^+, and w_{rms}^+ of RSM are basically the same as DNS.

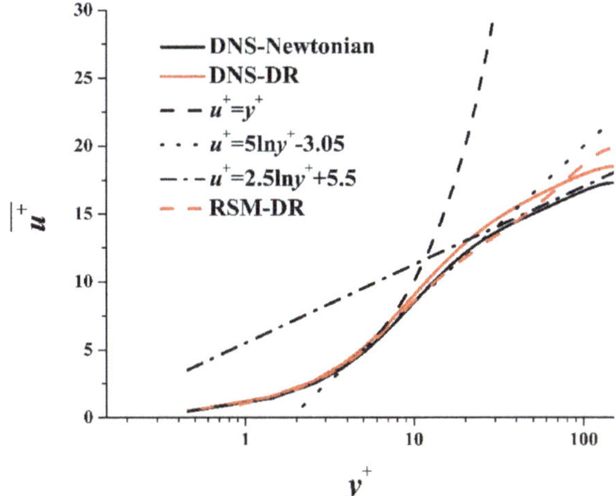

Figure 2. Time-average streamwise velocity profiles.

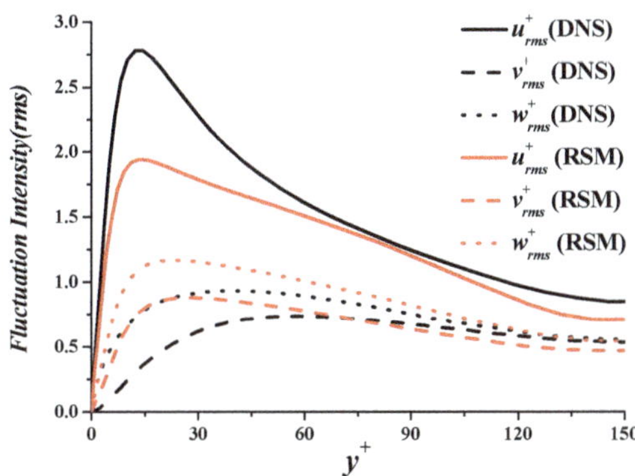

Figure 3. Fluctuation intensities.

Stress balance of Reynolds stress, viscous stress, and viscoelastic stress is shown in Figure 4. Prediction results of the three stresses agree well with those of DNS. Predicted Reynolds stress is smaller than DNS. Predicted viscoelastic stress is almost the same as that of DNS.

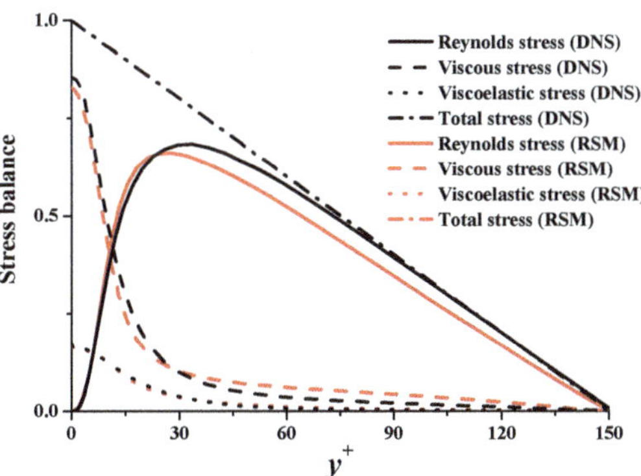

Figure 4. Stress balance.

C_{mij} represents the mean conformation tensor $\overline{c_{ij}^+}$. The four main components of the conformation tensor, with subscripts 1, 2, 3 representing streamwise, wall-normal, and spanwise directions, are compared in Figure 5. It shows that the four components agree well with DNS. The peak value of C_{m11} is 15% lower than the value in DNS. C_{m12} of RSM coincides well with DNS. It determines the value of viscoelastic stress, such that the deviation of viscoelastic stress is small in Figure 4.

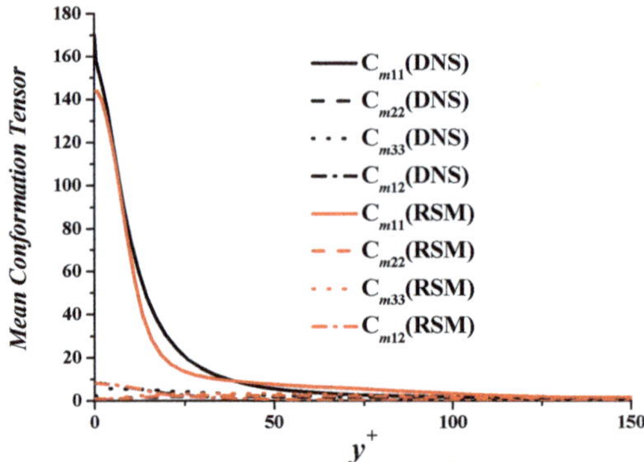

Figure 5. Mean conformation tensor.

The RSM for the above case is basically verified. We further verify the RSM using the DNS results in literature [30] at different Weissenberg numbers. Table 2 shows the bulk mean variables such as drag reduction. It is apparent that the RSM can also simulate the drag-reducing flow accurately at $We_\tau = 12.5$ and 30, which correspond to low and medium drag reduction, respectively. Figures 6 and 7 show that the critical features (e.g., average velocity profile and Reynolds stress distribution) can be simulated by the RSM in satisfied precision.

Table 2. Comparison of bulk mean variables with literature [30].

	$\overline{u_m^+}$	Re_m	C_f	DR%
DNS ($We_\tau = 12.5$) [30]	16.13	4838	0.00769	12.1%
RSM ($We_\tau = 12.5$)	16.10	4829	0.00772	11.9%
DNS ($We_\tau = 30$) [30]	20.6	6180	0.00471	42.8%
RSM ($We_\tau = 30$)	21.2	6373	0.00443	45.8%

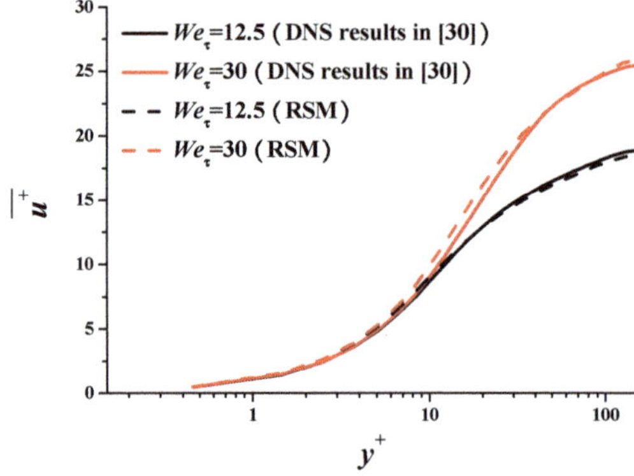

Figure 6. Mean velocity profile comparison with literature.

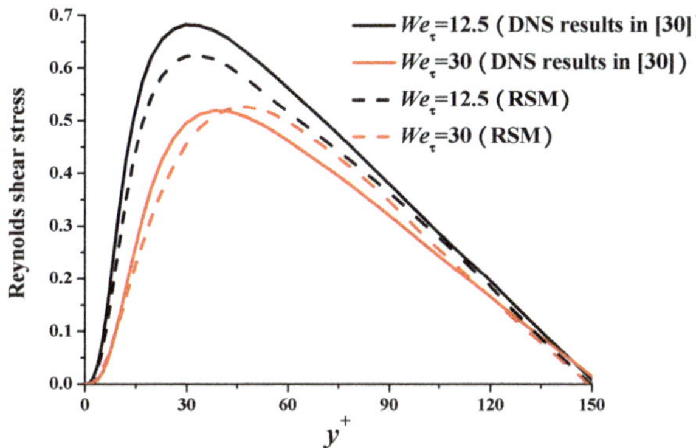

Figure 7. Reynolds shear stress comparison with literature.

RSM only concerns time-average variables so that it does not need to resolve the small eddies near the walls like DNS. The complex and time-consuming numerical methods, such as numerous iterations of pressure, can be avoided. Therefore, the computational time of RSM can be much smaller than that of DNS. Table 3 verifies that RSM can largely save the computational time because the acceleration ratio of computational time of DNS and RSM is as high as 120,960.

Table 3. Computational time.

DNS	RSM	Acceleration Ratio
604,800s	5s	120,960

To further examine the precision of the RSM at higher Reynolds numbers, it is compared with the typical DNS results made by Housiadas et al. [31] in Table 4. The drag reduction of the RSM agrees well with the DNS data in low ($Re_\tau = 125$) and medium ($Re_\tau = 180$) Reynolds numbers. The deviation becomes larger at higher Reynolds number ($Re_\tau = 395$). The model was also used to predict the effect of Weissenberg number at the higher Reynolds number (Table 5). Drag reduction increases with increasing We_τ and tends to a stable value about more than 40%.

Table 4. Drag reduction ($DR\%$) comparison with Housiadas' results [31] at different Re_τ.

Case	DNS	RSM	Deviation
$Re_\tau = 125, We_\tau = 50, \beta = 0.9$	29.8%	29.2%	2.0%
$Re_\tau = 180, We_\tau = 50, \beta = 0.9$	29.9%	28.6%	4.3%
$Re_\tau = 395, We_\tau = 50, \beta = 0.9$	29.6%	25.7%	13.2%

Table 5. $DR\%$ at different We_τ of the RSM for high Re_τ.

Case	DR%
$Re_\tau = 395, We_\tau = 25, \beta = 0.9$	24.7%
$Re_\tau = 395, We_\tau = 50, \beta = 0.9$	25.7%
$Re_\tau = 395, We_\tau = 100, \beta = 0.9$	28.0%
$Re_\tau = 395, We_\tau = 150, \beta = 0.9$	33.8%
$Re_\tau = 395, We_\tau = 200, \beta = 0.9$	41.1%
$Re_\tau = 395, We_\tau = 250, \beta = 0.9$	44.4%

5. Conclusions

Through the discussions, it is apparent that the newly established Reynolds stress turbulent model can predict bulk mean velocity and drag reduction accurately. The predictions of stresses and mean conformation tensor are also good. The time-average streamwise velocity outside the viscous sublayer is quite different from that in DNS. Streamwise fluctuation intensity is much smaller than that in DNS. This is probably because the modeling of the redistribution term ϕ_{ij} is the same as that for Newtonian fluid. This term may affect viscoelasticity and pressure in a complex way, which needs further study in future to find a more accurate modeling manner. The acceleration ratio is as high as 120,960, showing the extremely high speed of the new turbulent model of polymeric drag-reducing flow. The Reynolds stress model appears to have good accuracy at lower Reynolds numbers compared with typical DNS results. The model predictions significantly worsen as the Reynolds number and Weissenberg number increase, limiting the applicability of the model up to modest levels of drag reduction. This suggests (in association with the other limitations already stated before) that there is room for model improvement in future work.

Funding: This research was funded by National Natural Science Foundation of China (NSFC), grant number No.51576210.

Acknowledgments: The guidance and support of Bo Yu is highly appreciated.

Conflicts of Interest: The authors declare no conflict of interest. The funders had no role in the design of the study; in the collection, analyses, or interpretation of data; in the writing of the manuscript, or in the decision to publish the results.

References

1. Toms, B.A. Some observations on the flow of linear polymer solutions through straight tubes at large Reynolds numbers. *Proc. Int. Congr. Rheol.* **1948**, *135*, 1948.
2. Burger, E.D.; Munk, W.R.; Wahl, H.A. Flow increase in the Trans Alaska Pipeline through use of a polymeric drag reducing additive. *J. Petrol. Technol.* **1982**, *34*, 377–386. [CrossRef]
3. Motier, J.F.; Chou, L.C.; Kommareddi, N.S. Commercial drag reduction: past, present, and future. In *Proceedings of the ASME Fluids Engineering Division Summer Meeting*; ASME: San Diego, CA, USA, 1996; pp. 229–234.
4. Nijs, L. New generation drag reducer. *Pipeline Technology* **1995**, *2*, 143–149.
5. Dujmovich, T.; Gallegos, A. Drag reducers improve throughput cut costs. *Offshore* **2005**, *65*, 1–4.
6. Motier, J.F.; Carreir, A.M. Recent studies on polymer drag reduction in commercial pipelines. In *Drag Reduction in Fluid Flows: Techniques for Friction Control*; Sellin, R., Moses, R., Eds.; Ellis Horwood: West Sussex, UK, 1989; pp. 197–204.
7. Sureshkumar, R.; Beris, A.N.; Handler, R.A. Direct numerical simulation of the turbulent channel flow of a polymer solution. *Phys. Fluids* **1997**, *9*, 743–755. [CrossRef]
8. Dimitropoulos, C.D.; Sureshkumar, R.; Beris, A.N. Direct numerical simulation of viscoelastic turbulent channel flow exhibiting drag reduction: Effect of the variation of rheological parameters. *J. Non-Newtonian Fluid Mech.* **1998**, *78*, 433–468. [CrossRef]
9. Min, T.; Yoo, J.Y.; Choi, H.; Joseph, D.D. Drag reduction by polymer additives in a turbulent channel flow. *J. Fluid Mech.* **2003**, *486*, 213–238. [CrossRef]
10. Ptasinski, P.K.; Boersma, B.J.; Nieuwstadt, F.T.M.; Hulsen, M.A.; Van Den Brule, B.H.A.A.; Hunt, J.C.R. Turbulent channel flow near maximum drag reduction: Simulations, experiments and mechanisms. *J. Fluid Mech.* **2003**, *490*, 251–291. [CrossRef]
11. White, C.M.; Mungal, M.G. Mechanics and prediction of turbulent drag reduction with polymer additives. *Ann. Rev. Fluid Mech.* **2008**, *40*, 235–256. [CrossRef]
12. Dubief, Y.; White, C.; Terrapon, V.; Shaqfeh, E.; Moin, P.; Lele, S. On the coherent drag-reducing and turbulence-enhancing behaviour of polymers in wall flows. *J. Fluid Mech.* **2004**, *514*, 271–280. [CrossRef]
13. Graham, M.D. Drag reduction and the dynamics of turbulence in simple and complex fluids. *Phys. Fluids* **2014**, *26*, 625–656. [CrossRef]

14. Thais, L.; Mompean, G.; Gatski, T.B. Spectral analysis of turbulent viscoelastic and Newtonian channel flows. *J. Non-Newtonian. Fluid Mech.* **2013**, *200*, 165–176. [CrossRef]

15. Pereira, A.S.; Mompean, G.; Thais, L.; Thompson, R.L. Statistics and tensor analysis of polymer coil-stretch mechanism in turbulent drag reducing channel flow. *J. Fluid Mech.* **2017**, *824*, 135–173. [CrossRef]

16. Pereira, A.S.; Mompean, G.; Thais, L.; Soares, E.J.; Thompson, R.L. Active and hibernating turbulence in drag-reducing plane Couette flows. *Phys. Rev. Fluids* **2017**, *2*, 084605. [CrossRef]

17. Edwards, M.F.; Smith, R. The use of eddy viscosity expressions for predecting velocity profiles in Newtonian, non-Newtonian and drag-reducing turbulent pipe flow. *J. Non-Newtonian. Fluid Mech.* **1980**, *7*, 153–169. [CrossRef]

18. Azouz, I.; Shirazi, S.A. Numerical simulation of drag reducing turbulent flow in annular conduits. *ASME J. Fluids Eng.* **1997**, *119*, 838–846. [CrossRef]

19. Durst, F.; Rastogi, A.K. Calculations of turbulent boundary layer flows with drag reducing polymer additives. *Phys. Fluids* **1977**, *20*, 1975–1985. [CrossRef]

20. Hassid, S.; Poreh, M. A turbulent energy model for flows with drag reduction. *ASME J. Fluids Eng.* **1975**, *97*, 234–241. [CrossRef]

21. Hassid, S.; Poreh, M. A Turbulent energy dissipation model for flows with drag reduction. *Trans. ASME J. Fluids Eng.* **1978**, *100*, 107–112. [CrossRef]

22. Poreh, M.; Hassid, S. Mean velocity and turbulent energy closures for flows with drag reduction. *Phys. Fluids* **1977**, *20*, S193–S196. [CrossRef]

23. Cruz, D.O.A.; Pinho, F.T. Turbulent pipe flow predictions with a low Reynolds number k–ε model for drag reducing fluids. *J. Non-Newtonian. Fluid Mech.* **2003**, *114*, 109–148. [CrossRef]

24. Pinho, F.T. A GNF framework for turbulent flow models of drag reducing fluids and proposal for a k–ε type closure. *J. Non-Newtonian. Fluid Mech.* **2003**, *114*, 149–184. [CrossRef]

25. Wang, Y.; Yu, B.; Wu, X.; Wang, P. POD and wavelet analyses on the flow structures of a polymer drag-reducing flow based on DNS data. *Int. J. Heat Mass Transf.* **2012**, *55*, 4849–4861. [CrossRef]

26. Daly, B.J.; Harlow, F.H. Transport Equations in turbulence. *Phys. Fluids* **1970**, *13*, 2634–2649. [CrossRef]

27. Shima, N. Low-Reynolds-number second-moment closure without wall-reflection redistribution terms. *Int. J. Heat Fluid Flow* **1998**, *19*, 549–555. [CrossRef]

28. Dean, R.B. Reynolds number dependence of skin friction and other bulk flow variables in two-dimensional rectangular duct flow. *J. Fluids Eng.* **1978**, *100*, 215–223. [CrossRef]

29. Virk, P.S.; Mickley, H.S.; Smith, K.A. The ultimate asymptote and mean flow structure in Toms' phenomenon. *J. Appl. Mech.* **1970**, *37*, 488–493. [CrossRef]

30. Yu, B.; Kawaguchi, Y. Effect of Weissenberg number on the flow structure DNS study of drag-reducing flow with surfactant additives. *Int. J. Heat Fluid Flow* **2003**, *24*, 491–499. [CrossRef]

31. Housiadas, K.D.; Beris, A.N. On the skin friction coefficient in viscoelastic wall-bounded flows. *Int. J. Heat Fluid Flow* **2013**, *42*, 49–67. [CrossRef]

 *polymers*

Article

The Effect of Polydimethylsiloxane-Ethylcellulose Coating Blends on the Surface Characterization and Drug Release of Ciprofloxacin-Loaded Mesoporous Silica

Adrianna Skwira, Adrian Szewczyk and Magdalena Prokopowicz *

Department of Physical Chemistry, Faculty of Pharmacy, Medical University of Gdańsk, Hallera 107, Gdańsk 80-416, Poland
* Correspondence: magdalena.prokopowicz@gumed.edu.pl

Received: 12 August 2019; Accepted: 2 September 2019; Published: 4 September 2019

Abstract: In this study, we obtained novel solid films composed of ciprofloxacin-loaded mesoporous silica materials (CIP-loaded MCM-41) and polymer coating blends. Polymer coating blends were composed of ethylcellulose (EC) with various levels of polydimethylsiloxane (PDMS, 0, 1, 2% (v/v)). The solid films were prepared via the solvent-evaporation molding method and characterized by using scanning electron microscopy (SEM), optical profilometry, and wettability analyses. The solid-state of CIP present in the solid films was studied using X-ray diffraction (XRD) and differential scanning calorimetry (DSC). The release profiles of CIP were examined as a function of PDMS content in solid films. The surface morphology analysis of solid films indicated the progressive increase in surface heterogeneity and roughness with increasing PDMS content. The contact angle study confirmed the hydrophobicity of all solid films and significant impact of both PDMS and CIP-loaded MCM-41 on surface wettability. DSC and XRD analysis confirmed the presence of amorphous/semi-crystalline CIP in solid films. The Fickian diffusion-controlled drug release was observed for the CIP-loaded MCM-41 coated with PDMS-free polymer blend, whereas zero-order drug release was noticed for the CIP-loaded MCM-41 coated with polymer blends enriched with PDMS. Both the release rate and initial burst of CIP decreased with increasing PDMS content.

Keywords: coating; drug delivery; surface roughness; polymers; mesoporous silica

---

1. Introduction

Local drug delivery systems have been widely applied in bone tissue diseases [1,2]. They require two most pressing goals to be achieved: a controlled initial release and long-term delivery of drug [3]. At present, mesoporous silica materials, e.g., MCM-41, have been studied as local antibiotic/anticancer drug delivery systems for the bone treatment of osteomyelitis or osteosarcoma due to their desirable features such as large surface area (~800 m^2/g), uniform pore size (~2–6 nm), modifiable surface properties [4,5]. Moreover, mesoporous silica materials have been found as a nontoxic to the surrounding bone tissue and biocompatible with the osteoblast cells [6]. Furthermore, Braun et al. reported that mesoporous silica materials may release small amounts of Si (~60 ppm) through dissolution in simulated body fluid [7]. It has been confirmed that even relatively low concentration of Si has a positive impact on bone homeostasis and density [8]. However, another characteristic feature of many drug-loaded mesoporous silica materials is the high amount of drug released in a burst stage [9]. This can lead to a high local concentration of drug which may be cytotoxic to surrounding tissue. The reduction of burst release can be achieved by using polymeric coatings.

Controlled local drug release provides a suitable dosing of drug in a specific area in a prolonged manner. The use of bone drug delivery systems which provide controlled release is considered as an optimal and benign treatment characterized by increased effectiveness, extended duration of therapeutic effect, and reduced adverse effects, e.g., toxicity [10,11]. Nowadays, polymeric coatings have been emphasized as an interesting way to modify drug release from mesoporous silica matrices providing a physical barrier between the drug entrapped in the pores and the fluids [12]. Different polymers, such as poly(ethylene glycol) (PEG) [13], Eudragit RL and Eudragit S [14] were used for this purpose. Furthermore, blends of two types of polymers, which are known to be non-toxic and exhibit different physicochemical characteristics can be also used [15]. In the polymer coatings the release profiles of drug primarily depend on the polymer/polymer blend ratio. By simply modifying the polymer/polymer blend ratio, the properties of the obtained film can be effectively altered and thus broad ranges of drug release profiles can be provided [15,16].

Apart from the appropriate drug release, another requirement for mesoporous silica materials as local drug delivery systems to bone is their integration with the host tissue. Surface roughness has been considered as a crucial limiting factor for applied materials in the context of physical and biological compatibility with surrounding tissue [17]. In recent years, many materials intended for implantation with various surface structure were designed and examined for cellular response, activity, and bone formation [18,19]. It was proved for endosseous dental implants that a relatively rough surface improves implant stability, promotes tissue ingrowth, and enhances bone apposition [19–21]. Multiple methods of textural property modification are available [22,23]. However, two primary trends can be distinguished: control of the porosity of a material during its synthesis (1), and surface post-treatment of a synthesized material (2) among which a coating has been considered as one of the most effective methods [24].

The main features of appropriate implantable coating mixtures are biocompatibility, biostability and desired mechanical properties [25]. These characteristics are well documented for polydimethylsiloxane (PDMS) [26] and ethylcellulose (EC) [27]. EC is the most stable cellulose derivative and a water-insoluble polymer that has been used in many pharmaceutical applications as a coating or time-release agent. In addition, EC, due to its function as binder, flexible film former, water barrier and rheology modifier, has been increasingly applied in bone tissue engineering [27–29]. PDMS is characterized by a water-insoluble hydrophobic nature, good adhesive and plasticizing capacity, which characterize its influence on drug release and the physical properties of coated materials [30].

Therefore, the use of PDMS:EC combinations as new coating blends for ciprofloxacin-loaded MCM-41 (CIP-loaded MCM-41) was investigated in this paper. As the MCM-41 has been previously well described as a drug delivery system, we decided to modify the drug release rate by using a polymeric coating. Ciprofloxacin was chosen as the model drug due to its broad spectrum activity and common application in osteomyelitis [31]. The main idea of our studies was to prepare novel solid films of polymer-coated CIP-loaded MCM-41 which would provide the desired prolonged drug release with low burst release in vitro. Furthermore, the physicochemical properties of the obtained solid films were thoroughly investigated.

2. Materials and Methods

2.1. Materials

Tetraethyl orthosilicate (TEOS), cetyltrimethylammonium bromide (CTAB), ethanol, aqueous ammonia (25%), hydroxyl-terminated polydimethylsiloxane fluid (PDMS, 65 cSt, d = 0.97 mg/mL) were all obtained from Sigma-Aldrich. Ethylcellulose (EC, Ethocel® 20 cP) was obtained from Dow Chemical. Acidic solution of ciprofloxacin lactate (pH = 3.5, 10 mg/mL, Proxacin) was obtained from Polfa S.A. (Warsaw, Poland).

2.2. Synthesis of Mesoporous Silica Materials (MCM-41)

The synthesis of mesoporous silica materials, MCM-41, was carried out by a templating method using tetraethyl orthosilicate (TEOS) as a silica source and the cationic surfactant cetyltrimethylammonium bromide (CTAB) as a template [32]. Briefly, 125 g of water, 12.5 g of ethanol, 9.18 g of ammonia aq. (25 %), and 2.39 g of CTAB were stirred together in polypropylene beaker on a magnetic mixer (300 rpm) at 25 °C for approx. 15 min until a homogenous solution was formed. The pH of the solution was 10. Then, 10.03 g of TEOS was added and the resulting mixture was continuously stirred for 2 h. Next, hydrothermal treatment of the mixture was carried out at 90 °C for 5 days without stirring. The resulting solid product was recovered by vacuum filtration, washed with 100 mL of absolute ethanol and dried at 40 °C for 1 h. The CTAB template was removed from the product by calcination in air at 550 °C (with heating rate of 1 °C/min) for a period of 6 h in a muffle furnace (FCF 7SM, CZYLOK, Jastrzebie-Zdroj, Poland). The final MCM-41 samples in powders form with the fraction size range of 200–500 μm were obtained by micronisation (Mortar Grinder Pulverisette 2, Fritsch, Weimar, Germany). The micronisation lasted 10 min at 50 rpm.

2.3. Ciprofloxacin Adsorption onto MCM-41

Ciprofloxacin (CIP) adsorption onto MCM-41 was performed according to the following procedure (the optimal conditions of adsorption process were determined in preliminary studies). Briefly, the synthesized MCM-41 (particles fraction in the range of 200 μm–500 μm) was immersed in acidic solution (pH = 3.5) of ciprofloxacin lactate (10 mg/mL) at ratio 100 mg of silica per 2 mL of solution and incubated with stirring (300 rpm) at 25 ± 0.5 °C for 3 h to ensure the equilibrium state. Then, CIP-loaded MCM-41 was centrifuged, separated from supernatant, and freeze-dried (−52 °C, 0.1 mBar, 24 h). The absorbance of the CIP remaining in the supernatant was measured spectrophotometrically at 278 nm (model UV-1800 UV-Vis spectrophotometer, Shimadzu, Kyoto, Japan). The amount of CIP adsorbed onto MCM-41 and adsorption efficiency were calculated using Equations (1) and (2), respectively:

$$Q_e = \frac{(C_0 - C_e) \cdot V}{m} \tag{1}$$

$$\%Ads = \left(\frac{C_0 - C_e}{C_e}\right) \cdot 100\% \tag{2}$$

where Q_e (mg/g) is an amount of CIP adsorbed at the equilibrium state, %Ads (%) is an adsorption efficiency coefficient, C_0 (mg/L) is an initial CIP concentration, C_e (mg/L) is a CIP concentration at equilibrium state, V (L) is a volume of CIP solution, and m (g) is the mass of MCM-41 used.

2.4. Fabrication of Solid Films Composed of CIP-Loaded MCM-41 and Polymer Blends

The polymer coating blends were prepared by mixing 5% (*w/w*) ethylcellulose (EC) ethanolic solutions with 0, 1, 2% (*v/v*) of hydroxyl-terminated polydimethylsiloxane (PDMS, 65 cSt, *d* = 0.97 mg/mL), giving the various PDMS:EC ratios (Table 1). 95% (*v/v*) ethanol was used as a solvent in each formulation. The prepared coating blends were homogenized by sonication for 20 min using a cooling bath. The pure solid films of each formulation were prepared via solvent-evaporation molding method as follows. Equal volumes (250 μL) of F1, F2, and F3 coating blends (Table 1) were poured into polypropylene molds, and incubated till complete ethanol evaporation (30 ± 0.5 °C, 60 ± 10% relative humidity, 24 h). The end of the solidification of the polymer coating blends is regarded as the moment when no changes in weight (within the limits of instrumental error (± 0.01 g)) were detected. After complete evaporation of the ethanol, the PDMS:EC ratios in the pure solid films changed from 1:99 and 2:98 (*v/v*, before solvent evaporation) into 1:4 and 1:2 (*w/w*, after solvent evaporation) for F2 and F3, respectively (Table 1).

Table 1. Composition of each formulation (F1–F3) with PDMS:EC ratio before (*v/v*) and after solvent evaporation (*w/w*).

Before solvent evaporation			
Formulation	**PDMS:EC ratio**	**PDMS content [µL]**	**EC ethanolic solution content [µL]**
F1	0:100	-	250.0
F2	1:99	2.5	247.5
F3	2:98	5.0	245.0
After solvent evaporation			
Formulation	**PDMS:EC ratio**	**PDMS content [mg]**	**EC content [mg]**
F1	0:1	-	10.0
F2	1:4	2.4	9.9
F3	1:2	4.8	9.8

A constant amount of CIP-loaded MCM-41 (200 µm–500 µm) was immersed in each polymer blend at a ratio 6 mg per 250 µL. Solid films of CIP-loaded MCM-41 coated with F1, F2, and F3 (MCM-F1, MCM-F2, MCM-F3) were fabricated in the same manner as pure solid films. The obtained solid films of F1, F2, and F3 and MCM-F1, MCM-F2, MCM-F3 were then removed from the molds and stored in desiccators at room temperature for further analyses.

2.5. Physicochemical Characterization

Pure solid films of F1, F2, and F3 were characterized using the Fourier transform infrared spectroscopy (FTIR, Jasco FT/IR-4200, Jasco, Pfungstadt, Germany). Samples for analysis were prepared using the KBr tablet technique. For a better comparison, the spectra of F2 and F3 were normalized to a maximum absorption of the dominant peak at 804 cm^{-1}, attributed to the Si-C stretching in the Si–CH$_3$ group [33].

The morphological characterization of MCM-F1, MCM-F2, and MCM-F3 solid films was carried out by using scanning electron microscopy (SEM). For comparative purposes, SEM analysis of CIP-loaded MCM-41 (before polymer coating) and pure solid films of F1, F2, F3 was performed. In order to identify the components of pure solid films, the samples were analyzed using the energy-dispersive X-ray spectroscopy (EDX). The samples for SEM were fixed on carbon tape and sputter coated with gold, prior to the analysis. SEM images was obtained on electron microscope (Hitachi SU-70, Japan) at an acceleration voltage of 3 kV.

The images of surface roughness with quantitative data of MCM-F1, MCM-F2, MCM-F3 solid films were obtained using an optical profilometer (Contour GTK, Bruker, Billerica, MA, USA). To obtain quantitative characterization, mapping of 46.9 µm × 62.5 µm size surface was done for 10 samples of each formulation and the average values together with standard deviations of roughness parameters were calculated.

Wettability studies of solid films of F1, F2, F3, and MCM-F1, MCM-F2, MCM-F3 were performed by contact angle measurement at room temperature (DSA G10 goniometer, Kruss GmbH, Hamburg, Germany). The images of sessile drop of water and diiodomethane at the point of intersection (three-phase contact points) between the drop contour and the surface (baseline) were recorded for drop shape analysis (DSA) and determination of contact angle. The contact angle value was determined as an average of 10 measurements of each sample.

The X-ray diffraction (XRD) analysis of F3, MCM-F3, CIP-loaded MCM-41, and CIP (as reverence) and differential scanning calorimetry (DSC) of F3, MCM-F3, CIP-loaded MCM-41, CIP, and MCM-41 (as references) were performed to characterize solid-state of CIP present in the obtained solid films. XRD data were collected using diffractometer (Empyrean PANalytical, Malvern, UK) with a CuKα radiation beam operating at 40 kV and 40 mA, in the 2θ range between 5° and 70° with a step width of

0.02° and at a scanning rate of 0.5°/min. DSC measurements were carried out using a DSC instrument (822e Mettler Toledo, Columbus, Ohio, USA). Samples of about 5 mg were placed in an aluminum pan with a hole in the lid. The experiments were performed under an N_2 atmosphere (20 mL/min gas flow rate). The thermal behavior of the samples was investigated by heating the samples from 25 to 230 °C with 10 °C/min step. Calibration of the instrument was performed using an indium standard.

2.6. Drug Release Analysis

The release study was performed under "sink" conditions (considering approx. 100 mg/mL solubility of CIP in water [34]) according to the following procedure. Each solid film of MCM-F1, MCM-F2, MCM-F3 was immersed in 2 mL of purified water and shaken at 37 ± 0. 5°C (80 rpm) for 7 days. The release medium was exchanged every 24 h, and the amount of CIP released was measured spectrophotometrically at wavelength of 278 nm (Shimadzu model UV-1800 UV-Vis spectrophotometer). Quantitative determinations of the amount of released CIP were based on pre-calibration of the spectrometer using standard water solutions of the CIP.

The results were obtained from data groups of n = 7 and expressed as mean values ± standard deviation. The release kinetic parameters were calculated using linearized forms of Korsmeyer-Peppas, Higuchi models and zero order kinetics presented in Equations (3)–(5), respectively:

$$\log Q = \log k + n \log t \tag{3}$$

$$Q = k_H t \tag{4}$$

$$Q = Q_0 + k_0 t \tag{5}$$

where Q (%) denotes the fraction released over time t (h; days), Q_0 (%) is the initial fraction of released drug, n is a release exponent related to the drug release mechanism, k (h^{-n}; $days^{-n}$) is a rate constant, the k_H is a Higuchi dissolution constant ($h^{-1/2}$; $days^{-1/2}$), and k_0 is the zero order release constant (% of dose released per day). In Equation (3), n < 0.5 indicates quasi-Fickian diffusion, $n = 0.5$ indicates a Fickian diffusion, for n between 0.5 and 1, the drug release is considered as non-Fickian diffusion, and $n = 1$ corresponds to zero-order release of case II diffusion. To find out the mechanism of drug release, the data for first 60% of drug release fraction (Q) were fitted with Korsmeyer-Peppas and Higuchi models.

3. Results and Discussion

3.1. Ciprofloxacin Adsorption onto MCM-41

The CIP adsorption onto successfully synthesized MCM-41 was verified by the Fourier transform infrared spectroscopy technique (Jasco model 4700 FTIR). The characteristic bands of the CIP molecule were observed what confirmed the presence of drug onto MCM-41 after the loading procedure (Supplementary Materials). The adsorption efficiency was 56 ± 2% what corresponded to the mean amount of CIP-loaded of 112 ± 4 mg per each 1 g of MCM-41.

3.2. Solid Films Formation

The pure solid films of three formulations of polymer blends were obtained via solvent-evaporation molding method (F1, F2, F3). The solid films of CIP-loaded MCM-41 coated with polymer blends (MCM-F1, MCM-F2, MCM-F3) were obtained in the same manner. The final thickness and mass of the pure solid films of F1, F2, F3 were ranged from 30 to 40 μm, and 10.0 to 14.6 mg, respectively, whereas solid films of MCM-F1, MCM-F2, MCM-F3 were characterized by thicknesses ranging from 50 to 60 μm and masses ranging from 16.0 to 20.6 mg.

3.3. Physicochemical Characterization

3.3.1. FTIR Characterization

The bulk molecular structure of F1, F2, and F3 pure polymer solid films was characterized using FTIR technique (Figure 1). FTIR spectra of the precursors (EC powder and PDMS liquid) are presented for comparative purposes. The spectrum of F1 (the formulation containing 100% of EC) showed the characteristic broad peaks at ~1055 cm^{-1} and ~1095 cm^{-1} attributed to C–O–C stretching vibrations, and ~1377 cm^{-1} attributed to C–H bending vibrations in the EC molecule [35]. For the pure solid films of F2, F3, the characteristic peak at ~1024 cm^{-1} corresponding to Si–O–Si asymmetrical stretching [36] confirmed the presence of PDMS. In addition, sharp peaks at ~1261 cm^{-1} attributed to Si–C stretching and ~804 cm^{-1} attributed to CH$_3$ deformation in Si–CH$_3$ group of PDMS were noticed [33]. The intensity of the peak at ~1024 cm^{-1} slightly increased for F3 compared to F2, thus confirming a higher content of PDMS in the F3 sample. Additionally, a small peak at ~1377 cm^{-1} of EC is observed for F2, what confirmed the higher content of EC in F2 compared to F3. For both F2 and F3, a peak at ~1094 cm^{-1} was also observed due to the overlapping of Si–O–Si asymmetrical stretching at ~1089 cm^{-1} from the PDMS together with the C–O–C stretching mode at ~1095 cm^{-1} from the EC.

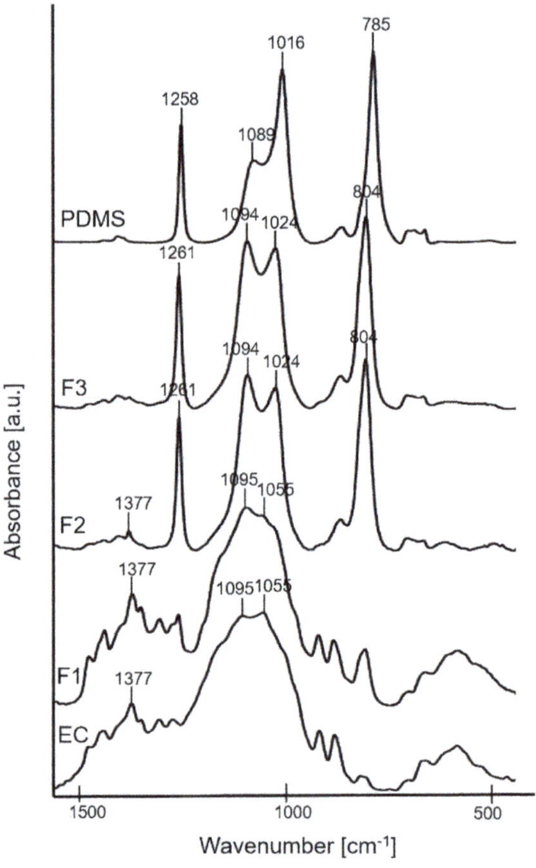

Figure 1. FTIR spectra of F1, F2, F3 solid films, and precursors: EC and PDMS.

3.3.2. Surface Morphology

The morphology of MCM-F1, MCM-F2, MCM-F3 solid films was characterized using SEM analysis (Figure 2). The SEM image of CIP-loaded MCM-41 powder before coating is also presented for comparative purposes (Figure 2A). The CIP-loaded MCM-41 powder was characterized by spherical particles loosely connected to each other. After coating with polymer blends the structure of the obtained solid films was more compact and heterogeneous. Moreover, the results suggested that the surface roughness increases with the increase of PDMS content in the solid films (Figure 2A). In order to confirm the observed differences in surface roughness the SEM-EDX analysis of pure solid films of F1, F2, F3 was also performed (Figure 2B). As it can be seen, the film of EC without PDMS (F1) was characterized by a smooth, gapless surface, whereas F2 and F3 present plain surface structure with an increasing number of cavities distributed in the solid films. The higher the PDMS content in polymer blend, the greater the number of cavities that was observed, indicating that they could be attributed to PDMS droplets (also confirmed by EDX profile in which the presence of silicon in F2 and F3 was noticed). It should be noted that no cavities attributed to PDMS were observed for MCM-F2, MCM-F3 solid films (Figure 2A) contrary to pure solid films of F2, F3 (Figure 2B). This may be related to occlusion of PDMS in the MCM-41 particles.

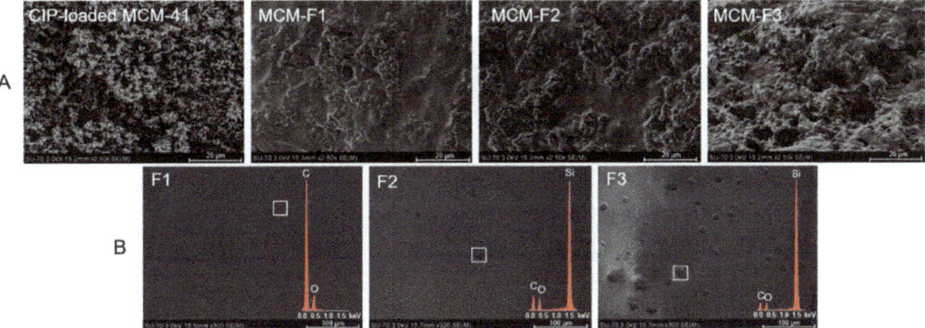

Figure 2. SEM images of CIP-loaded MCM-41, and solid films of: MCM-F1, MCM-F2, MCM-F3 (**A**); F1, F2, F3 (**B**).

The quantitative characterization of the surface of MCM-F1, MCM-F2, MCM-F3 solid films was performed by using optical profilometry. Table 2 shows the roughness parameters (R_a—average roughness, R_q—root-mean-square roughness, and R_t—peak-valley difference) calculated over the whole tested area (46.9 µm × 62.5 µm) as the means of 10 samples of each formulation together with standard deviations. As presented in Table 2, the quantitative results confirmed the increase of surface roughness in the following order: MCM-F1 < MCM-F2 < MCM-F3. The R_a value increased with the increase of PDMS content in solid film from 0.57 ± 0.10 µm for MCM-F1 to 2.49 ± 0.54 µm for MCM-F3. The highest values of roughness parameters were observed for MCM-F3 and correlated with the biggest cavities (Figure 3). These data indicate that the surface roughness of solid films is significantly dependent on the PDMS content in polymer blend what might be a promising feature in the context of tissue ingrowth. According to numerous in vitro experiments, implant roughness may affect osteoblast morphology, growth, proliferation and differentiation. It has been highlighted that the cells cultured on the rough surfaces show better morphology and proliferation in comparison to the controls cultured on flat surfaces [37].

Table 2. Mean roughness parameters with standard deviations of surface of MCM-F1, MCM-F2, MCM-F3 solid films.

Formulation	$R_a \pm$ SD [µm]	$R_q \pm$ SD [µm]	$R_t \pm$ SD [µm]
MCM-F1	0.57 ± 0.10	0.86 ± 0.17	11.22 ± 2.49
MCM-F2	1.86 ± 0.40	2.30 ± 0.45	19.12 ± 2.85
MCM-F3	2.49 ± 0.54	3.14 ± 0.70	23.67 ± 6.06

SD—standard deviation; R_a—average roughness; R_q—root-mean-square roughness; and R_t—peak-valley difference.

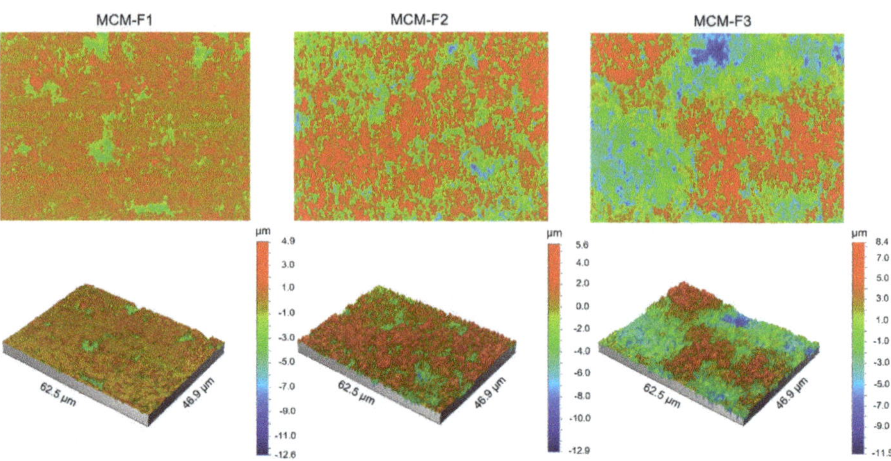

Figure 3. The optical profilometer images of 62.5 µm × 46.9 µm surfaces of MCM-F1, MCM-F2, MCM-F3 solid films.

3.3.3. Contact Angle and Wetting Properties

In order to verify the wettability of the solid films of F1, F2, F3, and MCM-F1, MCM-F2, MCM-F3 a contact angle analysis was performed. The average contact angles and surface free energies determined for all samples are shown in Table 3. All of the examined samples were characterized as hydrophobic. For the pure solid films of F1, F2, F3 (Section A in Table 3) the most hydrophilic surface was observed for F1. As expected, for F2 the addition of hydrophobic hydroxyl-terminated PDMS resulted in a higher water contact angle value, suggesting an increase in hydrophobicity, compared to F1. Conversely, for F3 a slight decrease of water contact angle and increase of the polar component of the surface free energy were noticed, compared to F2. This can be explained by gaining insight into the molecular structure of hydroxyl-terminated PDMS which contains hydrophobic PDMS-rich domains (polydimethylsiloxane chains) terminated with hydrophilic hydroxyl groups. For MCM-F1, MCM-F2, MCM-F3 solid films (Section B in Table 3) the water contact angle increased, compared to pure solid films, due to the presence of CIP-loaded MCM-41. Relatively the most hydrophobic surface was found for the solid film of MCM-F1 (without PDMS). The increase of polar component of surface energy as a function of PDMS content was observed. This can be explained by the high adhesion of polydimethylsiloxane chains to hydrophobic CIP-loaded MCM-41. This behaviour results in an increase of the number of hydrophilic hydroxyl groups of PDMS oriented to the sample surfaces and thus an increase of surface polar components.

Table 3. Average contact angle and surface free energy (γ_s) with its dispersive (γ_s^d) and polar (γ_s^P) components of solid films of F1, F2, F3 (**A**), and MCM-F1, MCM-F2, MCM-F3 (**B**).

	Sample	Average Contact Angle [θ, °]		Surface Free Energy [mJ/m^2]		
		Measuring Liquid				
		Water	Diiodomethane	Total (γ_s)	Dispersive (γ_s^d)	Polar (γ_s^P)
	F1	82.2	63.0	29.27	21.35	7.92
A	F2	86.3	54.2	32.03	27.81	4.22
	F3	83.5	54.4	32.50	26.99	5.51
	MCM-F1	112.9	81.9	16.40	16.07	0.33
B	MCM-F2	125.1	67.5	29.28	27.97	1.31
	MCM-F3	124.5	62.9	32.86	31.21	1.65

3.3.4. Solid-State Characterization of Ciprofloxacin

The XRD patterns of solid films of MCM-F3 and F3 (as representative examples), CIP-loaded MCM-41 and freeze-dried CIP reference are presented in Figure 4. The freeze-dried CIP reference was characterized by numerous well-defined sharp diffraction peaks demonstrating the crystalline nature of the drug. For the CIP-loaded MCM-41 sample the broad halo in the range of 15-35° derived from the amorphous silica with significant reduction of peaks characteristic for the CIP. Two diffraction peaks at 8 and 27° 2θ observed in CIP-loaded MCM-41 sample suggested the semi-crystalline nature of CIP molecules adsorbed onto the silica surface [38]. In the case of both the F3 and MCM-F3 the XRD patterns showed two broad amorphous halos at 10-15° and 15-30° 2θ which clearly revealed the amorphous nature of the obtained films.

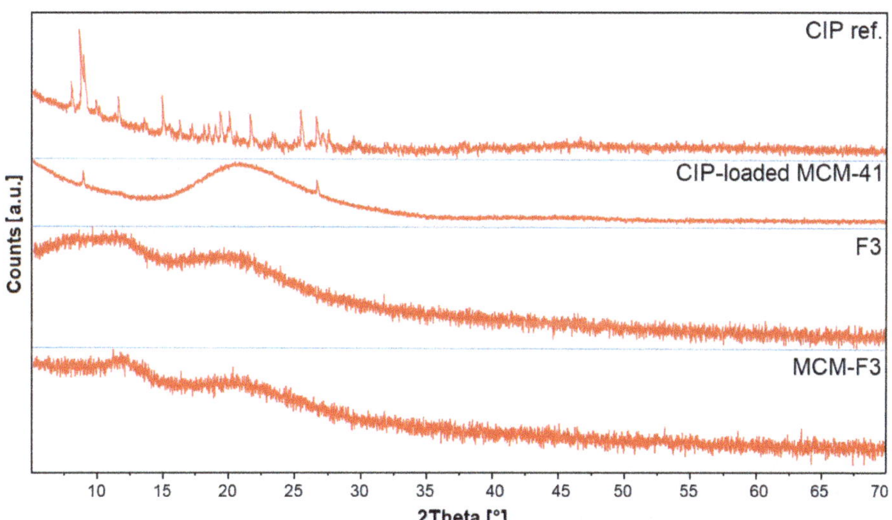

Figure 4. The XRD patterns of solid films of MCM-F3 and F3, CIP-loaded MCM-41, and CIP reference.

The DSC thermograms of solid films of MCM-F3, F3, unloaded MCM-41, freeze-dried CIP (as references) are presented in Figure 5. As selected in the frame the freeze-dried CIP was characterized by the endothermic peak of melting point at approx. 207 °C. MCM-F3 exhibited a small endothermic peak shifted to lower temperatures (approx. 199 °C) corresponding to the melting of drug, which suggested the presence of the CIP in the amorphous or semi-crystalline state what was also confirmed by the XRD results (Figure 4). The amorphous nature of drug loaded into the mesoporous silica is already known phenomenon [39]. Moreover, the obtained DSC results are in accordance with the observations of Mesallati et al. [34] who have claimed that solid dispersions of CIP with various polymers displayed a clear melting endotherm of the drug, which could be taken as confirmation of the amorphous nature of the CIP inside the polymer matrices.

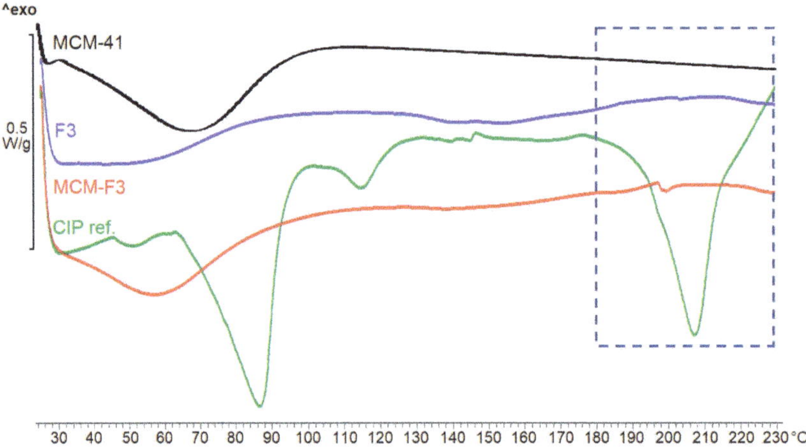

Figure 5. The DSC thermograms of solid films of MCM-F3, F3, unloaded MCM-41 and CIP reference.

3.4. Ciprofloxacin Release

The drug release was studied for MCM-F1, MCM-F2, MCM-F3 solid films to verify the impact of PDMS content on release profiles. Each solid film contained 6 mg of CIP-loaded MCM-41, corresponding to 0.67 ± 0.1 mg of CIP. CIP-elution profiles describing the cumulative percent amount of CIP released as a function of time (1–7 days), are presented in Figure 6A. Inset B shows the non-cumulative percent amount of released CIP at the sampling point as a function of the initial time period (0–2 days, Figure 6B). The CIP release from the CIP-loaded MCM-41 (6 mg, 200 µm–500 µm; in the powder form) and from solid film F3 with addition of pure CIP (CIP-F3, 0.67 ± 0.2 mg of CIP) are shown in Figure 6 for comparative purposes. As presented in Figure 6C, both CIP-loaded MCM-41 and CIP-F3 solid film were characterised by a relatively rapid release of CIP with a huge burst during the first 24 h (80 and 88%, respectively) with almost complete CIP release after 7 days (98 and 95%, respectively). In contrast, the initial amount of the drug eluted during the first day (initial burst) was reduced to 39% of the total amount of loaded CIP for the MCM-F1 solid film (PDMS-free, Figure 6B). Next, the release kinetics slowed-down and after four exchanges of the medium, the mean value of the recovered CIP from MCM-F1 corresponded to 50% of the total amount of loaded CIP (Figure 6A), whereas it was 75% after 39 daily exchanges (estimated from the release profile). In the case of the PDMS-containing solid films F2 and F3, the initial burst was reduced to 16% and 5% of the total amount of loaded CIP (corresponding to 107.5 and 33.6 µg of CIP released during the first day), respectively (Figure 6B). Then the release kinetics gently slowed down (Figure 6A). The $t_{25\%}$, $t_{50\%}$, and $t_{75\%}$ values (corresponding to 25%, 50% and 75% of the total amount of loaded CIP, respectively), calculated from the release profiles, were 15, 66, and 107 days or 46, 117, and 168 days for MCM-F2 and MCM-F3 solid films, respectively. Thus, the release of CIP from MCM-F1, MCM-F2, MCM-F3 solid films presented a significant modification compared to CIP-loaded MCM-41 and CIP-F3 solid film. The initial burst of CIP was reduced with the increase in content of PDMS in solid films. That is why the greatest reduction in initial burst was reported for MCM-F3 solid film characterized by a prolonged release rate similar to already exist antibiotic drug delivery systems [40].

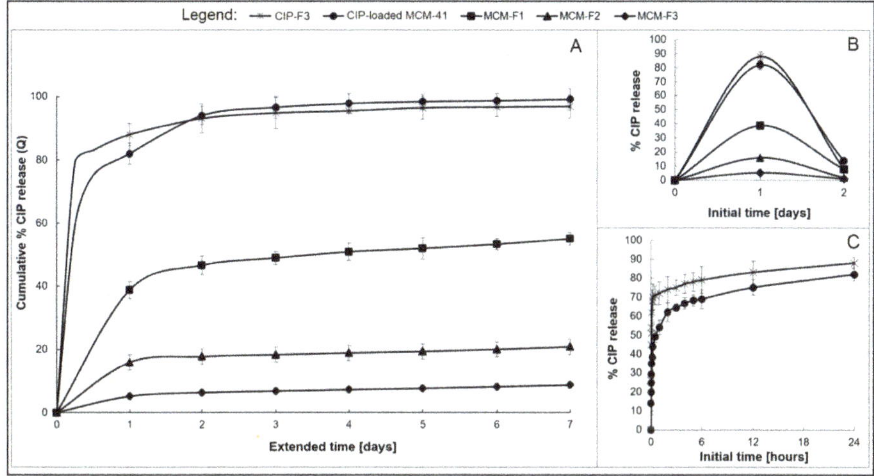

Figure 6. The overall CIP release profiles from MCM-F1, MCM-F2, MCM-F3, CIP-F3 solid films, and CIP-loaded MCM-41 (**A**); Insets: not-cumulative CIP release during the initial time: 0–2 days; (**B**) and initial 24 h CIP release from CIP-loaded MCM-41 and CIP-F3 (**C**).

The calculated kinetic parameters for the obtained release profiles are summarized in Table 4. It can be observed that all release profiles of solid films (excluding the first 24 h initial drug release) were characterized by zero order kinetics ($R^2 \geq 0.89$) with constant drug release rates per day. It should be noted that zero-order release CIP kinetics were observed for all PDMS-containing solid films, even after 7 days of release studies. Moreover, the release profiles of solid films were well described by both the Higuchi model ($R^2 \geq 0.93$) and the release exponent n values ≤ 0.55 suggesting the release of drug is controlled by diffusion of CIP from a non-disintegrating matrix [41]. However the use of CIP-loaded MCM-41 in solid films instead of pure CIP seemed to be essential in order to reduce the burst release, what was manifested in reduction of k_H values from 27.01 to 8.92 for CIP-F3 and MCM-F1 solid films, respectively.

Table 4. The kinetic parameters of fitted experimental data for MCM-F1, MCM-F2, MCM-F3, CIP-F3 solid films and CIP-loaded MCM-41.

Formulation	Korsmeyer-Peppas Model		Higuchi Model		Zero Order Kinetics *	
	n	R^2	k_H	R^2	k_0	R^2
CIP-F3	0.51	0.95	27.01	0.94	0.03	0.89
CIP-loaded MCM-41	0.34	0.92	4.93	0.85	0.02	0.82
MCM-F1	0.17	0.97	8.92	0.93	1.60	0.99
MCM-F2	0.33	0.87	6.80	0.98	0.61	0.99
MCM-F3	0.55	0.93	6.49	0.98	0.48	0.99

R^2—coefficient of determination; n—release exponent in Korsmeyer-Peppas model; k_H—Higuchi dissolution constant ($h^{-1/2}$; $days^{-1/2}$); k_0—zero order release constant (% of dose released per day); * Calculated excluding the first 24 h initial release of CIP.

Taking into account both the CIP release profile and the kinetic models the possible mechanism of drug release from CIP-loaded MCM-41 coated with polymer blends can be described as diffusion controlled [42]. In our case the water-soluble CIP molecules loaded into the mesoporous silica channels are entrapped in the EC-PDMS polymeric film which acts as a water-insoluble matrix. The structure of the formed EC film is dependent on the polymer-solvent interaction parameters. Jones et al. found that after ethanol evaporation, the structure of EC is heterogeneous, with numerous nanopores [43]. This confirms the release of CIP via simple diffusion from CIP-loaded MCM-41 coated with EC (MCM-F1 solid film) through the formed pores (release exponent $n = 0.51$, calculated from the Korsmeyer-Peppas model). However, the observed reduction of both the initial burst and the rate of CIP release from CIP-loaded MCM-41 coated with PDMS:EC blends may be attributed to the adhesion of PDMS chains onto the CIP-loaded MCM-41 surface, which is also in agreement with the obtained SEM results (Figure 2A) and wettability data (Table 3). Another reason can be related to a "locking" of the pore surface of EC due to the greater surface contact and thus easier penetration of the PDMS into the pores. The effect of PDMS on the release rate of water-soluble drugs from silica/PDMS composites has previously been reported [39]. It was found that an increase in PDMS content resulted in a decreasing rate of drug release, and as reported here, with a simultaneous increase in the values of the release exponent n. This phenomenon was correlated with a decrease in the porosity of the studied silica/PDMS composites, explained by the effect of the occlusion of hydrophobic PDMS-rich domains on the silica particles that may delay the penetration of water molecules into the composites, and hence dissolution of a loaded drug [44,45]. To sum up, the increase of PDMS content in solid films of MCM-F2, MCM-F3 may result in an increasing number of PDMS-filled nanopores, and also lock the CIP in the pores of MCM-41, thus slowing down the drug release [46]. Consequently, the zero-order CIP release could be explained by the very small magnitude of the interfacial partition coefficient of the drug and the small thickness of the drug depletion layer.

4. Conclusions

Novel solid films were obtained by coating CIP-loaded MCM-41 with PDMS:EC blends. Solid films with various levels of PDMS addition (0, 1, 2% (v/v)) were characterized in terms of their physicochemical properties. SEM and optical profilometry results indicated a progressive increase in the surface heterogeneity and roughness for the solid films of CIP-loaded MCM-41 coated with polymer blends as a function of PDMS content. The presence of amorphous/semi-crystalline CIP in the obtained films was confirmed by DSC and XRD analyses. Release studies showed a prolonged CIP release from all solid films for over 7 days with Fickian diffusion-controlled drug release for the EC formulation without PDMS and the zero-order drug release for formulations containing PDMS. This is probably related to the adhesion of hydrophobic chains of PDMS onto CIP-loaded MCM-41. The physicochemical characterization of the obtained solid films showed that by simple modification of the PDMS:EC ratio, the roughness and release profile of drug adsorbed into the MCM-41 pores can be easily altered. As the surface roughness has a fundamental significance for the integration with host bone tissue, the provided mesoporous silica coated with polymer blends could be a promising drug delivery system. The most beneficial physicochemical properties in the context of further biological evaluation, such as the greatest reduction of initial burst of CIP and the highest values of roughness parameters were observed for MCM-F3 solid film. The findings presented in this paper may be an excellent starting point for further investigations on the biocompatibility and material/cell interactions of the systems.

Supplementary Materials: The following are available online at http://www.mdpi.com/2073-4360/11/9/1450/s1.

Author Contributions: Conceptualization, M.P.; Formal analysis, M.P.; Funding acquisition, A.S. (Adrianna Skwira) and M.P.; Investigation, A.S. (Adrianna Skwira), A.S. (Adrian Szewczyk) and M.P.; Methodology, A.S. (Adrianna Skwira), A.S. (Adrian Szewczyk) and M.P.; Supervision, M.P.; Writing—original draft, A.S. (Adrianna Skwira), A.S. (Adrian Szewczyk) and M.P.

Funding: This research was supported by the project OPUS 15 (2018/29/B/NZ7/00533) co-financed by National Science Centre, and partially supported by the project POWR.03.02.00-00-I035/16-00 co-financed by the European Union through the European Social Fund under the Operational Programme Knowledge Education Development 2014–2020.

Acknowledgments: The authors wish to thank Marta Ziegler-Borowska from Nicolaus Copernicus University in Torun for experimental assistance.

Conflicts of Interest: The authors declare no conflict of interest. The funders had no role in the design of the study; in the collection, analyses, or interpretation of data; in the writing of the manuscript, or in the decision to publish the results.

References

1. Nandi, S.K.; Bandyopadhyay, S.; Das, P.; Samanta, I.; Mukherjee, P.; Roy, S.; Kundu, B. Understanding Osteomyelitis and Its Treatment through Local Drug Delivery System. *Biotechnol. Adv.* **2016**, *34*, 1305–1317. [CrossRef] [PubMed]
2. Mouriño, V.; Boccaccini, A.R. Bone Tissue Engineering Therapeutics: Controlled Drug Delivery in Three-Dimensional Scaffolds. *J. R. Soc. Interface* **2010**, *7*, 209–227. [CrossRef] [PubMed]
3. Park, K. Controlled Drug Delivery Systems: Past Forward and Future Back. *J. Control. Release* **2014**, *190*, 3–8. [CrossRef] [PubMed]
4. Wang, S. Ordered Mesoporous Materials for Drug Delivery. *Microporous Mesoporous Mater.* **2009**, *117*, 1–9. [CrossRef]
5. Prokopowicz, M.; Czarnobaj, K.; Szewczyk, A.; Sawicki, W. Preparation and in Vitro Characterisation of Bioactive Mesoporous Silica Microparticles for Drug Delivery Applications. *Mater. Sci. Eng. C* **2016**, *60*, 7–18. [CrossRef] [PubMed]
6. Beck, G.R.; Ha, S.W.; Camalier, C.E.; Yamaguchi, M.; Li, Y.; Lee, J.K.; Weitzmann, M.N. Bioactive Silica-Based Nanoparticles Stimulate Bone-forming Osteoblasts, Suppress Bone-Resorbing Osteoclasts, and Enhance Bone Mineral Density in Vivo. *Nanomed. Nanotechnol. Biol. Med.* **2012**, *8*, 793–803. [CrossRef]
7. Braun, K.; Pochert, A.; Beck, M.; Fiedler, R.; Gruber, J.; Lindén, M. Dissolution Kinetics of Mesoporous Silica Nanoparticles in Different Simulated Body Fluids. *J. Sol-Gel Sci. Technol.* **2016**, *79*, 319–327. [CrossRef]
8. Götz, W.; Tobiasch, E.; Witzleben, S.; Schulze, M. Effects of Silicon Compounds on Biomineralization, Osteogenesis, and Hard Tissue Formation. *Pharmaceutics* **2019**, *11*, 117. [CrossRef]
9. Pérez-Esteve, É.; Ruiz-Rico, M.; De La Torre, C.; Villaescusa, L.A.; Sancenón, F.; Marcos, M.D.; Amorós, P.; Martínez-Máñez, R.; Barat, J.M. Encapsulation of Folic Acid in Different Silica Porous Supports: A Comparative Study. *Food Chem.* **2016**, *196*, 66–75. [CrossRef]
10. Lee, J.H.; Yeo, Y. Controlled Drug Release from Pharmaceutical Nanocarriers. *Chem. Eng. Sci.* **2015**, *125*, 75–84. [CrossRef]
11. Siepmann, F.; Siepmann, J.; Walther, M.; MacRae, R.J.; Bodmeier, R. Polymer Blends for Controlled Release Coatings. *J. Control. Release* **2008**, *125*, 1–15. [CrossRef] [PubMed]
12. Duo, Y.; Li, Y.; Chen, C.; Liu, B.; Wang, X.; Zeng, X.; Chen, H. DOX-Loaded pH-Sensitive Mesoporous Silica Nanoparticles Coated with PDA and PEG Induce Pro-Death Autophagy in Breast Cancer. *RSC Adv.* **2017**, *7*, 39641–39650. [CrossRef]
13. Bhattacharyya, S.; Wang, H.; Ducheyne, P. Polymer-Coated Mesoporous Silica Nanoparticles for the Controlled Release of Macromolecules. *Acta Biomater.* **2012**, *8*, 3429–3435. [CrossRef] [PubMed]
14. Trendafilova, I.; Szegedi, Á.; Yoncheva, K.; Shestakova, P.; Mihály, J.; Ristić, A.; Konstantinov, S.; Popova, M. A pH Dependent Delivery of Mesalazine from Polymer Coated and Drug-Loaded SBA-16 Systems. *Eur. J. Pharm. Sci.* **2016**, *81*, 75–81. [CrossRef] [PubMed]
15. Smith, J.R.; Lamprou, D.A. Polymer Coatings for Biomedical Applications: A Review. *Trans. IMF* **2014**, *92*, 9–19. [CrossRef]
16. Lecomte, F.; Siepmann, J.; Walther, M.; MacRae, R.J.; Bodmeier, R. Polymer Blends used for the Coating of Multiparticulates: Comparison of Aqueous and Organic Coating Techniques. *Pharm. Res.* **2004**, *21*, 882–890. [CrossRef] [PubMed]
17. Ponche, A.; Bigerelle, M.; Anselme, K. Relative Influence of Surface Topography and Surface Chemistry on Cell Response to Bone Implant Materials. Part 1: Physico-Chemical Effects. *Proc. Inst. Mech. Eng. Part H J. Eng. Med.* **2010**, *224*, 1471–1486. [CrossRef] [PubMed]
18. Hokmabad, V.R.; Davaran, S.; Aghazadeh, M.; Rahbarghazi, R.; Salehi, R.; Ramazani, A. Fabrication and characterization of Novel ethyl Cellulose-Grafted-Poly (ε-caprolactone)/alginate Nanofibrous/Macroporous Scaffolds Incorporated with Nano-Hydroxyapatite for Bone Tissue Engineering. *J. Biomater. Appl.* **2019**, *33*, 1128–1144. [CrossRef]

19. Le Guéhennec, L.; Soueidan, A.; Layrolle, P.; Amouriq, Y. Surface Treatments of Titanium Dental Implants for Rapid Osseointegration. *Dent. Mater.* **2007**, *23*, 844–854. [CrossRef]

20. Krishna Alla, R.; Ginjupalli, K.; Upadhya, N.; Shammas, M.; Krishna Ravi, R.; Sekhar, R. Surface Roughness of Implants: A review. *Trends Biomater. Artif. Organs* **2011**, *25*, 112–118.

21. Linez-Bataillon, P.; Monchau, F.; Bigerelle, M.; Hildebrand, H.F. In Vitro MC3T3 Osteoblast Adhesion with Respect to Surface Roughness of Ti6Al4V Substrates. *Biomol. Eng.* **2002**, *19*, 133–141. [CrossRef]

22. Wirth, J.; Tahriri, M.; Khoshroo, K.; Rasoulianboroujeni, M.; Dentino, A.R.; Tayebi, L. Surface Modification of Dental Implants. In *Biomaterials for Oral and Dental Tissue Engineering*; Woodhead Publishing: Cambridge, MA, USA, 2017; pp. 85–96.

23. Garg, H.; Bedi, G.; Garg, A. Implant Surface Modifications: A review. *J. Clin. Diagn. Res.* **2012**, *6*, 319–324.

24. Bose, S.; Bogner, R.H. Solventless Pharmaceutical Coating Processes: A review. *Pharm. Dev. Technol.* **2007**, *12*, 115–131. [CrossRef] [PubMed]

25. Teo, A.J.T.; Mishra, A.; Park, I.; Kim, Y.J.; Park, W.T.; Yoon, Y.J. Polymeric Biomaterials for Medical Implants and Devices. *ACS Biomater. Sci. Eng.* **2016**, *2*, 454–472. [CrossRef]

26. Blanco, I. Polysiloxanes in Theranostics and Drug Delivery: A Review. *Polymers (Basel)* **2018**, *10*, 755. [CrossRef] [PubMed]

27. Murtaza, G. Ethylcellulose Microparticles: A review. *Acta Pol. Pharm.-Drug Res.* **2012**, *69*, 11–22.

28. Tian, B.; Tang, S.; Li, Y.; Long, T.; Qu, X.H.; Yu, D.G.; Guo, Y.J.; Guo, Y.P.; Zhu, Z.A. Fabrication, Characterization, and Biocompatibility of Ethyl Cellulose/Carbonated Hydroxyapatite Composite coatings on Ti6Al4V. *J. Mater. Sci. Mater. Med.* **2014**. [CrossRef] [PubMed]

29. Mehta, R.Y.; Missaghi, S.; Tiwari, S.B.; Rajabi-Siahboomi, A.R. Application of Ethylcellulose Coating to Hydrophilic Matrices: A Strategy to Modulate Drug Release Profile and Reduce Drug Release Variability. *AAPS Pharm. Sci. Tech.* **2014**, *15*, 1049–1059. [CrossRef]

30. Mata, A.; Fleischman, A.J.; Roy, S. Characterization of Polydimethylsiloxane (PDMS) Properties for Biomedical Micro/Nanosystems. *Biomed. Microdevices* **2005**, *7*, 281–293. [CrossRef] [PubMed]

31. Tsourvakas, S. Local Antibiotic Therapy in the Treatment of Bone and Soft Tissue Infections. *Sel. Top. Plast. Reconstr. Surg.* **2012**, 17–46.

32. Vallet-Regi, M.; Rámila, A.; Del Real, R.P.; Pérez-Pariente, J. A new property of MCM-41: Drug delivery system. *Chem. Mater.* **2001**, *13*, 308–311. [CrossRef]

33. Hanoosh, W.S.; Abdelrazaq, E.M. Polydimethyl Siloxane Toughened Epoxy Resins: Tensile Strength and Dynamic Mechanical Analysis. *Malaysian Polym. J.* **2009**, *4*, 52–61.

34. Mesallati, H.; Umerska, A.; Paluch, K.J.; Tajber, L. Amorphous Polymeric Drug Salts as Ionic Solid Dispersion Forms of Ciprofloxacin. *Mol. Pharm.* **2017**, *14*, 2209–2223. [CrossRef] [PubMed]

35. Trivedi, M.K.; Branton, A.; Trivedi, D.; Nayak, G.; Mishra, R.K.; Jana, S. Characterization of Physicochemical and Thermal Properties of Biofield Treated Ethyl Cellulose and Methyl Cellulose. *Int. J. Biomed. Mater. Res.* **2015**, *3*, 83–91.

36. Johnson, L.M.; Gao, L.; Shields, C.W.; Smith, M.; Efimenko, K.; Cushing, K.; Genzer, J.; López, G.P. Elastomeric microparticles for acoustic mediated bioseparations. *J. Nanobiotechnology* **2013**, *11*, 22. [CrossRef] [PubMed]

37. Yamashita, D.; Machigashira, M.; Miyamoto, M.; Takeuchi, H.; Noguchi, K.; Izumi, Y.; Ban, S. Effect of surface roughness on initial responses of osteoblast-like cells on two types of zirconia. *Dent. Mater. J.* **2009**, *28*, 461–470. [CrossRef] [PubMed]

38. Wang, Z.; Chen, B.; Quan, G.; Li, F.; Wu, Q.; Dian, L.; Dong, Y.; Li, G.; Wu, C. Increasing the oral bioavailability of poorly water-soluble carbamazepine using immediate-release pellets supported on SBA-15 mesoporous silica. *Int. J. Nanomedicine* **2012**, *7*, 5807–5818. [PubMed]

39. Prokopowicz, M. Correlation between physicochemical properties of doxorubicin-loaded silica/polydimethylsiloxane xerogel and in vitro release of drug. *Acta Biomater.* **2009**, *5*, 193–207. [CrossRef]

40. Nandi, S.K.; Mukherjee, P.; Roy, S.; Kundu, B.; De, D.K.; Basu, D. Local antibiotic delivery systems for the treatment of osteomyelitis - A review. *Mater. Sci. Eng. C* **2009**, *29*, 2478–2485. [CrossRef]

41. Costa, P.; Sousa Lobo, J.M. Modeling and comparison of dissolution profiles. *Eur. J. Pharm. Sci.* **2001**, *13*, 123–133. [CrossRef]

42. Huynh, C.T.; Lee, D.S. Controlled Release. *Encycl. Polym. Nanomater.* **2015**, 439–449. [CrossRef]

43. Jones, D.S.; Medlicott, N.J. Casting solvent controlled release of chlorhexidine from ethylcellulose films prepared by solvent evaporation. *Int. J. Pharm.* **1995**, *114*, 257–261. [CrossRef]
44. Prokopowicz, M. Atomic force microscopy technique for the surface characterization of sol–gel derived multi-component silica nanocomposites. *Colloids Surfaces A Physicochem. Eng. Asp.* **2016**, *504*, 350–357. [CrossRef]
45. Kim, H.J.; Matsuda, H.; Zhou, H.; Honma, I. Ultrasound-triggered smart drug release from a poly(dimethylsiloxane)- mesoporous silica composite. *Adv. Mater.* **2006**, *18*, 3083–3088. [CrossRef]
46. Popova, M.; Trendafilova, I.; Zgureva, D.; Kalvachev, Y.; Boycheva, S.; Novak Tušar, N.; Szegedi, A. Polymer-coated mesoporous silica nanoparticles for controlled release of the prodrug sulfasalazine. *J. Drug Deliv. Sci. Technol.* **2018**, *44*, 415–420. [CrossRef]

 *polymers*

Article

Novel Antibacterial and Toughened Carbon-Fibre/Epoxy Composites by the Incorporation of TiO$_2$ Nanoparticles Modified Electrospun Nanofibre Veils

Cristina Monteserín [1,*], Miren Blanco [1], Nieves Murillo [2], Ana Pérez-Márquez [2], Jon Maudes [2], Jorge Gayoso [2], Jose Manuel Laza [3], Estíbaliz Hernáez [3], Estíbaliz Aranzabe [1] and Jose Luis Vilas [3,4]

[1] Unidad de Química de superficies y Nanotecnología, Fundación Tekniker, Iñaki Goenaga 5, 20600 Eibar, Spain; miren.blanco@tekniker.es (M.B.); estibaliz.aranzabe@tekniker.es (E.A.)

[2] Division Industria y Transporte, TECNALIA, P Mikeletegi 7, E-20009 Donostia-San Sebastian, Spain; nieves.murillo@tecnalia.com (N.M.); ana.perez@tecnalia.com (A.P.-M.); jon.maudes@tecnalia.com (J.M.); jorge.gayoso@tecnalia.com (J.G.)

[3] Grupo de Química Macromolecular (LABQUIMAC) Dpto. Química-Física, Facultad de Ciencia y Tecnología, Universidad del País Vasco (UPV/EHU), 48940 Leioa, Bizkaia, Spain; josemanuel.laza@ehu.eus (J.M.L.); estibaliz.hernaez@ehu.eus (E.H.); joseluis.vilas@ehu.eus (J.L.V.)

[4] BCMaterials, Basque Center for Materials, Applications and Nanostructures, UPV/EHU Science Park, 48940 Leioa, Spain

[*] Correspondence: cristina.monteserin@tekniker.es; Tel.: +34-943-20-67-44

Received: 22 July 2019; Accepted: 13 September 2019; Published: 19 September 2019

Abstract: The inclusion of electrospun nanofiber veils was revealed as an effective method for enhancing the mechanical properties of fiber-reinforced epoxy resin composites. These veils will eventually allow the incorporation of nanomaterials not only for mechanical reinforcement but also in multifunctional applications. Therefore, this paper investigates the effect of electrospun nanofibrous veils made of polyamide 6 modified with TiO$_2$ nanoparticles on the mechanical properties of a carbon-fiber/epoxy composite. The nanofibers were included in the carbon-fiber/epoxy composite as a single structure. The effect of positioning these veils in different composite positions was investigated. Compared to the reference, the use of unmodified and TiO$_2$ modified veils increased the flexural stress at failure and the fracture toughness of composites. When TiO$_2$ modified veils were incorporated, new antibacterial properties were achieved due to the photocatalytic properties of the veils, widening the application area of these composites.

Keywords: carbon-fibers; multifunctional composites; nanocomposites; fracture toughness

---

1. Introduction

Carbon-fiber reinforced epoxy resin composites have gained popularity over the years as engineering materials mainly in the marine, aircraft, automotive, and civil engineering sectors. Among the reasons for their success should be highlighted their lightweight, high stiffness, and strength [1–4]. However, insufficient fracture toughness, due to the fragility of the epoxy matrix and poor delamination resistance which could contribute to sudden failure of the composite, affects the long-term reliability of thermosetting matrix composites [5,6].

In order to improve the delamination resistance of reinforced polymer composites, several approaches were tested. The most common way is the inclusion of some toughening elements or thermoplastic binders; nanoparticles, such as carbon nanotubes (CNTs) [7,8], nanoclays [9], and rubber/thermoplastic materials [10,11] are often incorporated in the matrix. Nanoparticles, as a result of their theoretical high stiffness and strength, might improve the fracture toughness and energy at crack growth initiation of epoxy systems [12–14]. However, the addition of toughener elements to the matrix resin could produce a decrease in mechanical properties such as compression and shear strength, also resulting in higher cost and weight. Even more, some disadvantages appear when adding CNTs and nanoclays, the main difficulty being obtaining a homogeneous dispersion of the nanoparticles in the resin [15]. At the same time, the increased viscosity arising from the addition of tougheners could also dramatically reduce the processing ability of matrix resin.

Recently, a method that could potentially overcome these drawbacks has raised interest. It involves the incorporation of thermoplastic electrospun nanofiber mats between the fiber reinforcing plies prior to processing. Reneker, Kim, and Dzenis [16,17] pioneered the work in this field to demonstrate toughness improvement; now, interlayer toughening is used increasingly in the composite industry [18–20]. Furthermore, due to the ease of embedding thermoplastic nanofibrous structures in resin and including them in the composite as a nanosized phase without an increase of resin viscosity, this is presented as a solution for the dispersion issue. Moreover, since their length is on the macro scale, no health hazards are associated with the production and handling of these nanofibers [21–24].

Recent literature indicates that nanofibers may contribute substantially to the ductility and fracture toughness of the composites [25–27]. The improvement is highly dependent on the type of thermoplastic and of their intrinsic properties [28]. The electrospinning technology also provides the possibility of including charges inside the nanofibers, which can improve their mechanical properties or give them different properties. The incorporation of metallic or electrically conductive carbonaceous nanoparticles such as silver nanowires or carbon nanofibers can also produce electrically conductive nanofiber veils. In recent years, major efforts have been dedicated to the manufacturing of antimicrobial membranes for ultrafiltration based on modified nanofibers. These antimicrobial functionalities can be introduced through different methods such as surface treatment or the incorporation of additives like antibiotics, biocides [29–31], or active agents. These agents include silver nanoparticles and nanoparticles of metals and metal oxides [32,33] such as Zn [34], Ti [28,35–37], Cu [38], Co [39], and combinations of Zn and Ti [40–44] among others [45,46]. The availability of functional nanofibers for the development of composite materials can be a great advantage in many sectors.

As mentioned above, the mechanical properties of thermoplastic electrospun nanofiber mats have an important effect on the mechanical properties of the reinforced composite. To increase the mechanical strength of thermoplastic electrospun nanofiber mats, several stronger nanoscale fillers were incorporated—such as graphene, carbon nanofibers (CNFs), nano clays, nano TiO_2, and carbon nanotubes (CNTs) [47–52]. However, the obtainability of functional composites based on the incorporation of functional nanofibers was poorly studied.

In this paper, the effect of the inclusion of electrospun TiO_2 modified polyamide 6 nanofiber veils on the mechanical and functional properties of carbon-fiber epoxy composites obtained by the infusion technique is explored. The nanofibrous veils are incorporated into the composite as single structures forming different configurations. The mechanical properties of the nanofiber veil-reinforced composites were investigated via flexural tests, dynamical mechanical analysis, and mode I and mode II fracture toughness tests against reference specimens. The crack propagation in the loaded sample was studied through scanning electron microscopy. Furthermore, new functionalities introduced by the presence of TiO_2 nanoparticles in the nanofiber were determined by the analysis of the antibacterial activity of the composites.

2. Materials and Methods

2.1. Materials

Diglycidyl ether of bisphenol A (DGEBA) was supplied by Momentive (Waterford, NY, USA), as Epikote 828, with an equivalent weight of 182–190 g/equivalent and a hydroxyl/epoxy ratio of

0.03. Epikote 828 crosslinking was obtained with 4,4′-diaminodiphenylmethane (DDM), supplied by Alfa Aesar (Lancashire, UK), which is characterized by a molecular weight of 198 g/mol and an amine equivalent weight of 49.5 g/mol. Electrospun nanofibers were prepared from PA6 Ultramid® B24 N 03 pellets supplied by BASF SE (Ludwigshafen, Germany). For the manufacturing of the veils modified with nanoparticles, AEROXIDE® TiO_2 P25 nanoparticles supplied by Evonik (Essen, Germany) were used. These nanoparticles have a primary particle size of 21 nm and a specific surface area (BET) of approximately 50 m^2/g. Due to the formation of aggregates and agglomerates between them, the density is approximately 130 g/cm^3. They have a predominant structure of anatase, with an anatase/rutile weight ratio of approximately 80/20. A bidirectional carbon-fiber fabric formed by HT3k fibers (200 g/m^2), supplied by SP Systems (Newport, UK) was employed for the production of the composite.

The development and characterization of electrospun PA6 and 25 wt.% TiO_2 nanoparticle PA6 modified nanofiber veils was reported in previous work [53]. The characterization showed that TiO_2 nanoparticles are widely dispersed on the surface of the nanofibers, providing the excellent photocatalytic activity of the veils under UV light irradiation (UV lamp Vl-6-L, from Vilber Lourmat (Collégien, France). Escherichia coli (E. coli) and other coliform bacteria were removed after 24 h of contact with the PA6/25 wt.% TiO_2 electrospun nanofiber veils when irradiated with UV light.

2.2. Sample Preparation

Theoretical composites with two carbon-fiber plies interlayered with PA6 and PA6/25 wt.% TiO_2 electrospun nanofibrous veils were prepared. The veils were included only in the interlaminar region (Figure 1, configuration 1) or in interlaminar and also in the external layers of the composite (Figure 1, configuration 2). A composite with two carbon-fiber plies was used as a reference for comparison purposes. The composites were fabricated by vacuum infusion; where the carbon-fiber is impregnated by the liquid resin with the catalyst due to the vacuum effect by means of a vacuum bagging film. The infusion methodology is lineal to induce straight resin flow. DGEBA-DDM stoichiometric formulations were mixed for 10 min at 80 °C under stirring [54–56]. The 80 °C temperature was maintained during the vacuum infusion process through the whole laminate, and a slight increase in the infusion time of the resin for the composites with PA6 veils modified with 25 wt.% TiO_2 was observed. Even though the veils were porous, the impregnation of all the layers of the composite required more processing time due to the increased areal weight in the veil modified with the nanoparticles—4.30 g/m^2 for the TiO_2 modified nanofibers versus 1.94 g/m^2 for the unmodified nanofibers, calculated by weighing a piece of 10 cm × 10 cm of each veil. The same behavior was observed for the configuration with three veils. The composites were cured at 90 °C for 4 h, ensuring that the veils did not melt during curing. The incorporation of polyamide nanofiber veils did not affect the curing kinetics of the composite. The thermal characterization by differential scanning calorimetry (DSC) was performed in a DSC1 module from Mettler-Toledo (Giessen, Germany) can be found in the Supporting Information (Figure S1 and Table S1). The resulting composites were cut to the dimensions required for the different experimental analysis. The thicknesses of the fabricated composites with two carbon-fiber layers were between 0.6–0.7 mm. Following the AITM 1.0005 standard for the fracture test analysis (mode I, mode II), composites with 14 layers of carbon-fiber with one veil in the interlaminar region were fabricated by vacuum infusion technique, obtaining a thickness between 3.1–3.6 mm.

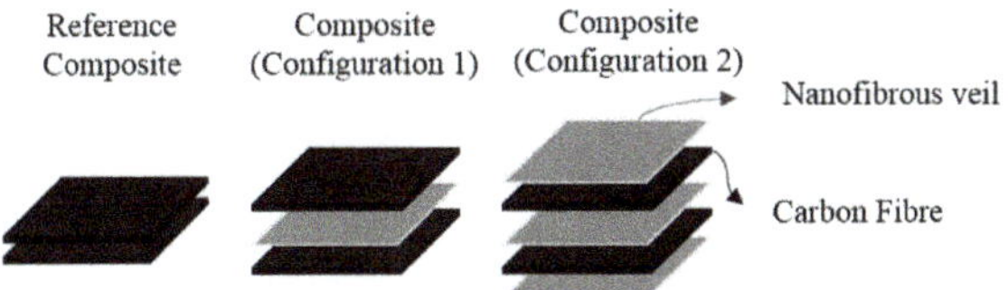

Figure 1. Configurations of the developed composites.

2.3. Characterization Techniques

The non-modified and 25 wt.% TiO_2 modified PA6 electrospun nanofiber veils were characterized by Fourier transform infrared spectroscopy (FTIR) (JASCO, Easton, MD, USA) scanning electron microscopy (SEM) (Carl Zeiss Microscopy, LLC, Thornwood, NY, USA), thermogravimetric analysis (TGA) (TA instruments, New Castle, DE, USA), and weight measurements in a previous paper [53]. Moreover, the photocatalytic decomposition of an organic pollutant, Remazol Black B (C.I. Reactive Black 5), and against E. coli and other coliform bacteria (ISO 7704: 1985) under UV radiation was also evaluated.

In this work, they were characterized by differential scanning calorimetry (DSC) to determine the TiO_2 nanoparticle influence on the crystallinity of the electrospun nanofibers. DSC measurements were performed with a Mettler-Toledo DSC1. Two successive temperature heating scans from 0 °C to 300 °C at 10 °C/min were performed. The mechanical properties of the composites were characterized by different techniques. Dynamic mechanical thermal analysis (DMTA) was performed using Polymer Laboratories Mark II DMTA equipment (Agilent Technologies, Santa Clara, CA, USA). All composite samples with PA6 and PA6/25 wt.% TiO_2, as well as the reference system, were performed in dual cantilever mode from −50 °C to 200 °C at a heating rate of 2 °C/min, 64 μm strain, and 1 Hz. Rectangular samples (10 mm × 35 mm) were directly cut from the composite sheets.

A three-point bending test was used to measure the flexural properties of the different composite samples using an electromechanical Instron 3369 machine with a load cell of 1 kN and a crosshead speed of 5 mm/min. Samples were cut to 60 mm × 20 mm, whilst avoiding the slipping of the specimens during the test. To calculate the flexural strength and deformation, the ISO 14125 methodology was followed. Mode I interlaminar fracture toughness studies were carried out using double cantilever beam (DCB) specimens following the AITM 1.0005 standard [57]. The DCB test specimens were prepared with dimensions of 250 mm × 25 mm and an initial crack length of 25 mm. Each experiment was repeated three times using a test speed of 10 mm/min; in each test, the specific thickness value was considered. Load, opening displacement, and crack length were recorded for the energy release rate (G_{IC}, G_{IIC}) calculation during the tests, as was explained in previous work [28]. The specimens employed in the analyses are the same three specimens that are tested in mode I. Scanning electron microscopy (SEM) and a Carl Zeiss SMT Ultra Gemini-II system were used to investigate the cross-sections after flexural analysis and Mode II fracture test of the composites.

A Microinstant® Coliforms Chromogenic Agar (CCA) kit (Scharlab, Barcelona, Spain) was employed to perform the antibacterial tests and for the detection of E. coli and coliform bacteria. The water sample was filtered through a 0.45 μm pore size membrane filter, according to ISO 7704:1985. The membrane was deposited face up on a plate containing the CCA medium. The plate was incubated under incubation conditions: 18 h at 36 ± 2 °C, followed by 6 h if red or colorless colonies appear to include possible late reactions of ß-galactosidase or ß-glucuronidase. After the incubation period, colonies colored from salmon pink to red were counted as Coliform bacteria other than E. coli, and colonies colored from dark blue to violet as E. coli. The total coliform count corresponds to the sum of the salmon pink to red colonies and the dark blue to violet colonies. The antibacterial capability of the TiO_2 modified PA6 nanofibers has previously been demonstrated [53].

3. Results and Discussion

3.1. Mechanical Characterization

To analyze the modified veil incorporation effect on the mechanical properties of composites, they were characterized by flexural, dynamic-mechanical, and fracture tests, along with SEM measurements.

DMTA is commonly used to measure viscoelastic properties like storage modulus (E′), which is related to the elastic stiffness; loss modulus (E″), which is related to the viscous response; and loss factor (tan δ), the ratio of the loss to the storage (i.e., the damping), which is related to the energy dissipation of the material. Figure 2 shows the storage modulus (E′) and tan δ change with respect to temperature of the composites reinforced with one and three nanofiber veils of PA6 and 25 wt.% TiO_2 modified PA6. The value of E′ in the glassy zone (low temperatures) is lower for composites modified with veils. Nevertheless, it must be taken into account that the curing degree achieved was around 96% at this temperature (Table S1); therefore, the resin was not fully cured. In regard to tan δ, the first peak observed, above 100 °C, was related to the curing process of composites at 90 °C. As the temperature increased, the crosslinking continued, and a second peak was obtained above 150 °C, which corresponds to the Tg∞ of the system. The presence of the veils did not seem to have a clear improvement in the values of Tg∞, but the systems modified with TiO_2 seemed to present a wider Tg∞ peak as a consequence of a more heterogeneous crosslinking due to the presence of nanoparticles.

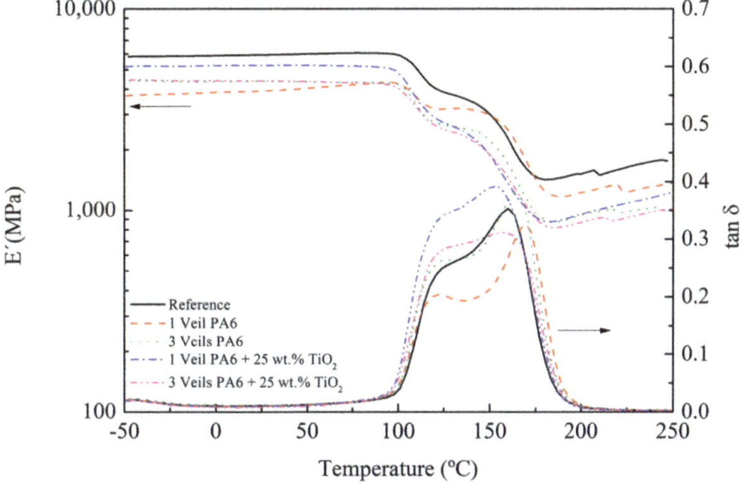

Figure 2. Storage modulus and tan δ of composites cured at a temperature of 90 °C.

The flexural strength and deformation of the carbon-fiber/epoxy composites with PA6 and 25 wt.% TiO_2 modified PA6 electrospun nanofiber veils were carried out, and the main results are shown in Table 1 along with the values for reference composites. The experimental curves can be found in the Supporting Information (Figure S2). Unsurprisingly, the composites did not present delamination during testing. All composites with both PA6 and 25 wt.% TiO_2 modified PA6 presented similar deformation values, but a clear increase in flexural strength against the reference composite was observed in all cases. For composites modified with one interlayered veil, an increase in flexural strength of 19.7% and 24.6% was observed for the composite with PA6 and 25 wt.% TiO_2 modified PA6 electrospun nanofiber veils, respectively. The flexural strength increase is lower for the three veils specimens with and without titanium dioxide nanoparticles, indicating that veils, on the exterior faces, did not contribute to the improvement of the flexural mechanical properties.

Table 1. Flexural strength (σ_{max}) and deformation at break (δ_{max}) for composites with PA6 [28] and 25 wt.% TiO_2 modified PA6 nanofibrous veils.

Sample	σ_{max} (MPa)	$\Delta\sigma_{max}$ (%)	δ_{max} (%)
Reference [28]	375.5 ± 33.2	-	2.2 ± 0.2
1veil PA6 [28]	449.5 ± 10.8	19.7	2.1 ± 0.0
3 veils PA6 [28]	415.4 ± 23.8	10.6	2.1 ± 0.2
1veil PA6 + 25 wt.% TiO_2	468.1 ± 32.5	24.6	2.2 ± 0.2
3 veils PA6 + 25 wt.% TiO_2	411.5 ± 25.4	9.6	2.1 ± 0.1

After flexural tests, the fracture surfaces of the specimens were characterized by SEM. Figure 3 shows the interlaminar section of the (a) composite with PA6 veil and (b) composite with PA6 modified with 25 wt.% TiO$_2$ veil. In both cases, fracture waves of the resin ended in the veil. The thicknesses of modified and non-modified veils were quite similar, 11 μm for unmodified and 20 μm for 25 wt.% TiO$_2$ modified PA6 electrospun nanofiber veils, so the increase in flexural strength was due to the presence of the veil which prevented the propagation of the crack through the epoxydic matrix [58]. As reported in previous work [28], the veils on the external and internal faces of the composite did not seem to contribute to an improvement of the properties. For composites with one or three veils, the obtained flexural strength values indicated that the presence of TiO$_2$ nanoparticles in the veils did not significantly affect the mechanical properties of the composites as initially expected [47–52]. On the one hand, TiO$_2$ nanoparticles could affect crystallization kinetics of the PA6 polymer, reducing the mechanical properties of the nanofibers themselves. In order to check if the presence of TiO$_2$ nanoparticles affected the crystallinity of PA6, modified and non-modified nanofiber veils were analyzed by DSC measurements. The PA6 pristine pellets used for obtaining these veils were also analyzed for a better result understanding. As can be observed in dynamic tests collected in Figure 4, the presence of a single melting peak almost between 224–225 °C in the first DSC scan of PA6 pellets reveals the presence of a higher proportion of α-form PA6 and a slight content of γ-form, whose melting temperatures appear at approximately 220 °C and 210 °C, respectively. For non-modified veils, apart from a slight peak at temperatures between 30–100 °C which was previously ascribed to the electrospinning process itself and is attributed to molecular rearrangement of PA6 chains within nanofibers [28,59,60], changes in PA6 nanofibers' crystallinity have not been observed, and these nanofibers present a similar higher amount of α-form crystals (even if a lower crystallization degree was observed). However, for the TiO$_2$ modified nanofibrous veils, the predominant crystalline phase observed was the γ-phase crystals of PA6. The γ-phase crystals, the predominant crystalline phase in the TiO$_2$ modified veil, were a lower stable phase than the α-phase crystals (prevalent in the unmodified veil). This could have negatively affected the mechanical properties of the nanofibers and, therefore, we did not observe significant variations in mechanical properties.

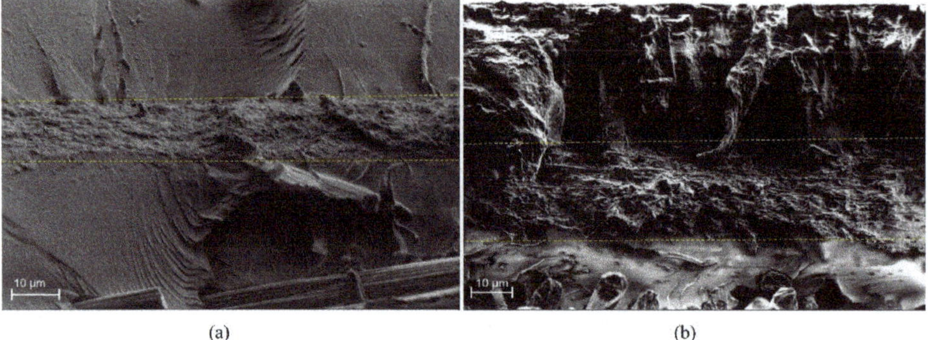

(a)   (b)

Figure 3. Scanning Electron Microscopy (SEM) images of (**a**) composite with PA6 veil and (**b**) composite with PA6 modified with 25 wt.% TiO$_2$ veil. The dotted lines show the veil presence on the composites.

On the other hand, the presence of the 25 wt.% TiO$_2$ modified PA6 nanofiber veils in the composite could affect the crosslinking of the epoxy-amine matrix, as was shown in DMTA measurements (Figure 2), where a widening in the Tg$_\infty$ peak for the composites with the TiO$_2$ modified nanofibers was observed, indicating a formation of a more heterogeneous structure.

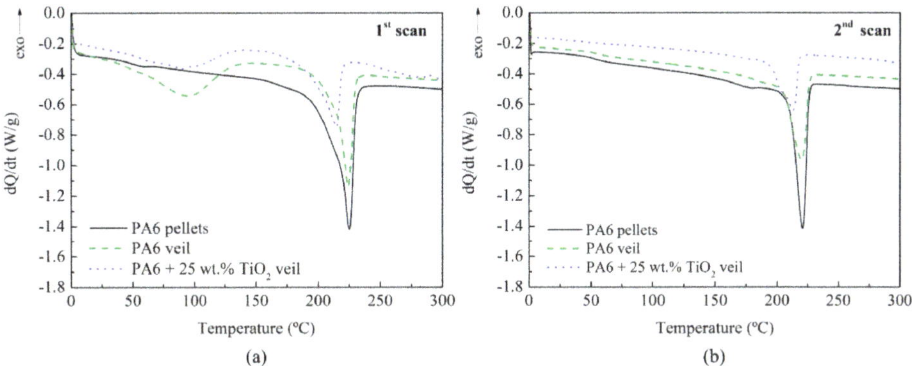

Figure 4. Dynamic differential scanning calorimetry (DSC) thermograms of the PA6 pellets, PA6 veil, and PA6 modified with 25 wt.% TiO_2 veil; (**a**) first scan and (**b**) second scan.

Mode I and Mode II fracture toughness tests were conducted on composites with 14 layers of carbon-fiber fabric and an interlaminar veil. The experimental results of the G_{IC} tests-load-displacement and mechanical energy-displacement graphs-obtained for the specimens of composites with PA6 and 25 wt.% TiO_2 modified PA6 nanofiber veils, as well as the results of the reference specimen, were presented in Figure 5, and the results were reported in Table 2. Both the force and the mechanical energy values were normalized to the specimen width. After an initial drop in load for low displacements in some of the composites, which was associated to the initiation of the crack propagation, composites with PA6 and 25 wt.% TiO_2 modified PA6 nanofibrous veils present higher load values compared to the reference composite. Via the load-displacement diagram integration, the mechanical energy propagation was calculated, confirming the trend. The mechanical energy absorbed during the crack propagation test was improved by the presence of the veil at the crack opening interface. Also, the G_{IC} values for both composites with veils were higher than for the reference; approximately 20% and 14% for the composites with PA6 and 25 wt.% TiO_2 modified PA6 nanofiber veils, respectively. However, the presence of the TiO_2 nanoparticles did not seem to contribute to an enhancement of the improvement already obtained with the unmodified PA6 veil but, rather, the properties were maintained.

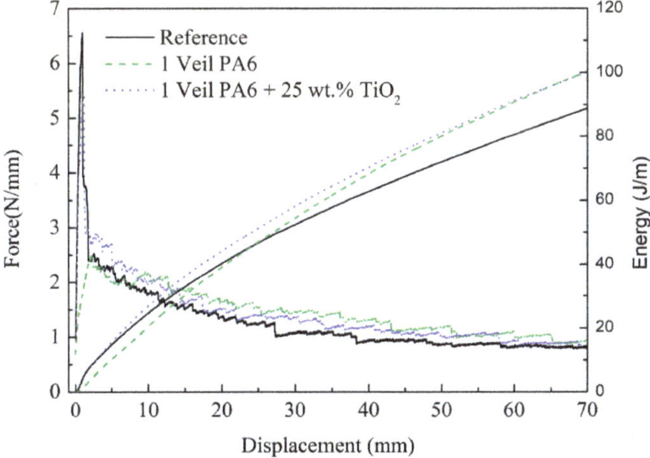

Figure 5. Mode I test results. Load and mechanical energy normalized to specimen width as a function of displacement.

Table 2. Mode I results for composites with PA6 [28] and 25 wt.% TiO$_2$ modified PA6 nanofibrous veils.

Sample	Energy (J/m)	ΔE%	G_{IC} (J/m^2)	ΔG_{IC}%
Reference [28]	62.7	-	389 ± 12.8	-
PA6 [28]	68.1	8.6	466 ± 72.9	20.0
PA6 + 25 wt.% TiO$_2$	70.2	12.0	444 ± 46.1	14.0

Figure 6 and Table 3 show the experimental results of the Mode II tests for composites with PA6 and 25 wt.% TiO$_2$ modified PA6 nanofiber veils, along with results for reference composites. In Figure 6, the force values were normalized to specimen width. The veils do not influence the stability of the composites; thus, the initial linear parts of the curves were similar. However, the effect of the veil on the load capacity was visible as soon as the crack began to propagate. Comparing the G$_{IIC}$ value obtained for the reference composite, a 4% of improvement was observed for the composite with the 25 wt.% TiO$_2$ modified PA6 nanofiber veils compared to the reference composite. TiO$_2$ nanoparticles added to the nanofiber polymer had a small influence on the fracture toughness of the composite, as the presence of the veil caused crack dispersion.

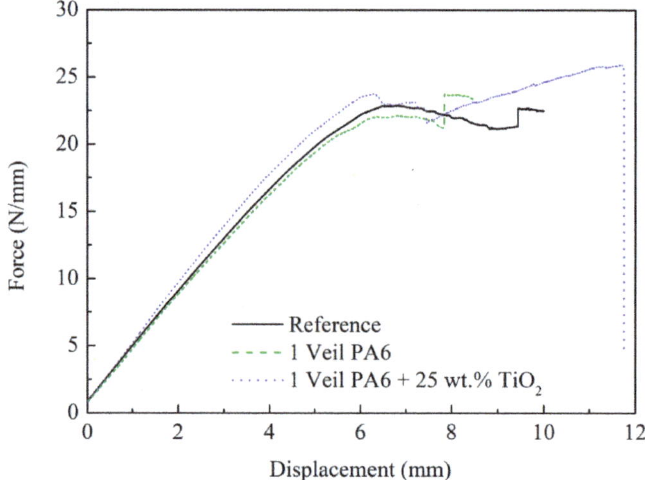

Figure 6. Mode II test results. Load normalized to specimen width as a function of displacement.

Table 3. Mode II results for composites with PA6 [28] and 25 wt.% TiO$_2$ modified PA6 nanofibrous veils.

Sample	G_{IIC} (J/m^2)	ΔG_{IIC} (%)
Reference [28]	2536 ± 257	-
PA6 [28]	2544 ± 304	0.28
PA6 + 25 wt.% TiO$_2$	2636 ± 124	3.95

3.2. Antibacterial Behavior Characterization

The efficiency of the 25 wt.% TiO$_2$ modified PA6 nanofiber veils in E. coli and other coliform bacteria elimination was previously demonstrated [53]. Therefore, in the present work, whether the veil maintains this capacity once incorporated into the composite was evaluated. For this purpose, the configuration 2 (Figure 1) of the composites was selected due to the presence of the veils in the external layers of the composite, which were in contact with the inoculated water. An area of 4 × 4 cm was cut from the reference composite, as well as from the composites with three veils of PA6 and 25 wt.% TiO$_2$ modified PA6 nanofiber veils. The same test procedure, as for the case of the veils, was followed.

The composites were placed in beakers in contact with 100 mL of artificially inoculated water for 24 h. In some tests, the beakers were irradiated with UV light, and in others, they were kept in the dark. After this period, the water in contact with the composites was filtered to determine the presence of E. coli and other coliform bacteria (ISO 7704: 1985). The plate was incubated with the membrane for different durations (24 and 96 h) at 36 ± 2 °C. Figure 7 shows the final appearance of the membranes that were used to filter the water in contact with the composites after 24 h and 96 h of incubation. After 96 h of incubation in an oven, all the membranes which were in contact with composites in dark environments presented purple-colored colonies of bacteria, indicating the simultaneous action of the enzymes present in E. coli and coliform bacteria. When the study was done under UV radiation, a significant reduction in the bacterial concentration was observed. However, the membrane that was used to filter the water in contact with composites with 25 wt.% TiO$_2$ modified PA6 nanofiber veils did not present any bacteria when the composite was irradiated with UV light. Therefore, the photocatalytic effect of the composites was confirmed and can be explained by the presence of part of the veils on the outer surface of the composites having direct contact with the bacteria. SEM images in Figure 8 confirm the presence of the TiO$_2$ modified PA6 nanofiber in the external part of the composites or covered by around 20 µm of resin.

Incubation time (h)	Reference composite (without UV)	Reference composite (with UV)	Composite + 3 veils PA6 (without UV)	Composite + 3 veils PA6 (with UV)	Composite + 3 veils PA6 /25wt.% TiO2 (without UV)	Composite + 3 veils PA6 /25wt.% TiO2 (with UV)
24						
96						

Figure 7. Antibacterial tests of composites with three veils of PA6, PA6 modified with 25% TiO$_2$, and reference.

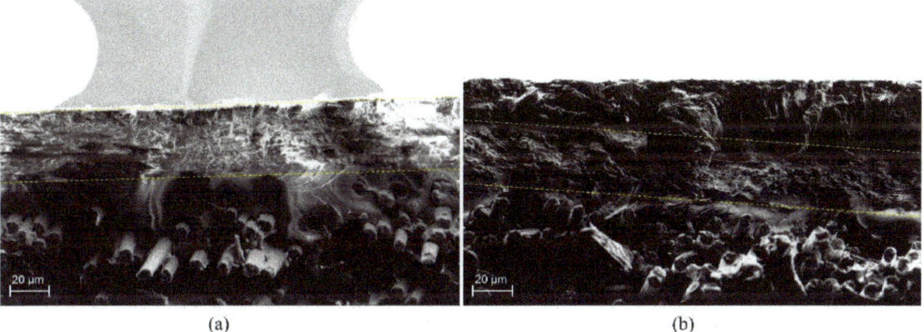

(a)   (b)

Figure 8. SEM images of the fracture surface of the composite specimens with PA6 modified with 25% TiO$_2$ veils, (**a**) external part and (**b**) covered with resin. The dotted lines show the veil presence on the composites.

4. Conclusions

The present study aimed to investigate the effect of incorporating 25 wt.% TiO$_2$ modified PA6 electrospun nanofiber veils on the mechanical and functional behavior of carbon-fiber/epoxy composites manufactured by a vacuum infusion process.

The mechanical performance of the composites was evaluated by dynamic-mechanical tests, flexural, and fracture (mode I and II) tests. The incorporation of PA6 electrospun nanofiber veils increased the flexural mechanical properties. For composites modified with one interleafed veil, an increase in flexural strength of 19.7% and 24.6% was observed for the composite with PA6 and 25 wt.% TiO_2 modified PA6 electrospun nanofiber veils, respectively. As the thickness of unmodified and modified veils was similar in the composites—11 μm for unmodified and 20 μm for 25 wt.% TiO_2 modified PA6 electrospun nanofiber veils—the increase in flexural strength was ascribed to the crack deflection capability of the veils. When 25 wt.% TiO_2 modified PA6 electrospun nanofiber veils were included in the composites, flexural strength values did not significantly increase compared to the results obtained for composites with unmodified veils. Initially, it was expected that TiO_2 modified veils would increase flexural strength to a greater extent than unmodified veils. However, TiO_2 nanoparticles could affect the crystallization of PA6, reducing the mechanical properties of the nanofibers themselves, as the γ-phase crystals (prevalent in the TiO_2 modified veil) were less stable than the α-phase crystals (prevalent in the unmodified veil). Also, a negative effect of modified veils in the crosslinking structure of epoxy-amine network could contribute to this fact, as more heterogeneous networks were observed by DMTA for composites with 25 wt.% TiO_2 modified PA6 electrospun nanofiber veils. For composites modified with three veils, the veils on the external faces of the conceptual composites did not contribute to an improvement of the flexural properties. Concerning the fracture behavior of composites, the G_{IC} values for composites with veils were higher than for the reference; approximately 20% and 14% for the composites with PA6 and 25 wt.% TiO_2 modified PA6 nanofiber veils, respectively. However, the G_{IIC} values of composites with veils were similar to values for reference composites. In every case, the presence of the TiO_2 nanoparticles had a small influence on the fracture toughness of the composites.

The incorporation of 25 wt.% TiO_2 modified PA6 nanofiber veils with already proven antibacterial properties in the external surface of composites resulted in composites with antibacterial properties when irradiated with UV light due to the photocatalytic effect of TiO_2 nanoparticles. This is due to the presence of the veils on some areas of the outer surface of the composites, as confirmed by SEM. The tested antibacterial capacity of composites against E. coli and other coliform bacteria could allow their use to be extended to applications where materials come into contact with microorganisms. For example, it could be used in marine applications, where the presence of fouling is a serious economic and safety problem; or in applications related to contact with people, such as the transport sector, sanitary sector, or food industry with applications in panels, containers, counters, or elements in contact with food in order to avoid the proliferation of dangerous bacteria.

Supplementary Materials: The following are available online at http://www.mdpi.com/2073-4360/11/9/1524/s1, Figure S1: DSC thermographs of the composites with PA6 and PA6 modified with with 25 % TiO_2 veils (a) 1st heating and (b) 2nd heating, Figure S2: Flexural strength and deformation curves, Table S1: DSC values of the composites with PA6 and PA6 modified with with 25 % TiO_2 veils.

Author Contributions: Conceptualization, C.M., J.L.V., E.A., N.M. and M.B.; methodology, C.M., M.B., A.P.-M., J.M., J.G. and E.H.; writing-review and editing, C.M., M.B., J.M.L.; supervision, J.L.V., M.B., E.A. and N.M.

Funding: This research is funded by the ELKARTEK Programme, "ACTIMAT", grupos de investigación del sistema universitario vasco (IT718-13), the Spanish government through the project TEC2015-63838-C3-1-R-OPTONANOSENS and from the Basque government through the project KK-2017/00089-μ4F.

Acknowledgments: The authors would like to acknowledge BASF for the useful support in the thermoplastic materials selection.

Conflicts of Interest: The authors declare no conflict of interest.

References

1. Peters, S.T. *Handbook of Composites*, 2nd ed.; Chapman and Hall: London, UK, 1998.
2. Beylergil, B.; Tanoglu, M.; Aktas, E. Enhancement of interlaminar fracture toughness of carbon fibre–epoxy composites using polyamide-6,6 electrospun nanofibers. *J. Appl. Polym. Sci.* **2017**, *134*, 45244. [CrossRef]

3. Duan, G.; Fang, H.; Huang, C.; Jiang, S.; Hou, H. Microstructures and mechanical properties of aligned electrospun carbon nanofibers from binary composites of polyacrylonitrile and polyamic acid. *J. Mater. Sci.* **2018**, *53*, 15096–15106. [CrossRef]

4. Zhou, S.; Zhou, G.; Jiang, S.; Fan, P.; Hou, H. Flexible and refractory tantalum carbide-carbon electrospun nanofibers with high modulus and electric conductivity. *Mater. Lett.* **2017**, *200*, 97–100. [CrossRef]

5. Kuwata, M.; Hogg, P.J. Interlaminar Toughness of Interleaved CFRP Using Non-Woven Veils: Part 1. Mode-I Testing. *Compos. Part A Appl. Sci. Manuf.* **2011**, *42*, 1551–1559. [CrossRef]

6. Jiang, S.; Chen, Y.; Duan, G.; Mei, C.; Greiner, A.; Agarwal, S. Electrospun nanofiber reinforced composites: A review. *Polym. Chem.* **2018**, *9*, 2685–2720. [CrossRef]

7. Gomez-del Río, T.; Salazar, A.; Pearson, R.A.; Rodríguez, J. Fracture behaviour of epoxy nanocomposites modified with triblock copolymers and carbon nanotubes. *Compos. Part B Eng.* **2016**, *87*, 343–349. [CrossRef]

8. Radue, M.S.; Odegard, G.M. Multiscale modeling of carbon fiber/carbon nanotube/epoxy hybrid composites: Comparison of epoxy matrices. *Compos. Sci. Technol.* **2018**, *166*, 20–26. [CrossRef]

9. Rafiq, A.; Merah, N.; Boukhili, R.; Al-Qadhi, M. Impact resistance of hybrid glass fiber reinforced epoxy/nanoclay composite. *Polym. Test.* **2017**, *57*, 1–11. [CrossRef]

10. Bahrami, A.; Cordenier, F.; Van Velthem, P.; Ballout, W.; Pardoen, T.; Nysten, B.; Bailly, C. Synergistic local toughening of high performance epoxy-matrix composites using blended block copolymer-thermoplastic thin films. *Compos. Part A Appl. Sci. Manuf.* **2016**, *91*, 398–405. [CrossRef]

11. Xu, F.; Du, X.S.; Liu, H.Y.; Guo, W.G.; Mai, Y.W. Temperature effect on nano-rubber toughening in epoxy and epoxy/carbon fiber laminated composites. *Compos. Part B Eng.* **2016**, *95*, 423–432. [CrossRef]

12. Coleman, J.N.N. Small but strong: A review of the mechanical properties of carbon nanotube-polymer composites. *Carbon* **2006**, *44*, 1624–1652. [CrossRef]

13. Tsu-Wei, C. An assessment of the science and technology of carbon nanotube-based fibers and composites. *Compos. Sci. Technol.* **2009**, *71*, 1–19.

14. Qian, H.; Greenhalgh, E.S.; Shaffer, M.S.P.; Bismarck, A. Carbon nanotube based hierarchical composites: A review. *J. Mater. Chem.* **2010**, *20*, 4751–4762. [CrossRef]

15. Pieters, R.; Miltner, H.E.; Van Assche, G.; Van Mele, B. Kinetics of temperature-induced and reaction-induced phase separation studied by modulated temperature DSC. *Macromol. Symp.* **2006**, *233*, 36–41. [CrossRef]

16. Dzenis, Y.A.; Reneker, D.H. Delamination Resistant Composites Prepared by Small Diameter Fiber Reinforcement at Ply Interfaces. U.S. Patent 6,265,333, 24 July 2001.

17. Kim, J.; Reneker, D.H. Mechanical properties of composites using ultrafine electrospun fibers. *Polym. Compos.* **1999**, *20*, 124–131. [CrossRef]

18. Chen, Y.; Sui, L.; Fanga, H.; Ding, C.H.; Li, Z.; Jiang, S.; Hou, H. Superior mechanical enhancement of epoxy composites reinforced by polyimide nanofibers via a vacuum-assisted hot-pressing. *Compos. Sci. Technol.* **2019**, *174*, 20–26. [CrossRef]

19. Duan, G.; Liu, S.; Jiang, S.; Hou, H. High-performance polyamide-imide films and electrospun aligned nanofibers from an amidecontaining diamine. *J. Mater. Sci.* **2019**, *54*, 6719–6727. [CrossRef]

20. Xu, H.; Jiang, S.; Ding, C.H.; Zhu, Y.; Li, J.; Hou, H. High strength and high breaking load of single electrospun polyimide microfiber from water soluble precursor. *Mater. Lett.* **2017**, *201*, 82–84. [CrossRef]

21. Ding, C.H.; Fang, H.; Duan, G.; Zou, Y.; Chen, S.; Hou, H. Investigating the draw ratio and velocity of an electrically charged liquid jet during electrospinning. *RSC Adv.* **2019**, *9*, 13608–13613. [CrossRef]

22. Yang, H.; Jiang, S.; Fang, H.; Hu, X.; Duan, G.; Hou, H. Molecular orientation in aligned electrospun polyimide nanofibers by polarized FT-IR spectroscopy. *Spectrochim. Acta A* **2018**, *200*, 339–344. [CrossRef]

23. Jian, S.; Zhu, J.; Jiang, S.; Chen, S.; Fang, H.; Song, Y.; Duan, G.; Zhang, Y.; Hou, H. Nanofibers with diameter below one nanometer from electrospinning. *RSC Adv.* **2018**, *8*, 4794–4802. [CrossRef]

24. Hou, H.; Xu, W.; Ding, Y. The recent progress on high-performance polymer nanofibers by electrospinning. *J. Jiangxi Norm. Univ. (Nat. Sci.)* **2018**, *42*, 551–564.

25. Zheng, N.; Huang, Y.; Liu, H.Y.; Gao, J.; Mai, Y.W. Improvement of interlaminar fracture toughness in carbon fiber/epoxy composites with carbon nanotubes/polysulfone interleaves. *Compos. Sci. Technol.* **2017**, *140*, 8–15. [CrossRef]

26. Beckermann, G.W. Nanofiber interleaving veils for improving the performance of composite laminates. *Reinf. Plast.* **2017**, *61*, 289–293. [CrossRef]

27. Brugo, T.; Palazzetti, R. The effect of thickness of Nylon 6,6 nanofibrous mat on Modes I-II fracture mechanics of UD and woven composite laminates. *Compos. Struct.* **2016**, *154*, 172–178. [CrossRef]

28. Monteserín, C.; Blanco, M.; Murillo, N.; Pérez-Márquez, A.; Maudes, J.; Gayoso, J.; Laza, J.M.; Aranzabe, E.; Vilas, J.L. Effect of Different Types of Electrospun Polyamide 6 Nanofibres on the Mechanical Properties of Carbon Fibre/Epoxy Composites. *Polymers* **2018**, *10*, 1190. [CrossRef] [PubMed]

29. Ignatova, M.; Rashkov, I.; Manolova, N. Drug-loaded electrospun materials in wound-dressing applications and in local cancer treatment. *Expert Opin. Drug Deliv.* **2013**, *10*, 469–483. [CrossRef]

30. Zhong, W.; Xing, M.M.Q.; Maibach, H.I. Nanofibrous materials for wound care. *Cutan. Ocul. Toxicol.* **2010**, *29*, 143–152. [CrossRef]

31. Zahedi, P.; Rezaeian, I.; Ranaei-Siadat, S.O.; Jafari, S.H.; Supaphol, P. A review on wound dressings with an emphasis on electrospun nanofibrous polymeric bandages. *Polym. Adv. Technol.* **2010**, *21*, 77–95. [CrossRef]

32. Liu, M.; Duan, X.P.; Li, Y.M.; Yang, D.P.; Long, Y.Z. Electrospun nanofibers for wound healing. *Mater. Sci. Technol.* **2017**, *76*, 1413–1423. [CrossRef]

33. Pant, H.R.; Pant, B.; Pokharel, P.; Kim, H.J.; Tijing, L.D.; Park, C.H.; Lee, D.S.; Kim, H.Y.; Kim, C.S. Photocatalytic TiO_2-RGO/nylon-6 spider-wave-like nano-nets via electrospinning and hydrothermal treatment. *J. Membr. Sci.* **2013**, *429*, 225–234. [CrossRef]

34. Shalumon, K.; Anulekha, K.; Nair, S.V.; Nair, S.; Chennazhi, K.; Jayakumar, R. Sodium alginate/poly (vinyl alcohol)/nano ZnO composite nanofibers for antibacterial wound dressings. *Int. J. Biol. Macromol.* **2011**, *49*, 247–254. [CrossRef] [PubMed]

35. Lee, K.; Lee, S. Multifunctionality of poly (vinyl alcohol) nanofiber webs containing titanium dioxide. *J. Appl. Polym. Sci.* **2012**, *124*, 4038–4046. [CrossRef]

36. Choi, J.; SomMoon, D.; Jang, J.U.; Yin, W.B.; Lee, B.; Lee, K.J. Synthesis of Highly Functionalized Thermoplastic Polyurethanes and Their Potential Applications. *Polymer* **2017**, *116*, 287–294. [CrossRef]

37. Han, K.Q.; Yu, M.H. Study of the preparation and properties of UV-blocking fabrics of a PET/TiO_2 nanocomposite prepared by in situ polycondensation. *J. Appl. Polym. Sci.* **2006**, *100*, 1588–1593. [CrossRef]

38. Haider, A.; Kwak, S.; Gupta, K.C.; Kang, I.K. Antibacterial activity and cytocompatibility of PLGA/CuO hybrid nanofiber scaffolds prepared by electrospinning. *J. Nanomater.* **2015**, *2015*, 1–10. [CrossRef]

39. Quirós, J.; Boltes, K.; Aguado, S.; de Villoria, R.G.; Vilatela, J.J.; Rosal, R. Antimicrobial metal–organic frameworks incorporated into electrospun fibers. *Chem. Eng. J.* **2015**, *262*, 189–197. [CrossRef]

40. Amna, T.; Hassan, M.S.; Barakat, N.A.; Pandeya, D.R.; Hong, S.T.; Khil, M.S.; Kim, H.Y. Antibacterial activity and interaction mechanism of electrospun zinc-doped titania nanofibers. *Appl. Microbiol. Biotechnol.* **2012**, *93*, 743–751. [CrossRef]

41. Hwang, S.H.; Song, J.; Jung, Y.; Kweon, O.Y.; Song, H.; Jang, J. Electrospun ZnO/TiO_2 composite nanofibers as a bactericidal agent. *Chem. Commun.* **2011**, *47*, 9164–9166. [CrossRef]

42. Ye, C.X.; Li, H.Q.; Cai, A.; Gao, Q.Z.; Zeng, X.R. Preparation and Characterization of Organic Nano-Titanium Dioxide/Acrylate Composite Emulsions by in-situ Emulsion Polymerization. *J. Macromol. Sci. A Pure Appl. Chem.* **2011**, *48*, 309–314. [CrossRef]

43. Yoshida, M.; Lal, M.; Kumar, N.D.; Prasad, P.N. TiO_2 nano-particle-dispersed polyimide composite optical waveguide materials through reverse micelles. *J. Mater. Sci.* **1997**, *32*, 4047–4051. [CrossRef]

44. Barnard, A.S.; Curtiss, L.A. Prediction of TiO_2 Nanoparticle Phase and Shape Transitions Controlled by Surface Chemistry. *Nano Lett.* **2005**, *5*, 1261–1266. [CrossRef] [PubMed]

45. Sundarrajan, S.; Ramakrishna, S. Fabrication of nanocomposite membranes from nanofibers and nanoparticles for protection against chemical warfare stimulants. *J. Mater. Sci.* **2007**, *42*, 8400–8407. [CrossRef]

46. Sundarrajan, S.; Luck, K.; Huat, S.; Ramakrishna, S. Electrospun nanofibers for air filtration applications. *Procedia Eng.* **2014**, *75*, 59–163. [CrossRef]

47. Danni, N.; Sasikumar, T.; Fazil, A.A. Mechanical properties of electrospun CNF/PVA nanofiber mats as reinforcement in polymer matrix composites. *Int. J. Appl. Chem.* **2016**, *12*, 107–119.

48. Li, B.; Yuan, H.; Zhang, Y. Transparent PMMA-based nanocomposite using electrospun graphene-incorporated PA-6 nanofibers as the reinforcement. *Compos. Sci. Technol.* **2013**, *89*, 134–141. [CrossRef]

49. Neppalli, R.; Causin, V.; Benetti, E.M.; Ray, S.S.; Esposito, A.; Wanjale, S.; Birajdar, M.; Saiter, J.M.; Marigo, A. Polystyrene/TiO_2 composite electrospun fibers as fillers for poly(butylene succinate-co-adipate): Structure, morpholog and properties. *Eur. Polym. J.* **2014**, *50*, 78–86. [CrossRef]

50. Marega, C.; Marigo, A. Effect of electrospun fibers of polyhydroxybutyrate filled with different organoclays on morphology, biodegradation, and thermal stability of poly(ε-caprolattone). *J. Appl. Polym. Sci.* **2015**, *132*, 42342–42354. [CrossRef]

51. Chang, Z.J. Development of a polyurethane nanocomposite reinforced with carbon nanotube composite nanofibers. *Mater. Sci. Forum* **2011**, *688*, 41–44. [CrossRef]

52. Liu, L.; Zhou, Y.; Pan, S. Experimental and analysis of the mechanical behaviors of multi-walled nanotubes/polyurethane nanoweb-reinforced epoxy composites. *J. Reinf. Plast. Compos.* **2013**, *32*, 823–834. [CrossRef]

53. Blanco, M.; Monteserín, C.; Angulo, A.; Pérez-Márquez, A.; Maudes, J.; Murillo, N.; Aranzabe, E.; Ruiz-Rubio, L.; Vilas, J.L. TiO_2 Doped Electrospun Nanofibrous Membrane for Photocatalytic Water Treatment. *Polymers* **2019**, *11*, 747. [CrossRef] [PubMed]

54. Daelemans, L.; Van der Heijden, S.; De Baere, I.; Rahier, H.; Van Paepegem, W.; De Clerck, K. Damage-Resistant Composites Using Electrospun Nanofibers: A Multiscale Analysis of the Toughening Mechanisms. *ACS Appl. Mater. Interfaces* **2016**, *818*, 11806–11818. [CrossRef] [PubMed]

55. Seyhan, A.T.; Tanoglu, M.; Schulte, K. Mode I and mode II fracture toughness of E-glass non-crimp fabric/carbon nanotubes (CNT) modified polymer based composites. *Eng. Fract. Mech.* **2008**, *75*, 5151–5162. [CrossRef]

56. Sager, R.J.; Klein, P.J.; Davis, D.C.; Davis, D.C.; Lagoudas, G.L.; Warren, G.L.; Sue, H.J. Interlaminar fracture toughness of woven fabric composite laminates with carbon nanotube/epoxy interleaf films. *J. Appl. Polym. Sci.* **2011**, *121*, 2394–2405. [CrossRef]

57. Brugemann, V.P.; Sinke, J.; de Boer, H. Fracture Toughness Testing in FML. In Proceedings of the 25th Structural Dynamics Conference, Orlando, FL, USA, 19–22 February 2007; pp. 141–151.

58. Wang, G.; Yu, D.; Kelkar, A.D.; Zhang, L. Electrospun nanofiber: Emerging reinforcing filler in polymer matrix composite materials. *Prog. Polym. Sci.* **2017**, *75*, 73–107. [CrossRef]

59. Malakhov, S.N.; Belousov, S.L.; Shcherbina, M.A.; Meshchankina, M.Y.; Chvalun, S.N.; Shepelev, A.D. Effect of Low Molecular Additives on the Electrospinning of Nonwoven Materials from a Polyamide-6 Melt. *Polym. Sci. Ser. A* **2016**, *58*, 236–245. [CrossRef]

60. Sinha-Ray, S.; Lee, M.W.; Sinha-Ray, S.; An, S.; Pourdeyhimi, B.; Yoon, S.S.; Yarin, A.L. Supersonic nanoblowing: A new ultra-stiff phase of nylon 6 in 20–50 nm confinement. *J. Mater. Chem. C* **2013**, *1*, 3491–3498. [CrossRef]

 *polymers*

Article

Hydrogel Small-Diameter Vascular Graft Reinforced with a Braided Fiber Strut with Improved Mechanical Properties

Guoping Guan [1], Chenglong Yu [1], Meiyi Xing [1], Yufen Wu [1], Xingyou Hu [1], Hongjun Wang [2] and Lu Wang [1,*]

[1] Engineering Research Center of Technical Textiles, Ministry of Education; Key laboratory of Textile Science and Technology, Ministry of Education; College of Textiles, Donghua University, Songjiang District, Shanghai 201620, China; ggp@dhu.edu.cn (G.G.); Belongdrew@163.com (C.Y.); 18363994765@163.com (M.X.); yufenwu1130@126.com (Y.W.); huxingyou@126.com (X.H.)

[2] Department of Biomedical Engineering, Stevens Institute of Technology, Hoboken, NJ 07030, USA; Hongjun.Wang@stevens.edu

* Correspondence: wanglu@dhu.edu.cn; Tel.: +86-21-6779-2637

Received: 18 April 2019; Accepted: 2 May 2019; Published: 6 May 2019

Abstract: Acute thrombosis remains the main limitation of small-diameter vascular grafts (inner diameter <6 mm) for bridging and bypassing of small arteries defects and occlusion. The use of hydrogel tubes represents a promising strategy. However, their low mechanical strength and high swelling tendency may limit their further application. In the present study, a hydrogel vascular graft of Ca alginate/polyacrylamide reinforced with a braided fiber strut was designed and fabricated with the assistance of a customized casting mold. Morphology, structure, swellability, mechanical properties, cyto- and hemocompatibility of the reinforced graft were characterized. The results showed that the reinforced graft was transparent and robust, with a smooth surface. Scanning electron microscopic examination confirmed a uniform porous structure throughout the hydrogel. The swelling of the reinforced grafts could be controlled to 100%, obtaining clinically satisfactory mechanical properties. In particular, the dynamic circumferential compliance reached (1.7 ± 0.1)%/100 mmHg for 50–90 mmHg, a value significantly higher than that of expanded polytetrafluoroethylene (ePTFE) vascular grafts. Biological tests revealed that the reinforced graft was non-cytotoxic and had a low hemolysis percentage (HP) corresponding to (0.9 ± 0.2)%. In summary, the braided fiber-reinforced hydrogel vascular grafts demonstrated both physical and biological superiority, suggesting their suitability for vascular grafts.

Keywords: hydrogel; vascular graft; braided fiber strut; swellability; mechanical property

---

1. Introduction

Cardiovascular diseases (CVDs) are the leading cause of human mortality worldwide, and the population of CVD patients has been increasing exponentially [1,2]. Polyethylene terephthalate (PET) and expanded polytetrafluoroethylene (ePTFE) are the most commonly used materials for large-diameter vascular grafts, such as an endovascular stent and graft [3,4]. However, small-diameter vascular grafts (inner diameter <6mm), which are highly demanded coronary artery bypasses for grafting or substituting lower limb arteries, are prone to causing thrombus formation, stenosis, pseudo-endothelial hyperplasia, and consequent luminal occlusion [5–7]. Thus, current commercial products are still far from ideal, and thrombosis is the primary issue limiting their intermediate and long-term utility [8].

Tissue-engineered small-diameter vascular grafts (TESDVGs) have demonstrated to be promising for CVD treatment and have been extensively explored in the past few decades [9]. Among the critical elements of TESDVG (e.g., scaffolds, cells, and growth factors (GFs)) [10], fabrication of a porous scaffold is relatively straightforward using established methods such as freeze-drying, phase separation, particulate leaching, gas foaming, and so on [11–13]. However, it remains a significant challenge to identify the appropriate cells for regenerating new extracellular matrix close to the native one, while preventing programmed cell death or apoptosis during implantation [14]. Furthermore, the involvement of multiple GFs in the spatiotemporal regulation of cellular activities also determines the necessity of synergistically incorporating these GFs in the creation of TESDVGs [15,16]. Up to date, it is not yet imaginable to design and construct a GF-induced, orchestrated cascade system. In vitro dynamic culture could help improve the functionality of TESDVGs [7]. However, deviation from in vivo circumstances in terms of mechanical and biological environments occurs. There is still a long way to go for TESDVGs before their broad adoption for clinical applications.

Because of their hydrophilicity and biocompatibility, hydrogels have found multifaceted applications in biomedicine [17,18], for example, as scaffolding materials for tissue formation, vehicles for drug and growth factor delivery [19–22], protection layer for encapsulated cells [23–25], and so on [21,26,27]. In particular, in consideration of their anti-coagulation and hemocompatibility [28], hydrogels may find great use in vascular grafts. However, some associated weaknesses such as poor mechanical properties and high swelling capacities should be addressed. Thus, increasing efforts have been made to improve the mechanical properties and control the swellability of hydrogels [29,30].

An ideal small-diameter vascular graft should have specific features concerning mechanical properties (suture retention, longitudinal and circumferential compliance, tensile and compressive strength, and so forth), biological performance (cyto- and hemocompatibility), blood permeability, biodegradability, morphological and structural stability, ease for processability, sterilization, packaging, storage, and transit performance. Therefore, in the present study, sodium alginate and acrylamide (AAm) were employed to fabricate an interpenetrating double-network hydrogel with a three-step process modified by a previous method [29,31–33]. By combination with a PET braided fiber strut, the mechanical properties of the hydrogel tube were enhanced significantly to meet the operational requirements in terms of tensile strength, tensile elongation, circumferential compliance, compressive strength, and suture retention. Furthermore, the swellability and inner diameter of the reinforced hydrogel vascular grafts could be maintained in an aqueous environment within the test time. As expected, the reinforced hydrogel tubes exhibited high cytocompatibility and meager hemolysis percentage (HP).

2. Materials and Methods

2.1. Design and Fabrication of a 3D Casting Mold

A stainless-steel casting mold was explicitly designed and fabricated to form a braided fiber strut reinforced hydrogel vascular graft. Briefly, the mold was composed of seven parts, i.e., base, base adaptor, lower adaptor, upper adaptor, rod, and two half pipes (Figure 1). The inner and outer diameters of the reinforced graft were designed to be 6 and 12 mm, respectively. Thus, the diameter of the rod was 6 mm. The outer diameters of the lower and upper adaptors and the inner diameter of the braided fiber strut were the same, i.e., 9 mm.

2.2. Preparation of the Ca Alginate/Polyacrylamide Hydrogel

The Ca alginate/polyacrylamide (PAAm) hydrogels were prepared as described previously [29] with some modifications. Three steps were involved. First, acrylamide (AAm) and sodium alginate (SA) were dissolved to form an aqueous solution. Ammonium persulfate (APS) as a photo-initiator, N, N'-methylenebisacrylamide (MBAA) as a crosslinker, and N, N, N', N'-tetramethylethylenediamine (TEMED) as an accelerator for polyacrylamide, were mixed into the aqueous solution to form a pre-gel. Second, the pre-gel was immersed into 1.0 M $CaCl_2$ aqueous solution to crosslink the alginate. Third, the crosslinked hydrogel was soaked in distilled water for 12 h to remove the residual monomers and finally obtain the Ca alginate/polyacrylamide hydrogel. The concentrations used were as follows: 16.8% (w/v) for AAm, 2.3% (w/v) for SA, 0.1% (w/v) for APS, 0.01% (w/v) for MBAA, and 0.09% (w/v) for TEMED. Various immersion times (3h, 4h, 5h, and 6h) of CaCl$_2$ aqueous solution were tested to identify the optimum crosslinking time, and the samples were correspondingly labeled as S1, S2, S3, and S4. All the raw materials were provided by Sinopharm Chemical Reagent Co., Ltd., Shanghai, China.

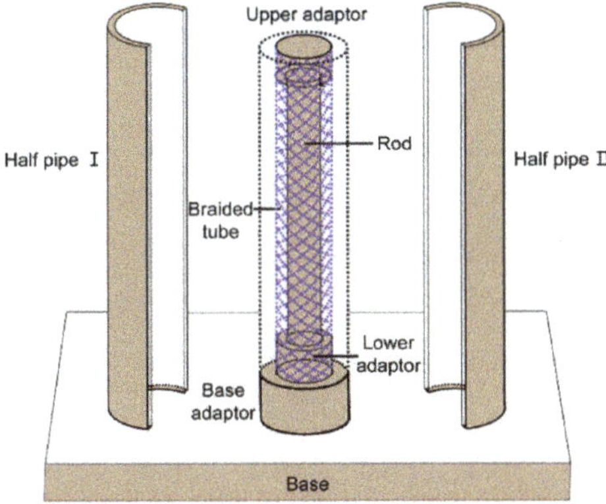

Figure 1. Schematic diagram of the stereo-casting mold.

2.3. Scanning Electron Microscopy (SEM) of the Hydrogels

The hydrogels with a size of 30 mm × 3 mm × 3 mm were freeze-dried by liquid nitrogen and broken into half for exposure of cross sections to SEM. Gold powder was sprayed on the samples prior to microscopy. A scanning electron microscope (TM 3000, Hitachi, Tokyo, Japan) was used to examine the microscopic morphology of the hydrogels. The accelerating voltage was 15 kV.

2.4. Volume Swellability of the Hydrogels

Swellability can be characterized by the swelling ratio in weight and in volume [30,34]. Since the focus of the present work was to attain a vascular graft with higher mechanical properties and lower swellability to maintain a constant diameter during implantation, the swelling ratio in volume was more important. The swelling ratio (SR) in volume was calculated according to Equation (1) [30]:

$$\mathrm{SR}(\%) = \frac{V}{V_0} \times 100 \tag{1}$$

where SR is the swelling ratio, V is the hydrogel volume upon swelling equilibrium, and V_0 is the initial hydrogel volume.

For measurement, the as-prepared hydrogels were cut into strips with a size of 50 mm × 20 mm × 3 mm. Volumes could be calculated by measuring the length, the width, and the height of the bulk hydrogels.

2.5. Tensile Strength and Elongation of the Hydrogels

The tensile strength of the hydrogels was tested on an automatic versatile mechanical property tester (YG(B)026G-500, Darong Textile Instrument Co., Ltd., Wenzhou, China) following the Chinese National standard GB/T 528-2009. The hydrogels reaching swelling equilibrium (swelling for 24 h) were cut into strips with a size of 50 mm × 20 mm × 3 mm, the gap was set to 25 mm, and the stretching speed was set to 500 mm/min.

2.6. Design and Fabrication of a Braided Fiber Strut

PET monofilaments (medical grade) with a diameter of 0.25 mm (Suzhou Suture Needle Company, Suzhou, China) were braided with a 32-bobbin braiding machine to form a tube with the inner diameter of 9 mm. The braided fiber strut on a rod was then heated at 190 °C for 10 min. Varying braiding angles (50°, 55°, and 60°) were chosen to adjust the strain of the braided fiber strut to match the strain of the hydrogel tube. Deformation of the braided fiber strut and of the hydrogel tube and their hybrid were evaluated using diameter reduction rate (DR) %, which was calculated following Equation (2):

$$DR(\%) = \frac{D_i - D_d}{D_i} \times 100 \tag{2}$$

where DR (%) is the diameter reduction rate (%), D_i is the initial inner diameter of a tube, and D_d is the inner diameter of the tube when strained at 50%.

2.7. Fabrication of Braided Fiber Strut-Reinforced Hydrogel Vascular Grafts

With the assistance of a mold, a braided fiber strut reinforced hydrogel vascular graft was fabricated from the Ca alginate/polyacrylamide hydrogel and PET strut (Figure 2). Briefly, the braided fiber strut was wrapped around the rod, and both ends of the braided fiber strut were tied to the upper adaptor and the lower adaptor. Two half pipes were combined to form a complete pipe outside the braided fiber strut, and then the pre-gel aqueous solution was poured into the cavity between the rod and the pipe. Upon pre-gelation, the two half pipes were removed, and the pre-gel structure was soaked into 1.0 M CaCl$_2$ aqueous solution for crosslinking, followed by water immersion to remove the residuals. Actually, the pre-gel was a network of PAAm embedding alginate chains. Afterward, the rod, the upper adaptor, and the lower adaptor were removed to release the braided fiber strut-reinforced hydrogel vascular grafts.

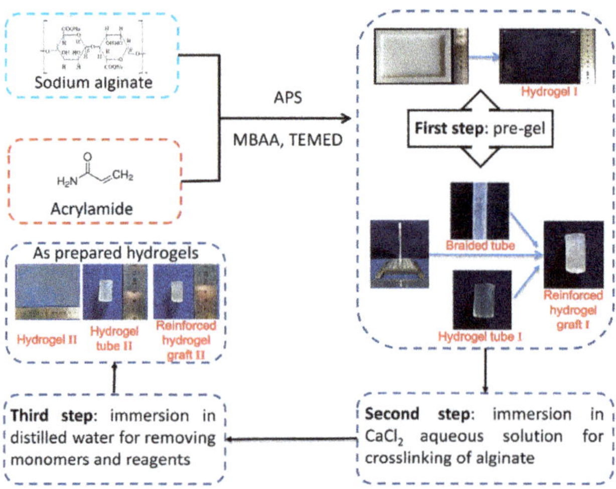

Figure 2. Schematic illustration of the preparation process of the reinforced hydrogel vascular graft. APS: ammonium persulfate, MBAA: N, N′-methylenebisacrylamide, TEMED: N, N, N′, N′-tetramethylethylenediamine.

2.8. Mechanical Properties of the Reinforced Grafts

The mechanical properties of the reinforced graft were similarly tested for tensile strength, dynamic circumferential compliance, compressive strength, and suture retention force as previously described [35,36]. Briefly, the tensile strength was longitudinally tested on an automatic versatile mechanical property tester (YG(B)026G-500, Darong Textile Instrument Co., Ltd., Wenzhou, China) by Chinese national standard GB/T 528-2009. The maximum strain was set to 50%. The dynamic circumferential compliance was tested under a sinusoidal wave formed at a 1 aHz frequency with a BioDynamicTM Test Instrument (Bose Corporation, The Mountain Framingham, MA, USA) using the following three pressure ranges: 50–90, 80–120, and 110–150 mmHg, in accordance to the standard ISO 7198:2016. The compressive strength was circumferentially tested on a specially designed radial compression measuring apparatus (Model LLY-06D, Laizhou Electronic Instrument Co., Ltd., Laizhou, China) according to ISO 25539-2:2012. The presser foot of the apparatus was 5 mm in diameter, and the reinforced graft was compressed to 50% of the outer diameter and then released at the same rate of 10 mm/min. The displacement of the presser foot was recorded to calculate elastic recovery. The suture retention force was measured on a universal mechanical property tester (Model YG-B 026H, Laizhou Electronic Instrument Co., Ltd., Laizhou, China) following the standard ISO 7198: 2016. A 2-0 braided polyester suture (Jinhuan Medical Supplies Co. Ltd., Shanghai, China) was used. The suture was passed through the reinforced graft at the site exactly 2 mm away from the edge of the end. Then, the suture was pulled at a speed of 50 mm/min until the specimen failed. Tubular hydrogels were also prepared and characterized as control samples.

For the swellability tests, the grafts were cut into segments with a size of 50 mm in length. Volumes could be calculated by measuring the inner and outer diameters of the tubes.

2.9. In Vitro Cytotoxicity Assay of the Reinforced Grafts

Porcine iliac artery endothelial cells (PIECs, Cell Bank, Shanghai Institutes for Biological Sciences, Chinese Academy of Sciences, Shanghai, China) were cultured in Dulbecco's Modified Eagle Medium (DMEM) with 10% fetal bovine serum and 1% penicillin/streptomycin (complete culture media, Thermo Fisher Scientific, Carlsbad, CA, USA) in a 37 °C, 5% CO_2 environment. Conditioned cell culture media (CM) was prepared by incubating the reinforced grafts in the media for 24 h to determine the effect of grafts eluates on PIECs. According to the standard ISO 10993-12:2012, 1 g of the reinforced graft was put into 10 mL cell culture media in a 37 °C, 5% CO_2 environment to prepare CM. The WST (water-soluble tetrazolium) assay was performed with CCK-8 (Yeasen Biotechnology Co., Ltd., Shanghai, China) according to the manufacturer's instructions. In total, 2000 cells/well were seeded in 100 µL complete culture media in 96-well plates for 4 h before changing to 100 µL CM. The absorbance was measured at 450 nm using a BioTek plate reader (BioTek, Winooski, VT, USA). Relative proliferation rate (RPR, %) was calculated on the basis of optical density values by equation (3):

$$\mathrm{RPR}(\%) = \frac{OD_s}{OD_c} \times 100 \tag{3}$$

where OD_s is the absorbance of the sample, and OD_c is the absorbance of the negative controls.

2.10. Hemolysis Test of the Graft

Fresh whole blood obtained from New Zealand white rabbits' ear veins was anticoagulated and utilized to test the hemolytic performance of the reinforced grafts. The blood was centrifuged (Biofuge Primo Model R centrifuge, Thermo Fisher Scientific, Carlsbad, CA, USA) and washed with PBS five times following standard procedures as reported [37,38]. A volume of 1 mL of red blood cells (RBCs) was suspended in 34 mL PBS. Then, 0.2 mL of the RBCs suspension was transferred to a 5 mL Eppendorf tube, which was filled with either 0.8 mL of deionized water as the positive control, or PBS buffer as the negative control. The reinforced graft samples were incubated in the suspension containing 0.2 mL of diluted RBCs and 0.8 mL of PBS buffer at 37 °C for 2 h, followed by centrifugation for 3 min at 10,000 rpm with an Eppendorf 5415 Model R centrifuge (Eppendorf, Hamburg, Germany). Then, the optical density of the supernatant was determined by a Perkin Elmer

Lambda 25 UV–visible spectrophotometer (Perkin Elmer, Waltham, MA, USA) at 540 nm. The HP was calculated using Equation (4) [39,40]:

$$HP(\%) = \frac{D_s - D_{nc}}{D_{pc} - D_{nc}} \times 100 \qquad (4)$$

where D_s is the absorbance of the sample, and D_{pc} and D_{nc} are the absorbances of the positive and negative controls, respectively.

2.11. Statistical Analysis

All data obtained in the present work were expressed as mean ± standard deviation (SD), and a significant difference was analyzed using ANOVA and unpaired student t-test; $p < 0.05$ was considered significant.

3. Results and Discussion

3.1. Morphology and Porous Structure of the Ca Alginate/Polyacrylamide Hydrogel

The Ca alginate/polyacrylamide hydrogel prepared in the present work was transparent and smooth (Figure 3a). The hydrogel was stretchable, flexible and elastic when tested for tensile and compressive strength. SEM examination revealed a porous structure of the hydrogel (Figure 3b). A wide range of pore size (pore area), i.e., 287–6150 μm^2 was measured, and the average pore size was 3556 ± 995 μm^2 (Figure 3c). In the present work, the as-prepared pre-gel was soaked in a $CaCl_2$ aqueous solution for gelation instead of adding the $CaCl_2$ solution into the mixed solution [41] or soaking in the $CaSO_4 \cdot 2H_2O$ aqueous solution [29], so the gelation rate could be tuned, which led to a homogeneous hydrogel.

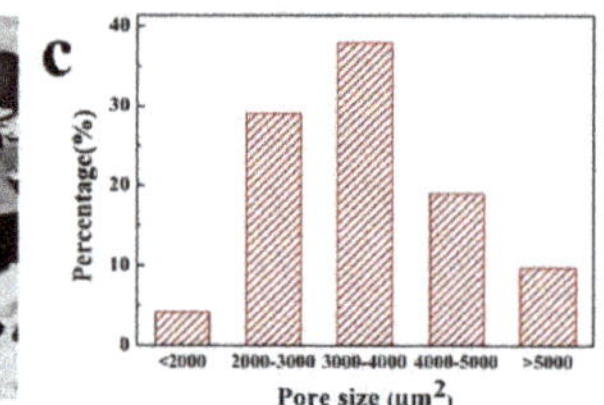

Figure 3. Images of the Ca alginate/polyacrylamide hydrogels (**a**), SEM microphotograph of a cross section (**b**) and distribution of the pore sizes (**c**).

3.2. Tensile Strength of the Ca Alginate/Polyacrylamide Hydrogel

A series of tensile tests were performed on the hydrogels with different immersion times (S1: 3h, S2: 4h, S3: 5h, and S4: 6h) to determine the optimal crosslinking time. Figure 4 shows the results of the tensile breaking strength and tensile breaking elongation of four types of hydrogel (S1–S4) (n = 3). The tensile breaking strengths of S1 and S2 decreased significantly after reaching the swelling equilibrium compared to those of the as-prepared hydrogels (Figure 4a). Meanwhile, a significant increase in water content and volume were seen for S1 and S2 hydrogels upon reaching the swelling equilibrium. While no apparent changes were observed for S3 and S4 before and after swelling equilibrium, the tensile breaking strength of S3 was much higher than those of S1 and S2. This phenomenon could be explained by the fact that the ionic crosslinking increased over prolonged immersion times till 5 h, and then the crosslinking of G blocks of alginate with Ca^{2+} reached its plateau by 5 h; as a result, the tensile strength increased with the increase of the crosslinking degree.

Figure 4b shows that the tensile breaking elongations of the four hydrogels decreased significantly because of swelling compared to those of the as-prepared hydrogels. However, no significant difference was identified among four hydrogels either before or after swelling, suggesting that the immersion time in $CaCl_2$ solution had a minimal effect on the elongation performance of the hydrogels. On the one hand, the amount of Ca^{2+} was meager for crosslinking; on the other hand, the influence of the difference of immersion time on crosslinking saturation could not be shown, since 3 h might be enough for almost complete crosslinking. Combined with the above results, the optimal immersion time selected was 5 h.

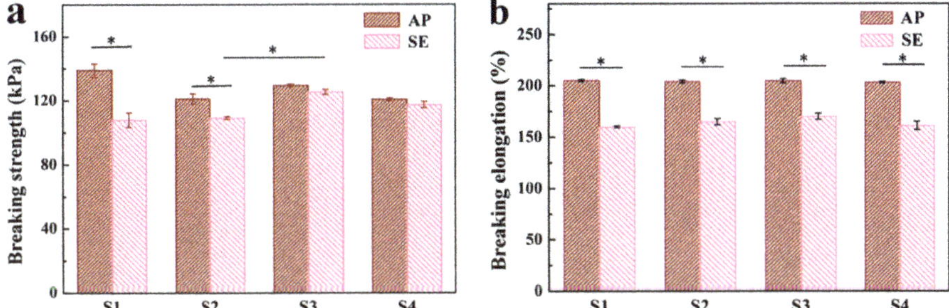

Figure 4. Results of the tensile breaking strength and tensile breaking elongation tests of four hydrogels (S1: 3h; S2: 4h; S3: 5h; S4: 6h). (**a**) Tensile breaking strength. (**b**) Tensile breaking elongation; * means significant difference at $p < 0.05$, AP: as-prepared hydrogels, SE: hydrogels at swelling equilibrium.

3.3. Volume Swellability of the Ca Alginate/Polyacrylamide Hydrogel

Representative swelling properties of the hydrogels as a function of time in deionized water are presented in Figure 5a. The results showed that the swelling ratios of all four types of hydrogels increased as a function of the immersion time (3 to 6 h) in deionized water until equilibrium. Meanwhile, the equilibrium times for hydrogels S1, S2, S3, and S4 were determined as 12, 12, 0, and 0 h, respectively. In other words, no volume changes were found for S3 and S4; that is, the swelling ratios were 100%. Figure 5b showed that the swelling ratios of S1 and S2 were significantly higher than those of S3 and S4 when they reached the swelling equilibrium. This observation indicates that vascular grafts derived from both S3 and S4 would have satisfying structural stability, and the volume and configuration (e.g., the inner diameter) would remain constant. This point is of critical importance for hydrogel-based vascular grafts, especially for small-diameter vascular grafts, as any subtle change to the diameter of vascular grafts may evoke enormous turbulence of the blood flow, which would induce acute thrombosis and, consequently, luminal occlusion [41]. These results are in good agreement with those of the mechanical properties. In combination, S3 would be an excellent choice to fabricate small-diameter vascular grafts possessing not only robust mechanical properties but also constant structural stability, suitable for in vivo implantation. Hence, S3 was selected for the following experiments.

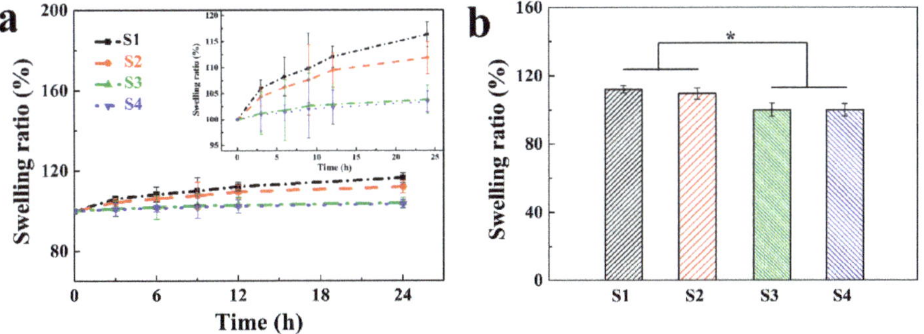

Figure 5. Swellability of four hydrogels. Processes of approaching the swelling equilibrium (**a**), equilibrium swelling ratios (**b**) (* $p < 0.05$).

3.4. Fabrication of Strut-Reinforced Hydrogel Vascular Grafts

With the assistance of a customized stainless-steel casting mold (Figure 6a), hydrogel-based vascular grafts (Figure 6d) reinforced with a braided fiber strut (Figure 6b) could be fabricated. To maintain the integrity between hydrogel and fiber strut, it became necessary to assure a similar deformation behavior of hydrogel tubes and fiber struts. In this regard, several fiber struts with different braiding angles (Table 1) were chosen and compared to S3 hydrogel tubes. In consideration of the typical shrinking rate of a native blood vessel of about 33% [42], an elongation rate of 50% was used for the tensile test. As shown in Table 1, the braided fiber strut with a braiding angle 55° had a deformation comparable to that of the S3 hydrogel tube (Table 1, n = 3). Therefore, this braided fiber strut was selected to fabricate reinforced hydrogel vascular grafts for the following experiments.

Table 1. Diameter reduction rate (%) of the braided fiber struts and S3 hydrogel tubes.

Number	Braiding Angle (°)	Diameter Reduction Rate (%)
1	50 ± 0.5	26.52 ± 0.64
2	55 ± 0.5	18.73 ± 0.68
3	60 ± 0.5	7.98 ± 0.22
S3	/	18.14 ± 0.94

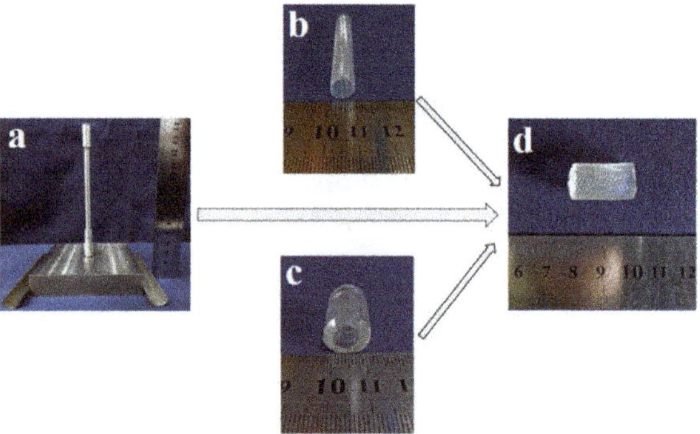

Figure 6. Images of the stereo-casting mold (**a**), braided polyethylene terephthalate (PET) tube (**b**), hydrogel tube (**c**), and reinforced hydrogel graft (**d**).

With the desired hydrogel and braided fiber struts, strut-reinforced hydrogel vascular grafts could be readily fabricated. During the fabrication, the inner diameter of the vascular graft and the thickness of the wall could be precisely adjusted and controlled by adjusting the mold parameters. Moreover, the braided fiber strut could be positioned in the middle, inner, or outer layer of the wall. Compared to other crimp small-diameter vascular grafts made of PET and ePTFE, a prominent advantage of the vascular graft described in the present work is that the lumen would not close while bending even without a crimp wall, because its morphology is similar to that of a native blood vessel possessing a layered thick and elastic wall. Hydrogel tubes without a braided fiber strut (Figure 6c) were also fabricated as controls.

3.5. Mechanical Properties of the Strut-Reinforced Hydrogel Grafts

Native blood vessels are strained inside the body and they are shortened upon cut-off from the primary site, e.g., the length of a swine carotid artery is reduced to 2/3 of its original length upon cut-off from the body [42]. Therefore, the length of an implanted vascular graft should refer to the original length of a target native blood vessel, and, ideally, the graft should have the same elongation rate as the native vessels. The tensile properties of the reinforced grafts were evaluated, setting the strain to 50%. The results showed that the incorporation of a braided fiber strut significantly improved the mechanical properties of the grafts (Figure 7), and the reinforced graft could almost completely recover to its original shape upon relief of the force. More importantly, a simultaneous response of the braided fiber strut and the hydrogel tube was invariably found, without any mismatch or interfacial separations. The measured stress of the reinforced grafts was almost the sum of those of the braided fiber strut and of the hydrogel tube and reached 201.3 ± 6.91 kPa with a set strain of 50%, close to but still higher than that of a swine carotid artery ($p < 0.05$) (Figure 7 inset). This observation implies that the strut-reinforced grafts can meet the tensile strength requirements for clinical application.

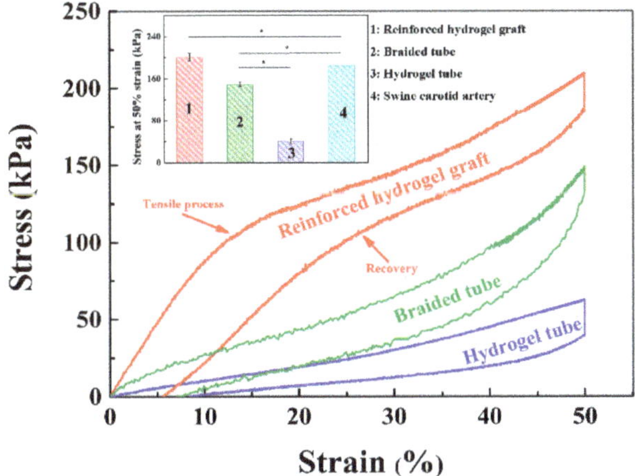

Figure 7. Strain—tress tensile curves of the strut-reinforced hydrogel grafts, hydrogel tubes, and braided fiber struts. Inset: results of the quantitative analysis.

The compliance of a vascular graft plays a pivotal role in maintaining a healthy blood flow, avoiding abnormal tissue hyperplasia and thrombosis [43,44]. However, the compliance of the existing vascular grafts made of either PET or ePTFE was low, corresponding to less than 1% [41], which partly accounts for the unsuitability of such grafts for small-diameter vessels [45]. The circumferential compliances of the reinforced graft, hydrogel tube, and a commercial ePTFE vascular graft were respectively tested according to the standard ISO 7198: 2016 on the BOSE platform under three spans of pressures, i.e., 50–90, 80–120, and 110–150 mmHg. The measurement revealed that the compliances of the hydrogel tubes and the reinforced grafts were much higher than that of the ePTFE vascular grafts ($p < 0.05$) in all three pressure conditions, despite a decreased compliance of the reinforced grafts by the braided fiber strut (Figure 8). The compliance of reinforced hydrogel grafts under a pressure of 50–90 mmHg reached (1.7 ± 0.1)%/100 mmHg, significantly higher than that of the ePTFE vascular grafts, indicating the superiority of reinforced hydrogel grafts for the substitution or bypass of small-diameter blood vessels.

In addition to the tensile strength, the compressive strength of both the hydrogel tube and the reinforced graft was tested. As shown in Figure 9, the incorporation of the braided fiber strut in the reinforced grafts significantly increased the compressive strength of the grafts, demonstrating the vital role of the braided fiber strut in strengthening the radial mechanical properties. Upon relief after compression to 50% of the outer diameter, the radial elastic recovery could reach (90.1 ± 1.3)%.

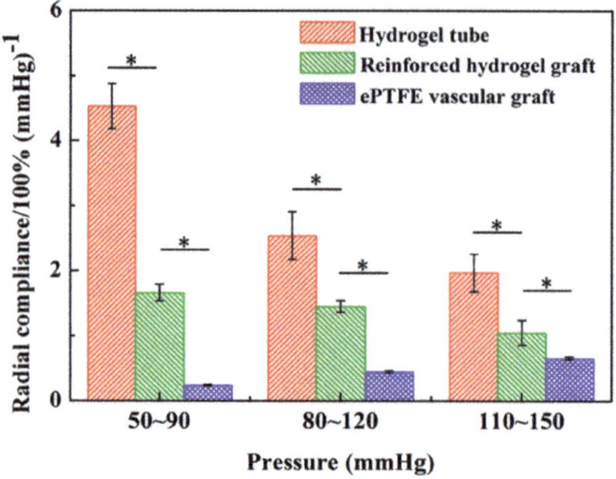

Figure 8. Circumferential compliance of the hydrogel tubes, strut-reinforced hydrogel grafts, and an approved expanded polytetrafluoroethylene (ePTFE) vascular graft (n=3).

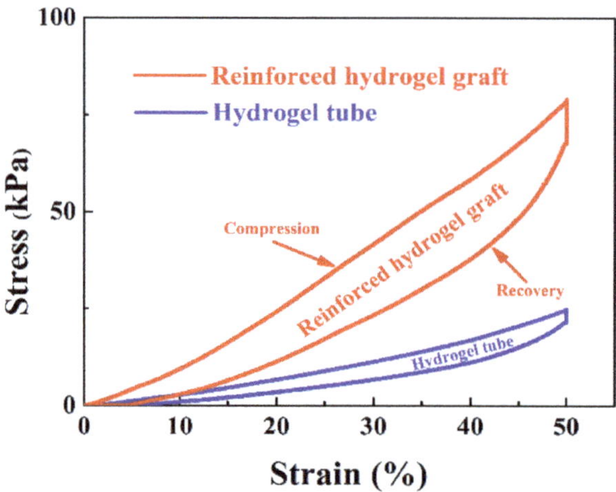

Figure 9. Compressive strength of the strut-reinforced grafts and hydrogel tubes.

Considering suturing is the first step of surgery, suture retention force is another essential parameter for the mechanical characterization of vascular grafts. Thus, a vascular graft should be able to suture to the native blood vessels. The suture retention force measurement showed that the reinforced graft yielded 8.4 ± 0.5 N retention force, remarkably higher than those of a canine femur artery (7.9 ± 0.4 N, $p < 0.05$) [46] or of hydrogel tubes (6.8 ± 1.3 N, $p < 0.05$). This finding implies that the braided fiber strut would be the dominant contributor to strengthen the suture retention force. Taken together, it becomes clear that the reinforced grafts can not only meet the mechanical need for surgery but also provide satisfactory structural flexibility.

3.6. Swellability of the Strut-Reinforced Vascular Grafts

The swellability of grafts, especially upon implantation, is critical to graft stability [47]. Thus, the swellability and inner diameter of hydrogel (S3) tubes and braided fiber strut-reinforced vascular grafts were evaluated for 24 h. As shown in Figure 10, the swellability and inner diameters of both samples kept constant. The swellability remained 100% for the entire testing period, with no volume changes. The inner diameters also remained regularly at 6.5 mm. These observations indicate that the strut-reinforced vascular grafts would maintain their structural stability during implantation.

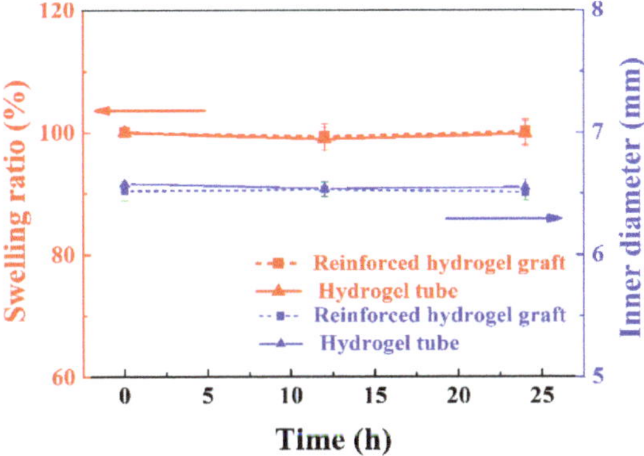

Figure 10. Evolution of swellability in volume and inner diameters of the reinforced hydrogel grafts and hydrogel tubes over time.

3.7. In Vitro Cellular Viability of the Strut-Reinforced Vascular Grafts

Strut-reinforced vascular grafts sterilized by autoclave were incubated in cell culture media for 24 h to prepare the conditioned cell culture media (CM). PIECs were divided into two groups, i.e., graft-incubated group (CM) and control group (CG, complete culture media). The cells cultured in either CM or CG media for 1, 3, and 5 days were analyzed with the WST assay. As shown in Figure 11, the optical density (OD), proportional to the cell number, increased over the culture time, and the relative proliferation ratio (RPR, %) in CM and CG for 1, 3, and 5 days was 81.3%, 90.6%, and 93.7%, respectively. On the basis of the criteria for cytocompatibility [48], the strut-reinforced vascular graft can be defined as non-cytotoxicity, consistently with previously reported results [31].

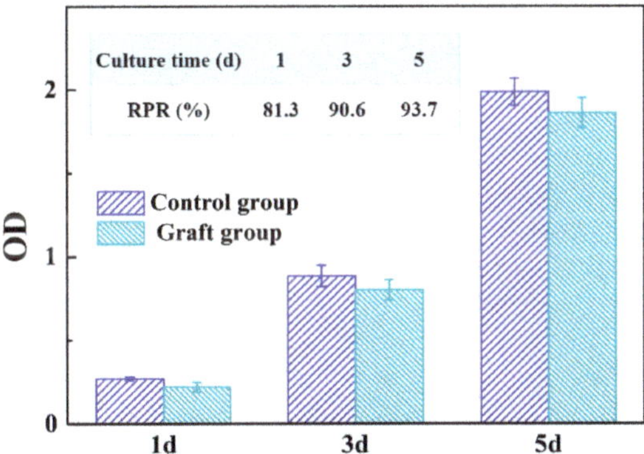

Figure 11. Results of cell culture in conditioned cell culture media (CM) and complete culture media. RPR: relative proliferation ratio.

3.8. Hemolysis Percentage of the Reinforced Graft

Vascular grafts are also required to be hemocompatible. HP is usually used to evaluate the destructive properties of biomaterials with respect to erythrocytes, especially for blood contact biomaterials [49]. The results of the hemolysis tests for the strut-reinforced graft showed that its HP was as low as (0.9 ± 0.2)%. Thus, it can be considered as a non-hemolytic material according to the standard ASTM F756-00, where the non-hemolytic range is defined as 0–2% [50,51]. On the basis of both biological test results, the strut-reinforced vascular graft showed satisfactory cyto- and hemocompatibilities. In summary, the newly developed vascular grafts have not only robust mechanical properties but also stable structure and excellent biocompatibility. In particular, they have a much higher circumferential compliance and lower swellability in volume.

4. Conclusions

A braided PET tubular strut-reinforced Ca alginate/polyacrylamide hydrogel vascular graft was fabricated with the assistance of a customized 3D casting mold. The mechanical properties concerning tensile strength, tensile elongation, dynamic circumferential compliance, compressive strength, and suture retention performance of the strut-reinforced grafts were improved significantly in comparison to native blood vessels and ePTFE prostheses. Furthermore, the shape of strut-reinforced grafts could be maintained upon prolonged incubation in an aqueous environment. Thus, this strut-reinforced graft holds great promise to meet the necessary requisites for clinical operations. More importantly, it has a satisfying cytocompatibility, hemocompatibility, suturing retention, and dynamic circumferential compliance. The present work paves the way to develop a new product-oriented small-diameter vascular graft.

Author Contributions: Formal analysis: G.G., C.Y., M.X., and Y.W.; investigation: G.G., C.Y., M.X., X.H., and L.W.; methodology: G.G., H.W., and L.W.; writing-original draft: G.G., C.Y., Y.W.; writing-review and editing: G.G., C.Y., Y.W., and L.W.

Funding: This research was funded by the Fundamental Research Funds of Central Universities (Grant No.2232019G-06), Innovation Program of Shanghai Municipal Education Commission (No. ZX201503000017, 15ZZ032) and "111 project" (B07024).

Acknowledgments: The authors would like to thank Deep Parikh for his technical support of the manuscript. Thanks also to China Scholarship Council for the support of joint research.

Conflicts of Interest: The authors declare no conflict of interest.

References

1. Chou, R.; Dana, T.; Blazina, I.; Daeges, M.; Jeanne, T.L. Statins for Prevention of Cardiovascular Disease in Adults Evidence Report and Systematic Review for the US Preventive Services Task Force. *J. Am. Med. Assoc.* **2016**, *316*, 2008–2024. [CrossRef] [PubMed]

2. Roger, V.L.; Go, A.S.; Lloyd-Jones, D.M.; Adams, R.J.; Berry, J.D.; Brown, T.M.; Carnethon, M.R.; Dai, S.F.; de Simone, G.; Ford, E.S.; et al. Executive Summary: Heart Disease and Stroke Statistics-2011 Update a Report from the American Heart Association. *Circulation* **2011**, *123*, 459–463. [CrossRef]

3. Seifu, D.G.; Purnama, A.; Mequanint, K.; Mantovani, D. Small-diameter vascular tissue engineering. *Nat. Rev. Cardiol.* **2013**, *10*, 410–421. [CrossRef] [PubMed]

4. Veith, F.J.; Gupta, S.K.; Ascer, E.; Whiteflores, S.; Samson, R.H.; Scher, L.A.; Towne, J.B.; Bernhard, V.M.; Bonier, P.; Flinn, W.R.; et al. Six-year prospective multicenter randomized comparison of autologous saphenous vein and expanded polytetrafluoroethylene grafts in infrainguinal arterial reconstructions. *J. Vasc. Surg.* **1986**, *3*, 104–114. [CrossRef]

5. Zilla, P.; Bezuidenhout, D.; Human, P. Prosthetic vascular grafts: Wrong models, wrong questions and no healing. *Biomaterials* **2007**, *28*, 5009–5027. [CrossRef] [PubMed]

6. Nomi, M.; Atala, A.; De Coppi, P.; Soker, S. Principals of neovascularization for tissue engineering. *Mol. Asp. Med.* **2002**, *23*, 463–483. [CrossRef]

7. Alexandre, N.; Amorim, I.; Caseiro, A.R.; Pereira, T.; Alvites, R.; Rema, A.; Goncalves, A.; Valadares, G.; Costa, E.; Santos-Silva, A.; et al. Long term performance evaluation of small-diameter vascular grafts based on polyvinyl alcohol hydrogel and dextran and MSCs-based therapies using the ovine pre-clinical animal model. *Int. J. Pharm.* **2016**, *513*, 332–346. [CrossRef]

8. Heath, D.E. Promoting Endothelialization of Polymeric Cardiovascular Biomaterials. *Macromol. Chem. Phys.* **2017**, *218*, 10. [CrossRef]

9.	Schmedlen, R.H.; Elbjeirami, W.M.; Gobin, A.S.; West, J.L. Tissue engineered small-diameter vascular grafts. *Clin. Plast. Surg.* **2003**, *30*, 507–517. [CrossRef]

10.	Langer, R.; Vacanti, J.P. Tissue engineering. *Science* **1993**, *260*, 920–926. [CrossRef]

11.	Cai, Q.; Yang, J.A.; Bei, J.Z.; Wang, S.G. A novel porous cells scaffold made of polylactide-dextran blend by combining phase-separation and particle-leaching techniques. *Biomaterials* **2002**, *23*, 4483–4492. [CrossRef]

12.	Lee, S.J.; Lee, D.; Yoon, T.R.; Kim, H.K.; Jo, H.H.; Park, J.S.; Lee, J.H.; Kim, W.D.; Kwon, I.K.; Park, S.A. Surface modification of 3D-printed porous scaffolds via mussel-inspired polydopamine and effective immobilization of rhBMP-2 to promote osteogenic differentiation for bone tissue engineering. *Acta Biomater.* **2016**, *40*, 182–191. [CrossRef]

13.	Moghadam, M.Z.; Hassanajili, S.; Esmaeilzadeh, F.; Ayatollahi, M.; Ahmadi, M. Formation of porous HPCL/LPCL/HA scaffolds with supercritical CO_2 gas foaming method. *J. Mech. Behav. Biomed.* **2017**, *69*, 115–127. [CrossRef] [PubMed]

14.	Badylak, S.F. The extracellular matrix as a biologic scaffold material. *Biomaterials* **2007**, *28*, 3587–3593. [CrossRef]

15.	Shin, Y.M.; Lee, Y.B.; Kim, S.J.; Kang, J.K.; Park, J.C.; Jang, W.; Shin, H. Mussel-Inspired Immobilization of Vascular Endothelial Growth Factor (VEGF) for Enhanced Endothelialization of Vascular Grafts. *Biomacromolecules* **2012**, *13*, 2020–2028. [CrossRef] [PubMed]

16.	Antonova, L.V.; Sevostyanova, V.V.; Kutikhin, A.G.; Mironov, A.V.; Krivkina, E.O.; Shabaev, A.R.; Matveeva, V.G.; Velikanova, E.A.; Sergeeva, E.A.; Burago, A.Y.; et al. Vascular Endothelial Growth Factor Improves Physico-Mechanical Properties and Enhances Endothelialization of Poly(3-hydroxybutyrate-co-3-hydroxyvalerate)/Poly(epsilon-caprolactone) Small-Diameter Vascular Grafts In vivo. *Front. Pharmacol.* **2016**, *7*, 8. [CrossRef] [PubMed]

17.	Calvert, P. Hydrogels for Soft Machines. *Adv. Mater.* **2009**, *21*, 743–756. [CrossRef]

18.	Kwon, H.J.; Yasuda, K.; Gong, J.P.; Ohmiya, Y. Polyelectrolyte hydrogels for replacement and regeneration of biological tissues. *Macromol. Res.* **2014**, *22*, 227–235. [CrossRef]

19.	Drury, J.L.; Mooney, D.J. Hydrogels for tissue engineering: Scaffold design variables and applications. *Biomaterials* **2003**, *24*, 4337–4351. [CrossRef]

20.	Lee, K.Y.; Mooney, D.J. Hydrogels for tissue engineering. *Chem. Rev.* **2001**, *101*, 1869–1879. [CrossRef]

21.	Ghobril, C.; Grinstaff, M.W. The chemistry and engineering of polymeric hydrogel adhesives for wound closure: A tutorial. *Chem. Soc. Rev.* **2015**, *44*, 1820–1835. [CrossRef]

22.	Billiet, T.; Vandenhaute, M.; Schelfhout, J.; Van Vlierberghe, S.; Dubruel, P. A review of trends and limitations in hydrogel-rapid prototyping for tissue engineering. *Biomaterials* **2012**, *33*, 6020–6041. [CrossRef]

23.	Bryant, S.J.; Vernerey, F.J. Programmable Hydrogels for Cell Encapsulation and Neo-Tissue Growth to Enable Personalized Tissue Engineering. *Adv. Healthc. Mater.* **2018**, *7*, 13. [CrossRef]

24.	Dou, X.Q.; Feng, C.L. Amino Acids and Peptide-Based Supramolecular Hydrogels for Three-Dimensional Cell Culture. *Adv. Mater.* **2017**, *29*, 21. [CrossRef]

25.	Seliktar, D. Designing Cell-Compatible Hydrogels for Biomedical Applications. *Science* **2012**, *336*, 1124–1128. [CrossRef]

26.	Browning, M.B.; Guiza, V.; Russell, B.; Rivera, J.; Cereceres, S.; Hook, M.; Hahn, M.S.; Cosgriff-Hernandez, E.M. Endothelial Cell Response to Chemical, Biological, and Physical Cues in Bioactive Hydrogels. *Tissue Eng. Part A* **2014**, *20*, 3130–3141. [CrossRef]

27.	Fan, M.; Ma, Y.; Tan, H.P.; Jia, Y.; Zou, S.Y.; Guo, S.X.; Zhao, M.; Huang, H.; Ling, Z.H.; Chen, Y.; et al. Covalent and injectable chitosan-chondroitin sulfate hydrogels embedded with chitosan microspheres for drug delivery and tissue engineering. *Mat. Sci. Eng. C-Mater.* **2017**, *71*, 67–74. [CrossRef]

28.	Alexandre, N.; Ribeiro, J.; Gartner, A.; Pereira, T.; Amorim, I.; Fragoso, J.; Lopes, A.; Fernandes, J.; Costa, E.; Santos-Silva, A.; et al. Biocompatibility and hemocompatibility of polyvinyl alcohol hydrogel used for vascular grafting—In vitro and in vivo studies. *J. Biomed. Mater. Res. A* **2014**, *102*, 4262–4275. [CrossRef]

29.	Sun, J.Y.; Zhao, X.H.; Illeperuma, W.R.K.; Chaudhuri, O.; Oh, K.H.; Mooney, D.J.; Vlassak, J.J.; Suo, Z.G. Highly stretchable and tough hydrogels. *Nature* **2012**, *489*, 133–136. [CrossRef]

30.	Kamata, H.; Akagi, Y.; Kayasuga-Kariya, Y.; Chung, U.; Sakai, T. "Nonswellable" Hydrogel without Mechanical Hysteresis. *Science* **2014**, *343*, 873–875. [CrossRef]

31. Darnell, M.C.; Sun, J.Y.; Mehta, M.; Johnson, C.; Arany, P.R.; Suo, Z.G.; Mooney, D.J. Performance and biocompatibility of extremely tough alginate/polyacrylamide hydrogels. *Biomaterials* **2013**, *34*, 8042–8048. [CrossRef]
32. Yang, C.H.; Wang, M.X.; Haider, H.; Yang, J.H.; Sun, J.Y.; Chen, Y.M.; Zhou, J.X.; Suo, Z.G. Strengthening Alginate/Polyacrylamide Hydrogels Using Various Multivalent Cations. *ACS Appl. Mater. Interfaces* **2013**, *5*, 10418–10422. [CrossRef] [PubMed]
33. Li, G.; Zhang, G.P.; Sun, R.; Wong, C.P. Mechanical strengthened alginate/polyacrylamide hydrogel crosslinked by barium and ferric dual ions. *J. Mater. Sci.* **2017**, *52*, 8538–8545. [CrossRef]
34. Li, Y.L.; Wang, C.X.; Zhang, W.; Yin, Y.J.; Rao, Q.Q. Preparation and Characterization of PAM/SA Tough Hydrogels Reinforced by IPN Technique Based on Covalent/Ionic Crosslinking. *J. Appl. Polym. Sci.* **2015**, *132*, 7. [CrossRef]
35. Yang, X.Y.; Wang, L.; Guan, G.P.; Zhang, H.Q.; Shen, G.T.; Guan, Y.; Peng, L.; Li, Y.L.; King, M.W. Mechanical and biocompatibility performance of bicomponent polyester/silk fibroin small-diameter arterial prostheses. *J. Appl. Biomater. Funct.* **2015**, *13*, E201–E209. [CrossRef]
36. Zou, T.; Wang, L.; Li, W.C.; Wang, W.Z.; Chen, F.; King, M.W. A resorbable bicomponent braided ureteral stent with improved mechanical performance. *J. Mech. Behav. Biomed.* **2014**, *38*, 17–25. [CrossRef] [PubMed]
37. Ren, Z.C.; Chen, G.Y.; Wei, Z.Y.; Sang, L.; Qi, M. Hemocompatibility evaluation of polyurethane film with surface-grafted poly(ethylene glycol) and carboxymethyl-chitosan. *J. Appl. Polym. Sci.* **2013**, *127*, 308–315. [CrossRef]
38. Zhao, Y.L.; Wang, S.G.; Guo, Q.S.; Shen, M.W.; Shi, X.Y. Hemocompatibility of electrospun halloysite nanotube- and carbon nanotube-doped composite poly(lactic-co-glycolic acid) nanofibers. *J. Appl. Polym. Sci.* **2013**, *127*, 4825–4832. [CrossRef]
39. He, Q.J.; Zhang, J.M.; Shi, J.L.; Zhu, Z.Y.; Zhang, L.X.; Bu, W.B.; Guo, L.M.; Chen, Y. The effect of PEGylation of mesoporous silica nanoparticles on nonspecific binding of serum proteins and cellular responses. *Biomaterials* **2010**, *31*, 1085–1092. [CrossRef]
40. Lin, Y.S.; Haynes, C.L. Synthesis and Characterization of Biocompatible and Size-Tunable Multifunctional Porous Silica Nanoparticles. *Chem. Mater.* **2009**, *21*, 3979–3986. [CrossRef]
41. Hiob, M.A.; She, S.; Muiznieks, L.D.; Weiss, A.S. Biomaterials and Modifications in the Development of Small-Diameter Vascular Grafts. *ACS Biomater. Sci. Eng.* **2017**, *3*, 712–723. [CrossRef]
42. Shen, G.; Zhang, W.; Lin, J.; Wang, L. Effects of testing conditions on radial compliance of textile vascular prostheses. *J. Med. Biomech.* **2014**, *29*, 113–118.
43. Zeugolis, D.I.; Khew, S.T.; Yew, E.S.Y.; Ekaputra, A.K.; Tong, Y.W.; Yung, L.Y.L.; Hutmacher, D.W.; Sheppard, C.; Raghunath, M. Electro-spinning of pure collagen nano-fibres—Just an expensive way to make gelatin? *Biomaterials* **2008**, *29*, 2293–2305. [CrossRef]
44. Xie, Y.; Guan, Y.; Kim, S.H.; King, M.W. The mechanical performance of weft-knitted/electrospun bilayer small diameter vascular prostheses. *J. Mech. Behav. Biomed.* **2016**, *61*, 410–418. [CrossRef] [PubMed]
45. Nottelet, B.; Pektok, E.; Mandracchia, D.; Tille, J.C.; Walpoth, B.; Gurny, R.; Moller, M. Factorial design optimization and in vivo feasibility of poly(epsilon-caprolactone)-micro- and narofiber-based small diameter vascular grafts. *J. Biomed. Mater. Res. A* **2009**, *89A*, 865–875. [CrossRef] [PubMed]
46. Kong, X.Y.; Han, B.Q.; Wang, H.X.; Li, H.; Xu, W.H.; Liu, W.S. Mechanical properties of biodegradable small-diameter chitosan artificial vascular prosthesis. *J. Biomed. Mater. Res. A* **2012**, *100A*, 1938–1945. [CrossRef] [PubMed]
47. Tang, J.Y.; Bao, L.H.; Li, X.; Chen, L.; Hong, F.F. Potential of PVA-doped bacterial nano-cellulose tubular composites for artificial blood vessels. *J. Mater. Chem. B* **2015**, *3*, 8537–8547. [CrossRef]
48. Liu, J.; Zhang, X.X.; Wang, H.Y.; Li, F.B.; Li, M.Q.; Yang, K.; Zhang, E.L. The antibacterial properties and biocompatibility of a Ti-Cu sintered alloy for biomedical application. *Biomed. Mater.* **2014**, *9*, 11. [CrossRef] [PubMed]
49. Deng, J.; Cheng, C.; Teng, Y.Y.; Nie, C.X.; Zhao, C.S. Mussel-inspired post-heparinization of a stretchable hollow hydrogel tube and its potential application as an artificial blood vessel. *Polym. Chem.* **2017**, *8*, 2266–2275. [CrossRef]

50. Li, L.H.; Tu, M.; Mou, S.S.; Zhou, C.G. Preparation and blood compatibility of polysiloxane/liquid-crystal composite membranes. *Biomaterials* **2001**, *22*, 2595–2599. [CrossRef]
51. Mercado-Pagan, A.E.; Stahl, A.M.; Shanjani, Y.; Yang, Y.Z. Vascularization in Bone Tissue Engineering Constructs. *Ann. Biomed. Eng.* **2015**, *43*, 718–729. [CrossRef] [PubMed]

Article

Use of Polyphenol Tannic Acid to Functionalize Titanium with Strontium for Enhancement of Osteoblast Differentiation and Reduction of Osteoclast Activity

Chris Steffi, Zhilong Shi, Chee Hoe Kong, Sue Wee Chong, Dong Wang and Wilson Wang *

Department of Orthopaedic Surgery, National University of Singapore, NUHS Tower Block Level 11, 1E Kent Ridge Road, Singapore 119228, Singapore
* Correspondence: doswangw@nus.edu.sg

Received: 14 June 2019; Accepted: 25 July 2019; Published: 29 July 2019

Abstract: Implant anchorage remains a challenge, especially in porous osteoporotic bone with high osteoclast activity. The implant surface is modified with osteogenic molecules to stimulate osseointegration. Strontium (Sr) is known for its osteogenic and anti-osteoclastogenic effects. In this study, Sr was immobilized on a titanium (Ti) surface using bioinspired polyphenol tannic acid (pTAN) coating as an ad-layer (Ti-pTAN). Two separate coating techniques were employed for comparative analysis. In the first technique, Ti was coated with a tannic acid solution containing Sr (Ti-pTAN-1Stp). In the second method, Ti was first coated with pTAN, before being immersed in a $SrCl_2$ solution to immobilize Sr on Ti-pTAN (Ti-pTAN-2Stp). Ti-pTAN-1Stp and Ti-pTAN-2Stp augmented the alkaline phosphatase activity, collagen secretion, osteocalcin production and calcium deposition of MC3T3-E1 cells as compared to those of Ti and Ti-pTAN. However, osteoclast differentiation of RAW 264.7, as studied by TRAP activity, total DNA, and multinucleated cell formation, were decreased on Ti-pTAN, Ti-pTAN-1Stp and Ti-pTAN-2Stp as compared to Ti. Of all the substrates, osteoclast activity on Ti-pTAN-2Stp was the lowest. Hence, an economical and simple coating technique using pTAN as an adlayer preserved the dual biological effects of Sr. These results indicate a promising new approach to tailoring the cellular responses of implant surfaces.

Keywords: osteoporosis; strontium; polyphenol tannic acid; titanium; osteoblasts; osteoclasts

---

1. Introduction

Low bone mineral density and altered microarchitecture are prominent characteristics of osteoporotic bone, especially in post-menopausal, elderly women [1]. Apart from being susceptible to fractures, the porous architecture of fragile osteoporotic bones affects the anchorage of screws, leading to post-surgical complications and implant failures [2]. Balanced regulation of osteoblast and osteoclast activity in the peri-implant region is crucial for successful osseointegration [3,4]. Thus, the surfaces of implants were functionalized with osteogenic molecules to promote bone-implant integration [5,6]. The physiochemical properties of the implant surface, such as roughness, topography and chemical modifications, play a crucial role in modulating the activity of bone cells at the bone-implant interface, and hence, influence osseointegration [7–9]. Surfaces were also functionalized with bone morphogenetic protein-2 (BMP-2), BMP-7, vascular endothelial growth factor and hydroxyapatite to regulate the cellular behavior on the implant surface and augment ossteointegration [6,10–12]. Anti-osteoporotic drugs, such as bisphosphonates, have gained research interest in surface science by demonstrating an improvement of bone-implant integration in osteoporotic rats [13]. Yet, atypical fractures caused by bisphosphonates remain a concern [14].

Strontium ranelate was shown to augment bone mineral density and reduce the risk of fragility fractures in post-menopausal, elderly women [15,16]. Along with long-term beneficial effects, the medication was cost-effective and was approved in Europe for osteoporosis treatment [16]. Oral administration of strontium ranelate to ovariectomized rats reduced osteoporosis-induced bone loss, while bone quality was preserved [17]. At a cellular level, the effects of Sr are cell specific. For instance, the differentiation of osteoblasts improved with strontium ranelate, whereas osteoclast activity was impeded by strontium ranelate treatment [18]. These established the suitability of the dual actions of strontium (Sr) for osteoporotic bone. Apart from oral administration, research has been carried out to immobilize Sr on implant surfaces [19]. A titanium surface was functionalized with strontium-substituted hydroxyapatite (HA) by electrochemical deposition, which boosted implant integration in osteopenic rats [20]. In another study, a Ti surface modified with Sr-loaded nanotubules promoted new bone formation and improved bone-implant contact in an in vivo rat model [21]. In a comparative analysis, HA, substituted with three different osteogenic metals, i.e., zinc, magnesium and strontium, was coated onto a Ti surface by electrochemical deposition [22]. All metal-HA substitutions revealed improved osseointegration in osteopenic rats as compared to that of unsubstituted-HA, with the osteopromotive effect of Sr being the highest of all. Because of the osteopromotive effects of Sr, in this study, Sr was shortlisted to functionalize Ti via an economical and facile coating technique using a bioinspired polyphenol coating.

Surface modification of bone implants is evolving from coating with passive materials having a composition similar to bone to the immobilization of active compounds via coatings which trigger desired biological responses and result in rapid and permanent implant fixations within native bone. Recently, bioinspired coatings were achieved using natural polyphenols such as polyphenol tannic acid (pTAN) [23]. These molecules oligomerized by successive autooxidation reactions. Thus, the formed oligomers have reduced solubility and a high affinity for surfaces on which to be deposited. Polyphenols form complexes with metal ions, providing the stability that is dependent on pH and valency states of the metal ions [24]. For example, metals ions (sodium) from a reaction buffer used for the coating were detected in the surface coatings of pTAN. The concentration of these metal ions influences the kinetics of pTAN coating, suggesting that apart from oxidation reactions, metal ion coordination bonds are involved in the coating process [25]. In a separate study, pTAN was shown to coat different surfaces rapidly by forming pH-dependent coordination bonds with Fe(III) ions [26]. Apart from the ability of pTAN to interact with metal ions, pTAN coatings possess anti-bacterial and anti-oxidant properties [23]. Moreover, U.S. Food and Drug Administration classifies tannic acid as generally regarded as safe (GRAS). These findings outline the advantages of using pTAN as an adlayer.

Consequently, this study exploited the metal binding capability of pTAN coating to immobilize Sr^{2+} on Ti. The potency of the immobilized Sr was monitored by studying osteoblast and osteoclast differentiation. Sr immobilized substrates were studied for osteoblasts markers such as alkaline phosphatase (ALP) activity, collagen type 1 (COLL 1), osteocalcin and calcium deposition. Osteoclast markers such as tartarate resistance acid phosphatase (TRAP) activity, total DNA, and multinucleated cell formation were also assessed on the surface coatings.

2. Materials and Methods

2.1. Materials

All chemicals obtained from Sigma-Aldrich (Darmstadt, Germany) were utilized as received. Chemicals and reagents purchased elsewhere are mentioned in the text. All experiments used milli-Q water (>18.2 MΩ·cm) unless otherwise stated (ultrapure water system, Arium 611UF, Sartorius Stedim Biotech GmbH, Göttingen, Germany).

2.2. Material Preparation

Sandpaper (600- and 1200-grit) was used to polish titanium alloy (Ti-6Al-4V foils, 10 mm × 10 mm, Goodfellow Cambridge Ltd. Huntingdon, England). The titanium (Ti) foils were then rinsed in water-bath ultrasonicator for 10 min. Sonication was further carried out in Kroll's reagent (4.0% HF, 7.2% HNO_3, 88.8% water) to remove the accumulated carbide resulted from polishing [10]. Next, the reaction was stopped by 1 N sodium hydroxide addition, and subsequently, the substrates were sonicated 10 min each in dichloromethane, acetone, and water. The substrates were immersed for 30 min in 40% nitric acid for surface passivation and thoroughly washed with water to remove nitric acid.

The procedure for polyphenol tannic acid (pTAN) coating for Ti (named as Ti-pTAN) was followed as previously reported with slight modification [23]. In brief, Ti foils were immersed for 24 h in 2 mg/mL solution of pTAN prepared in bicine buffer (100 mM) with 0.6 M NaCl at 7.8 pH. The procedure was carried out in dark with gentle shaking. The substrates were washed thoroughly in water and further submerged in water for 2 days at 37 °C; the water was changed every 24 h.

Ti-pTAN substrates were functionalized with Sr by two different methods (Figure 1). In the first one-step method, Ti foils were incubated in pTAN solution (2 mg/mL, 100 mM bicine buffer, pH = 7.8) containing $SrCl_2$ (0.3 M Sr) for 24 h with gentle shaking in dark. 0.3 M $SrCl_2$ was used to maintain the similar ionic stoichiometry as 0.6 M NaCl in bicine buffer. These substrates were named Ti-pTAN-1Stp. The second method was a 48 h-long, two-step process wherein pTAN coating and Sr coating were carried out separately. In the first 24 h, Ti was coated with pTAN, as discussed above. Subsequently, Ti-pTAN substrates were immersed in a $SrCl_2$ solution (0.1 M Sr in milli-Q water) for another 24 h in the dark. In this case, the Sr solution comprising minimum Sr concentration with maximum immobilisation efficiency was used. These substrates were denoted as Ti-pTAN-2Stp. After the Sr coating, Ti-pTAN-1Stp and Ti-pTAN-2Stp were sonicated for 5 min and washed thoroughly with milli-Q water to remove unbound Sr. The substrates were irradiated with UV for 30 min and transferred to a 24-well cell culture plate prior to cell seeding.

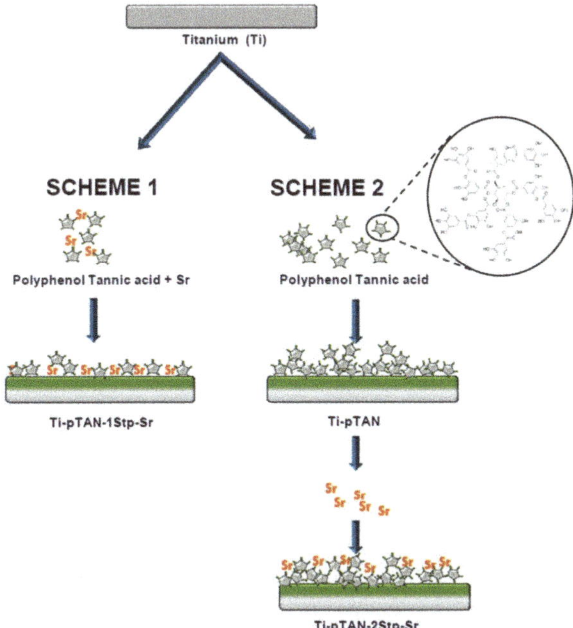

Figure 1. Schematic representation of the workflow of Ti-pTAN-1Stp-Sr and Ti-pTAN-2tp-Sr coatings.

2.3. Substrate Characterization

Surface chemical composition of all the substrates were analyzed using X-ray photoelectron spectroscopy (XPS) (Kratos AXIS Ultra DLD spectrometer, Kratos Analytical Ltd, Manchester, UK). The binding energies of all the elements were referenced to C 1s signal at 284.6 eV. The surface roughness was studied using a contact based Nanomap-LS 3D profilometer (AEP Technologies, Santa Clara, CA, USA). Water contact angle (WCA) analysis for all the substrates were performed using contact angle analyzer Phoenix 300 touch (Surface Electro Optics, Gyeonggi-do, Korea). The measurements were carried out according to the sessile drop technique and a video camera. Ten microliters of sessile drop of distilled water was transferred onto the substrate using a micro-syringe. The contact angle values on both sides of the drop were measured using tangent rule to ensure symmetry and horizontal level. To estimate the amount of bound Sr ions, Ti-pTAN-1Stp and Ti-pTAN-2Stp substrates were soaked in a 2% nitric acid solution overnight with shaking. Sr ions dissolved in the nitric acid solution were quantified using inductively coupled plasma mass spectrometry (ICP-MS, CX 7700x Agilent, Santa Clara, CA, USA).

2.4. Strontium Ions Release Study

Ti-pTAN-1Stp and Ti-pTAN-2Stp substrates were immersed in 1 mL of phosphate buffer saline (PBS) at 37 °C for 14 days. PBS was collected at regular intervals and replenished with fresh PBS. Sr ions in released in PBS were measured using inductively coupled plasma mass spectrometry (ICP-MS, CX 7700x Agilent, CA, USA).

2.5. Cell Culture

Mouse pre-osteoblasts MC3T3-E1 (ATCC, Manassa, VA, USA) were grown in complete Minimum Essential Medium Alpha (MEM-α, 10% fetal bovine serum (FBS), 100 U/mL penicillin and 100 µg/mL streptomycin (Thermo Fisher Scientific, Waltham, MA, USA). Osteoblast differentiation media was prepared by adding 10 mM sodium β-glycerophosphate and 50 µg/mL ascorbic acid to complete phenol red free MEM-α. Then, 40,000 cells were seeded per substrate placed in a 24-well plate.

RAW 264.7 (ATCC, Manassa, VA, USA) murine pre-osteoclasts were cultured in complete Dulbecco's Modified Eagle Medium (10% heat inactivated FBS, 100 U/mL penicillin and 100 µg/mL streptomycin (Invitrogen, Waltham, MA, USA). For osteoclast differentiation studies, RAW 264.7 were cultured in presence of 50 ng/mL murine RANKL (R&D Systems, Minneapolis, MN, USA) in complete phenol red free MEM-α (10% heat inactivated FBS, 100 U/mL penicillin, 100 µg/mL streptomycin) at cell density of $2000/cm^2$ on all the substrates. 100 µL of cell suspension was loaded on the substrates. After 6 h of incubation, an additional 400 µL of cell culture media was added to the wells. Cells were maintained in a humid environment at 37 °C with 5% CO_2, and the spent media were changed every 2–3 days.

2.6. MC3T3-E1 Proliferation

Cell proliferation of MC3T3-E1 on all the substrates was assessed on day 1, day 4, day 7 and day 14. Cells were incubated in Cell Counting Kit-8 (CCK-8, Dojindo Laboratories, Kumamoto, Japan) working reagent, prepared by mixing CCK-8 in complete MEM-α (1:10 *v/v*). After 4 h, the absorbance was measured at 450 nm by microplate reader (Synergy H1, BioTek Instruments, Inc. Winooski, VT, USA).

2.7. ALP Activity, Collagen-1 and Osteocalcin Production of MC3T3-E1 Cells

ALP activity of osteoblasts was estimated in the cell culture supernatant at various time points with QuantiChrom Alkaline Phosphatase Assay Kit (BioAssay Systems, Hayward, CA, USA). In brief, 50 µL of sample was incubated with 150 µL ALP working reagent for 4 min. Absorbance of 405 nm was measured at 0 and 4 min using microplate reader (Synergy H1). The ALP activity in terms of pNP produced was estimated as per the formula provided in the kit, and normalized with the total protein.

The unit for ALP was μM of pNP produced per minute per mg of protein. Collagen 1 and osteocalcin were also measured in cell culture supernatant on day 14 by using Pro-Collagen I alpha 1 SimpleStep ELISA kit (Abcam, Cambridge, UK) and Osteocalcin ELISA kit (Cloud-Clone Corp., Katy, TX, USA) respectively, and the values were normalized to total protein.

2.8. Collagen Staining of Osteoblasts

MC3T3-E1 was cultured on various substrates for 14 days. After washing with phosphate buffer saline (PBS), cells were fixed for 15 min in 4% paraformaldehyde which was followed by 5 min treatment with 0.1% triton X-100 to permeabilise cell membrane. Non-specific binding was blocked by overnight incubation of cells in 3% bovine serum albumin (BSA) at 4 °C. Cells were incubated with anti-mouse collage type I primary antibody (Merck, Darmstadt, Germany) for 90 min. Secondary antibody Alexa Fluor 546 tagged anti-rabbit IgG (H + L) (Invitrogen) and alexa Fluor 488 phalloidin (Invitrogen) for actin staining were added to cells for 30 min followed by nuclei labelling using DAPI. Cells images were acquired using Olympus FV1000 confocal laser scanning microscope (CSLM, Olympus, Tokyo, Japan).

2.9. Alizarin Staining and Calcium Estimation

After 14 days of culture MC3T3-E1 cells in osteogenic medium, the cells were washed with PBS and fixed with 70% ethanol at 4 °C for 1 h. Alizarin red solution (2% in water) was added to the cells for 1 h. The cells were washed with copious amount of water to remove unbound alizarin and observed under microscope (Leica Microsystems, Wetzlar, Germany).

Calcium deposited by MC3T3-E1 cells was also quantified after 14 days of culture. The substrates were submerged in 500 μL of 2% nitric acid overnight with gentle shaking, to dissolve the deposited calcium. The eluted calcium was then quantified using QuantiChrom Calcium Assay Kit (BioAssay Systems, Hayward, CA, USA) according to manufacturer's instructions. In brief, 5 μL of the sample was incubated with 200 μL of kit's working solution for 5 min. Absorbance was measured at 612 nm using a microplate reader (Synergy H1).

2.10. TRAP Activity

To trigger osteoclastogenesis, RAW 264.7 cells were exposed to 50 ng/mL of RANKL for 3 and 5 days. TRAP activity of the cells was estimated in cell culture supernatant at day 3 and day 5 of culture using TRAP activity assay kit (Takara Bio Inc, Shiga, Japan). Sample was mixed with acid phosphatase buffer (1:1 *v/v*) provided with the kit and incubated for 60 min at 37 °C. Stop solution, 0.5 N sodium hydroxide, was added and absorbance was taken at 405 nm by microplate reader (Synergy H1). The calculation of TRAP activity was based on pNP generated, which were normalized to total protein. Hence, the unit of TRAP activity was μM of pNP produced per minute per mg of protein.

2.11. DNA Quantification

RAW 264.7 cells were lysed by three freeze thaw cycles in milli-Q water following 5 days of culture with RANKL. After centrifugation for 20 min at 12,000 g (4 °C), the samples were mixed with picogreen reagent (Thermo Fisher Scientific, Waltham, MA, USA) and incubated for 5 min at RT. A microplate reader (Synergy H1) was used to excite the samples at 480 nm, and emission was measured at 520 nm.

2.12. Actin and TRAP Staining of Osteoclasts

RAW 264.7 cells were cultured on various substrates. After 5 days of RANKL exposure, the cells were fixed with 4% paraformaldehyde for 15 min and incubated for 5 min in triton 0.1% X-100 to increase cell membrane permeability. To prevent non-specific binding, the cells were immersed in 3% BSA (overnight, 4 °C). 90 min incubation in anti-TRAP primary antibody (Takara Bio Inc, Shiga, Japan) followed by 30 min incubation in premixed solution of secondary antibody (Alexa Fluor 546 tagged-Goat anti-Mouse IgG (H + L), Thermo Fisher Scientific) and phalloidin (Alexa fluor 488 tagged,

Thermo Fisher Scientific) was carried out to stain TRAP and actin in the cells. After nuclei staining with DAPI, the cells were imaged with Olympus FV1000 CLSM (Olympus, Japan). Multinucleated cells with more than three nuclei (blue), and positive for TRAP (red) were classified as osteoclasts.

2.13. Statistics

All experiments were replicated more than three times and the data were presented as mean ± standard deviation (SD). The statistical analysis was carried out by one-way analysis-of-variance and Tukey post hoc test. $p < 0.05$ was considered as statistically significant.

3. Results

3.1. Surface Characterization

In the current study, polyphenol tannic acid chemistry is utilized to immobilize Sr on a Ti surface. The elemental compositions of the surfaces were examined by XPS. Signal peaks for elements such as O 1s (530 eV), Ti 2p (460 eV) and C 1s (285 eV) were observed on pristine Ti (Figure 2A and Table 1). C 1s signal on Ti was present due to inevitable hydrocarbon contamination. Therefore, C 1s signals are commonly used as reference for signal calibrations in XPS scans [27]. Ti-pTAN substrates displayed a decrease in Ti 2P signal signifying successful deposition of pTAN on Ti (Figure 2B and Table 1). Ti-pTAN-1Stp and Ti-pTAN-2Stp also displayed a drop in Ti 2P signals. However, Ti 2P signals on Ti-pTAN-1Stp was higher than the Ti 2P signals on Ti-pTAN and Ti-pTAN-2Stp. This suggested that the substitution of Na with Sr in bicine buffer for 1 step coating method resulted in a comparatively thinner pTAN coating.

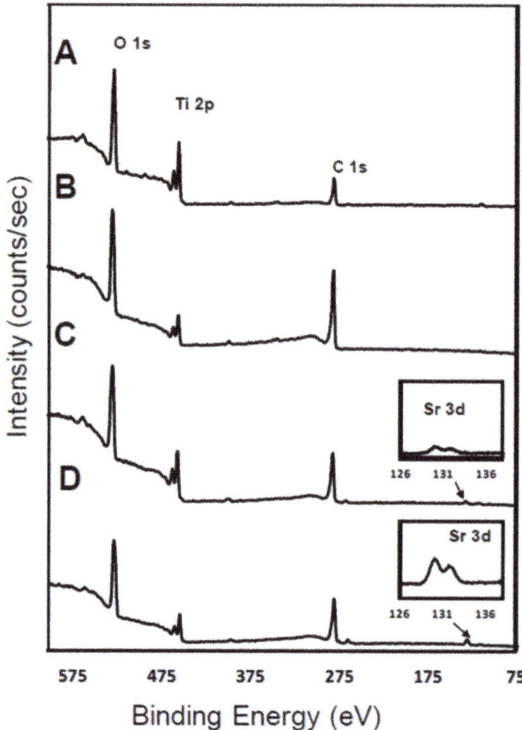

Figure 2. XPS wide spectra scan of (A) Ti; (B) Ti-pTAN; (C) Ti-pTAN-1Stp-Sr; (D) Ti-pTAN-2Stp-Sr.

Table 1. Elemental composition as percentages [a] assessed by XPS.

Substrates	C%	O%	Ti%	Sr%
Ti	38.11	50.75	11.14	0
Ti-pTAN	66.03	30.43	3.54	0
Ti-pTAN-1Stp	55.49	36.38	7.68	0.45
Ti-pTAN-2Stp	58.84	35.8	4.47	0.89

[a] Percentage estimation of elements were based on C, O, Ti and Sr only.

An increased C 1s signal was detected on Ti-pTAN substrate. The observed carbon to oxygen ratio (C/O, 2.17) was slightly higher than theoretical C/O ratio of pTAN (1.65), which may be the result of carbon contamination. Nonetheless, for Ti-pTAN-1Stp (Figure 2C and Table 1) and Ti-pTAN-2Stp (Figure 2D and Table 1), the C/O ratios were 1.52 and 1.64 respectively. Moreover, distinct Sr 3d signals were observed on Ti-pTAN-1Stp and Ti-pTAN-2Stp, suggesting the successful immobilization of Sr by pTAN coating.

The surface roughness of the coatings was comparable to Ti. The Ra values of Ti, Ti-pTAN, pTAN-1Stp-Sr and Ti-pTAN-2Stp-Sr were 0.44 ± 0.10 μm, 0.46 ± 0.05 μm, 0.46 ± 0.09 μm and 0.42 ± 0.06 μm, respectively. Water contact angle of the surface were examined. The contact angle of pristine Ti was $82° \pm 0.9°$ (Figure 3). Hydrophilicity was conferred by pTAN coating, as contact was decreased to $48° \pm 2.2°$ on Ti-pTAN. This is expected as multiple hydrophilic phenol moieties in pTAN increase wettability of the surface [28]. However, the water contact angles of Ti-pTAN-1Stp-Sr and Ti-pTAN-2Stp-Sr were $49° \pm 1.7°$ and $51° \pm 2.6°$ respectively, which were comparable to Ti-pTAN. Hence, Sr immobilization did not have any significant effect on contact angle of pTAN coating. Surface density of Sr on Ti-pTAN-1Stp-Sr and Ti-pTAN-2Stp-Sr determined ICP-MS were 0.32 ± 0.02 μg/cm^2 and 5.8 ± 0.18 μg/cm^2 respectively.

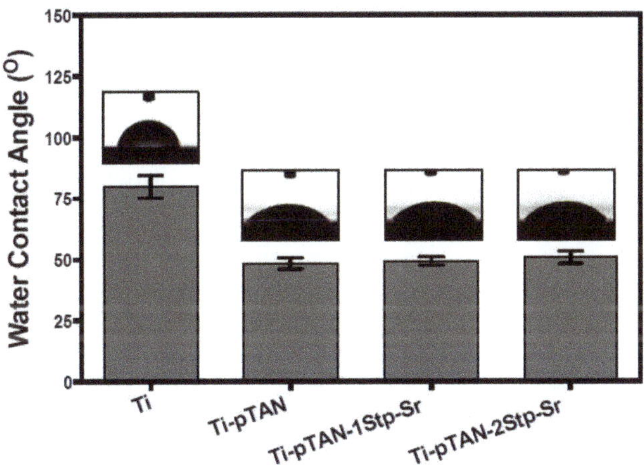

Figure 3. Contact angle analysis of Ti, Ti-pTAN, Ti-pTAN-1Stp-Sr and Ti-pTAN-2Stp-Sr.

3.2. Strontium Release

The release of Sr from Ti-pTAN-1Stp-Sr and Ti-pTAN-2Stp-Sr was estimated by ICP-MS. A burst release was observed after 24 h (Figure 4). However, subsequent time points revealed minimal release, with about 20% remaining on Ti-pTAN-1Stp-Sr and about 70% of the immobilized Sr remaining on Ti-pTAN-2Stp-Sr for 2 weeks. This shows that the binding of Sr^{2+} ions on Ti-pTAN-2Stp-Sr were stronger, compared to Ti-pTAN-1Stp-Sr.

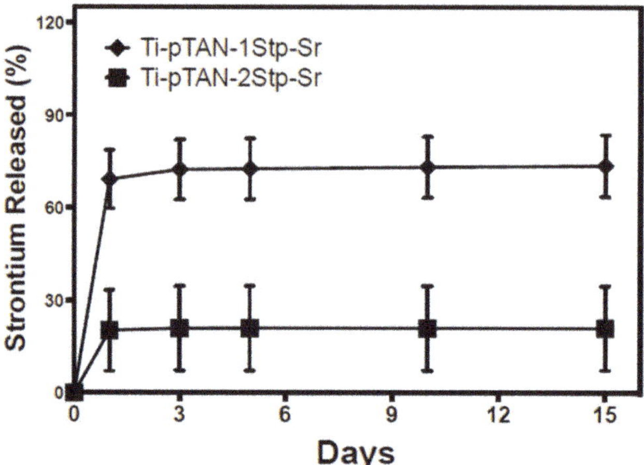

Figure 4. Sr release from Ti-pTAN-1Stp-Sr and Ti-pTAN-2Stp-Sr after incubation in PBS at 37 °C.

3.3. MC3T3-E1 Cell Viability

The cell viability of MC3T3-E1 cells on Ti, Ti-pTAN, Ti-pTAN-1Stp-Sr and Ti-pTAN-2Stp-Sr were monitored by CCK-8 assay kit. No difference in cell viability was observed for day 1, day 4 and day 7 of cell culture (Figure 5). However, a statistically significant decrease in cell metabolism was observed on day 14 for the MC3T3-E1 cells cultured on Ti-pTAN-1Stp-Sr and Ti-pTAN-2Stp-Sr as compared to Ti ($p < 0.01$) and Ti-pTAN ($p < 0.01$).

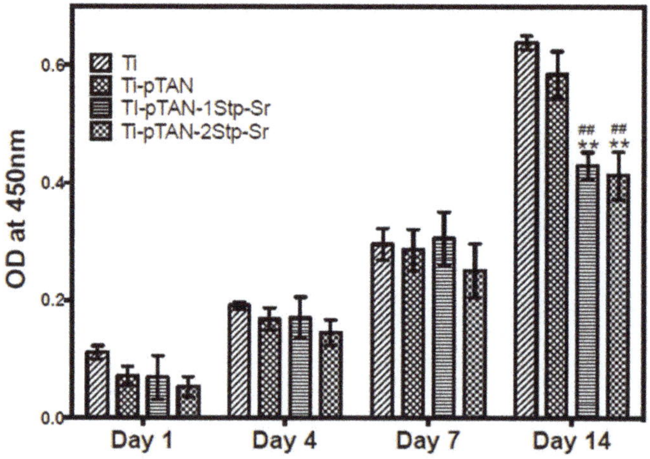

Figure 5. Proliferation study: CCK8 assay of MC3T3-E1 cells cultured on Ti, Ti-pTAN, Ti-pTAN-1Stp-Sr and Ti-pTAN-2Stp-Sr at various time points. (**) represents a statistically significant difference with $p < 0.01$ as compared to Ti and (##) represents the $p < 0.01$ as compared to Ti-pTAN for the same time point.

3.4. Osteoblast Differentiation

The effects of the substrates on ALP activity of MC3T3-E1 cells were monitored at day 3, day 5, day 7 and day 14 of culture (Figure 6A). For day 3 cultures, ALP activity of osteoblasts on Ti was at a similar level as those on the other substrates. ALP values reached peak at day 5. Osteoblasts on Ti-pTAN-1Stp-Sr and Ti-pTAN-2Stp-Sr had significantly higher ALP activity as compared to Ti ($p < 0.01$) and pTAN ($p < 0.05$). From day 7 onwards, decrease in ALP activity was observed. Nevertheless, ALP activity was significantly higher on Ti-pTAN-1Stp-Sr as compared to Ti ($p < 0.01$), and for Ti-pTAN-2Stp-Sr, the activity was higher as compared to both Ti ($p < 0.01$) and Ti-pTAN ($p < 0.01$) at the same time point. ALP activity on day 14 was the lowest for all the substrates, but the activity was still higher on Ti-pTAN-2Stp-Sr ($p < 0.01$) and Ti-pTAN ($p < 0.05$) as compared to Ti.

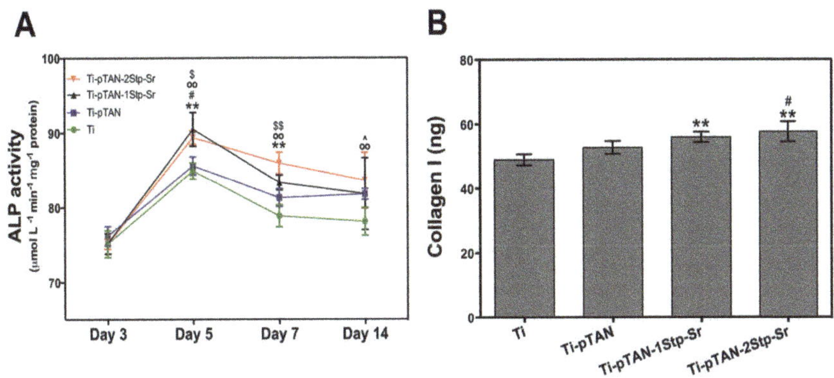

Figure 6. Early differentiation marker of MC3T3-E1: (**A**) ALP activity of MC3T3-E1 cells cultured on Ti, Ti-pTAN, Ti-pTAN-1Stp-Sr and Ti-pTAN-2Stp-Sr were studied on day 3, day 5, day 7 and day 14 of cell culture. Statistically significant differences are denoted as (**) $p < 0.01$ for Ti-pTAN-1Stp compared to Ti; (#) $p < 0.05$ for Ti-pTAN-1Stp compared to Ti-pTAN; (oo) for Ti-pTAN-2Stp compared to Ti; ($) $p < 0.05$ or ($$) $p < 0.01$ for Ti-pTAN-2Stp compared to Ti-pTAN and (^) $p < 0.05$ for Ti-pTAN compared to Ti. All the statistical comparisons were performed within the same time point; (**B**) Collagen 1 estimation was performed after 14 days of MC3T3-E1 culture on various substrates. Statistical significance is marked as (**) $p < 0.01$ as compared to Ti and (#) $p < 0.05$ as compare to Ti-pTAN.

COLL 1 production of osteoblast cultured on Ti-pTAN-1Stp-Sr and Ti-pTAN-2Stp-Sr was statistically higher as compared to osteoblast cultured on Ti ($p < 0.01$, Figure 6B). Unlike the osteoblasts cultured on Ti-pTAN-1Stp-Sr, the osteoblasts cultured on Ti-pTAN-2Stp-Sr had higher COLL1 secretion as compared to Ti-pTAN ($p < 0.05$), suggesting superior osteogenic effects of Ti-pTAN-2Stp-Sr over Ti-pTAN-1Stp-Sr. The COLL 1 amounts were statistically similar for osteoblast cultured on Ti and Ti-pTAN. Immunofluorescence labelling of cells further confirmed these results for COLL 1. After 21 days of culture, MC3T3-E1 cells were stained for COLL1 using anti-COLL 1 antibody (Figure 7). MC3T3-E1 cells cultured on Ti-pTAN-1Stp-Sr (Figure 7c3) and Ti-pTAN-2Stp-Sr (Figure 6d3) displayed higher COLL 1 production as compared to cells cultured on Ti (Figure 7a3) and Ti-pTAN (Figure 7b3). Actin staining was also performed for the cells (Figure 7a2,b2,c2,d2) to monitor cell morphology and density. The cells cultured on all the substrates were confluent with uniform cell density. Cuboidal morphology was observed for the cells cultured on Ti-pTAN-1Stp-Sr and Ti-pTAN-2Stp-Sr. But the cells on Ti and Ti-pTAN had fusiform morphologies.

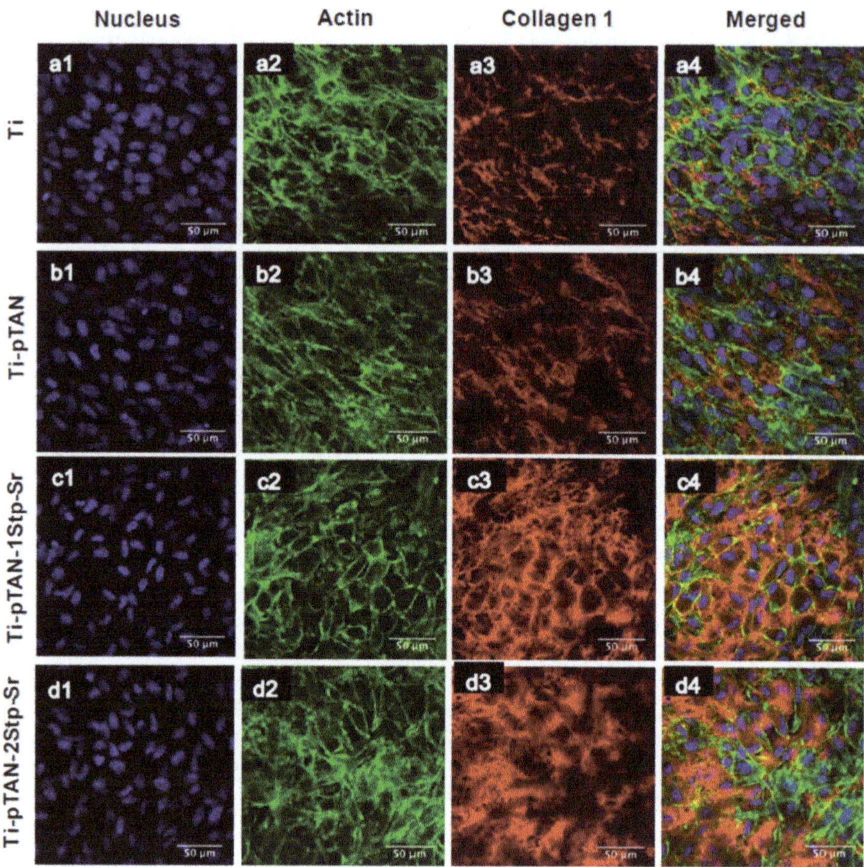

Figure 7. CSLM images of MC3T3-E1 cells after 21 days of culture on Ti (**a1–a4**); Ti-pTAN (**b1–b4**); Ti-pTAN-1Stp-Sr (**c1–c4**); Ti-pTAN-2Stp-Sr (**d1–d4**). MC3T3-E1 cells were stained for nucleus (**blue**, a1, b1, c1and d1); actin (**green**, a2, b2, c2and d2); collagen 1 (**red**, a3, b3, c3 and d3).

The effect of immobilized Sr on late differentiation markers such as osteocalcin [29] and matrix mineralization were investigated after 14 days of culture. Alizarin S stain and calcium estimation were performed to monitor the effects of Sr immobilized substrates on osteoblast mineralization. Stronger alizarin stain was observed on Ti-pTAN-1Stp-Sr (Figure 8Ac) and Ti-pTAN-2Stp-Sr (Figure 8Ad) as compared to Ti (Figure 7Aa) and Ti-pTAN (Figure 7Ab). Similarly, calcium estimated on Ti-pTAN-1Stp-Sr and Ti-pTAN-2Stp-Sr was higher than on Ti ($p < 0.01$) and Ti-pTAN ($p < 0.01$) (Figure 8B). Moreover, calcium amounts on Ti-pTAN were elevated as compared to Ti ($p < 0.05$), suggesting that pTAN coatings have a positive influence on osteoblast mineralization. Osteocalcin production was also higher on Ti-pTAN-1Stp-Sr and Ti-pTAN-2Stp-Sr as compared to Ti ($p < 0.01$) and Ti-pTAN ($p < 0.05$ and $p < 0.01$, respectively).

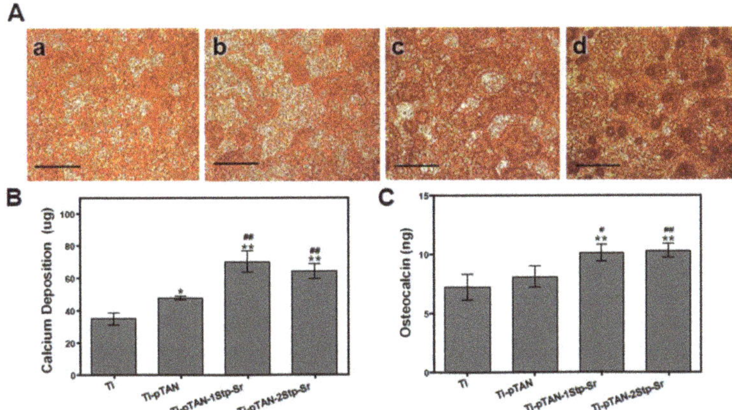

Figure 8. Calcium deposition and osteocalcin production of MC3T3-E1 cells after 14 days of culture on various substrates. (**A**) Alizarin S stain of calcium deposited by MC3T3-E1 cell on Ti (**a**); Ti-pTAN (**b**); Ti-pTAN-1Stp-Sr (**c**) and Ti-pTAN-2Stp-Sr (**d**); Scale bar 200 μm. (**B**) Estimation of calcium deposition by MC3T3-E1 cells; (**C**) Osteocalcin production by MC3T3-E1 cells cultured on the substrates. Statistical significance is represented as (*) $p < 0.05$ or (**) $p < 0.01$ as compared to Ti and (#) $p < 0.05$ or (##) $p < 0.01$ as compared to Ti-pTAN.

3.5. Osteoclast Differentiation

TRAP activity was estimated after day 3 and day 5 of culture in cell culture supernatant. For day 3 cultures, TRAP activity was reduced on Ti-pTAN, Ti-pTAN-1Stp-Sr and Ti-pTAN-2Stp-Sr as compared to Ti ($p < 0.01$) (Figure 9A). This suggested that the decrease of TRAP on all the substrates was due to pTAN coating. On day 5, cells on Ti-pTAN, Ti-pTAN-1Stp-Sr and Ti-pTAN-2Stp-Sr substrates also showed a reduction in TRAP as compared to Ti ($p < 0.01$). TRAP activity on Ti-pTAN-2Stp-Sr was even lower than cells on Ti-pTAN ($p < 0.01$), Ti-pTAN-1Stp-Sr ($p < 0.01$). This suggested that Sr immobilized on Ti-pTAN-2Stp-Sr with higher surface density has superior inhibitory effects towards osteoclast differentiation as compared to Ti-pTAN-1Stp-Sr. Total cell DNA estimation was used to study cell proliferation as per previous reports [30,31]. Cell proliferation after 5 days of RANKL stimulation of RAW 264.7 was also monitored. Total DNA was significantly reduced on Ti-pTAN, Ti-pTAN-1Stp-Sr and Ti-pTAN-2Stp-Sr as compared to Ti ($p < 0.01$) (Figure 9B).

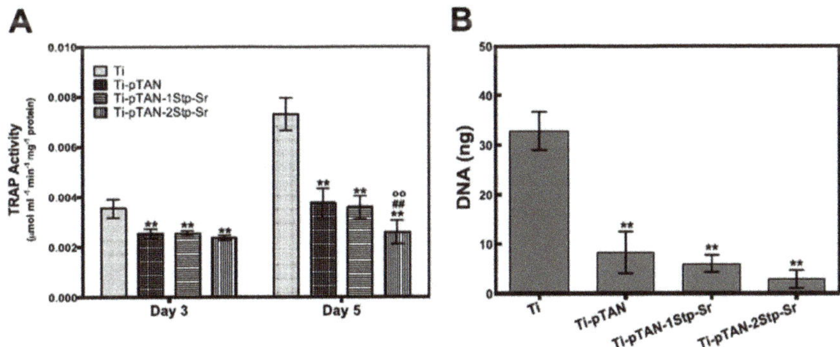

Figure 9. Osteoclast TRAP activity (**A**) and total DNA; (**B**) after 5 days of culture on Ti, Ti-pTAN, Ti-pTAN-1Stp-Sr, Ti-pTAN-2Stp-Sr.

Cells after 5 days of culture on Ti, Ti-pTAN, Ti-pTAN-1Stp-Sr and Ti-pTAN-2Stp-Sr were stained for actin and TRAP (Figure 10). TRAP positive multi-nucleated cells with more than 3 nuclei were considered as osteoclasts. Unlike Ti (Figure 10a1–4), which displayed large multinucleated TRAP positive osteoclasts, smaller and reduced number of multinucleated TRAP positive cells were observed on Ti-pTAN (Figure 9b1–4), Ti-pTAN-1Stp-Sr (Figure 10c1–4) and Ti-pTAN-2Stp-Sr (Figure 10d1–4). Similar to the TRAP activity estimation data of cell culture supernatant, intracellular TRAP production was decreased for cells on Ti-pTAN (Figure 10b3), Ti-pTAN-1Stp-Sr (Figure 9c3) and Ti-pTAN-2Stp-Sr (Figure 9d3) as compared to Ti (Figure 10a3). Cells especially on Ti-pTAN-2Stp-Sr depicted the lowest TRAP production of all the substrates. This suggested that the osteoclast inhibitory effects of Ti-pTAN-2Stp-Sr were superior to both Ti-pTAN-1Stp-Sr and Ti-pTAN.

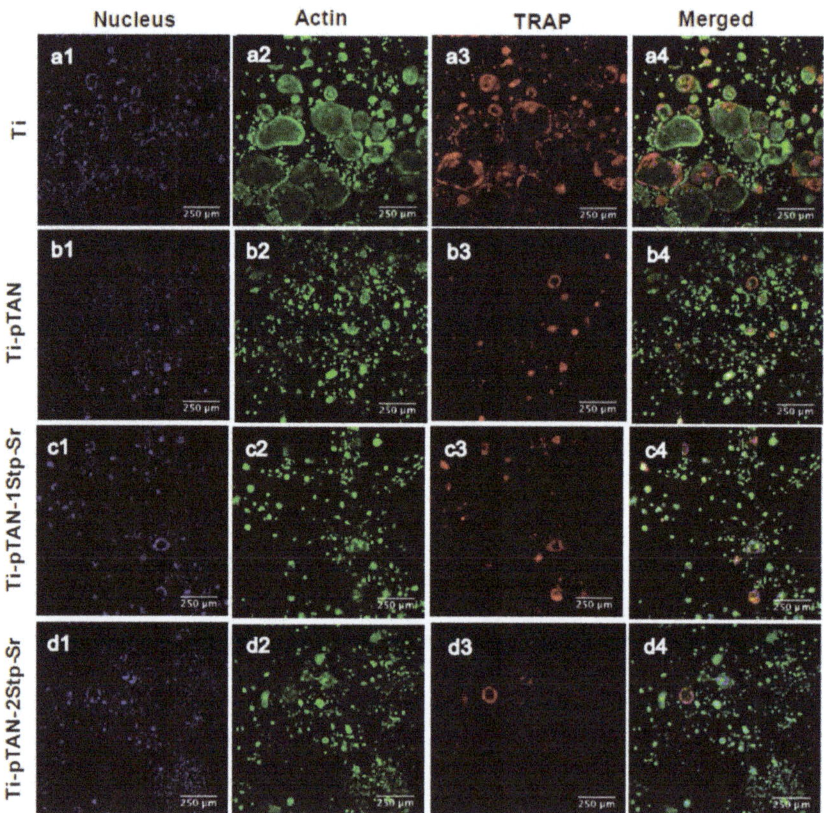

Figure 10. CSLM images of RAW 264.7 cells after 5 days of culture in presence of RANKL. Cells were cultured on Ti (**a1–a4**), Ti-pTAN (**b1–b4**), Ti-pTAN-1Stp-Sr (**c1–c4**), Ti-pTAN-2Stp-Sr (**d1–d4**) and were stained for nucleus (**blue**, a1, b1, c1and d1), actin (**green**, a2, b2, c2and d2), TRAP (**red**, a3, b3, c3 and d3).

4. Discussion

Since the imbalance in osteoclast and osteoblast activity in patients with bone metabolic disease can be an obstacle to bone implant success, coatings that actively aim at correcting this imbalance are required to recover bone turnover and improve the osseointegration of bone implants. In this study, a pTAN coating was used to immobilize Sr on a Ti surface. Two separate schemes were employed (Figure 1); the first scheme was a one-step process, wherein Ti-pTAN-1Stp-Sr substrates were prepared by replacing sodium with Sr in tannic acid solution in bicine buffer. In the second scheme, Ti-pTAN-2Stp-Sr substrates were prepared by a two-step method, wherein a pTAN coating was carried out in a bicine buffer containing sodium, before coating the substrates with Sr. The Sr metal ions were immobilized on pTAN-coated Ti (Figure 2 and Table 1). The surface immobilization efficiency was higher for the two-step coating method. Hence, unlike the one-step coating method, the two-step coating method not only displayed higher surface density of immobilized Sr, but also showed stronger binding, as 70% of the Sr remained bound and was not released for two weeks (Figure 4). The thicker tannic acid coating conferred by the two-step method could contribute to more phenol groups increasing their interactions with Sr^{2+} ions to increase immobilization efficiency. pTAN coating kinetics and mechanisms have been described previously [32]. It is proposed that the first layer of pTAN on the Ti surface is though the formation of TiO(OH)pTAN hydroxo complexes [25,32]. Subsequent layers are built up by the heterogeneous polymers formed by the autooxidation of pTAN [25]. pTAN is also known to form polymers through the formation of multivalent coordination bonding with metal ions [24,26]. Hence, it is assumed that the pTAN-Sr coatings on Ti-pTAN-1Stp-Sr and Ti-pTAN-2Stp-Sr are formed by auto-oxidation of pTAN and multivalent coordination bonding with Sr.

A decrease in osteoblast proliferation was observed for cells cultured on Ti-pTAN-1Stp-Sr and Ti-pTAN-2Stp-Sr (Figure 5). Joseph Caverzasio observed that the addition of Sr to a cell culture medium enhanced osteoblast proliferation [3]. Almeida et al. reported that a Sr salt supplemented in media did not affect MC3T3-E1 proliferation at lower concentrations (0.05 and 1 mM), but at higher concentrations such as 0.5 mM, a decrease in proliferation was observed as compared to that of untreated controls [33,34]. Furthermore, Ti functionalized by Sr via magnetron co-sputtering technology also decreased proliferation of human dental pulp stem cells. [19] Yet, in both studies, osteoblast differentiation was augmented due to Sr introduction, which was consistent with our results, as described below. It is known that a decrease in osteoblast proliferation and growth arrest are associated with the elevated production of differentiation markers such as ALP, COLL 1 etc. [35,36] As such, the effects of Ti-pTAN-1Stp-Sr and Ti-pTAN-2Stp-Sr on osteoblast differentiation markers were investigated.

ALP and COLL 1 are early differentiation markers for osteoblasts, whereas osteocalcin is a late differentiation marker [29]. By secreting ALP and collagen 1 in the early phases of differentiation and mineralizing matrix in the later stages, MC3T3-E1 exhibits successive differentiation sequence which is comparable to osteoblast in bone [35]. Hence, this cell line is widely used as in vitro model for osteoblasts research. Generally, the ALP expression of differentiating osteoblasts augments in the early phases of differentiation, leading to matrix maturation, while the level drops before matrix mineralization [37]. As presented, ALP production was enhanced on day 5 for all the substrates with a decline in the activity on day 7 and day 14, suggesting the progression of the cells to the mineralization phase (Figure 6A). Yet, compared to Ti, Sr-immobilized substrates such as Ti-pTAN-1Stp-Sr and Ti-pTAN-2Stp-Sr boosted ALP activity on day 5 and day 7. COLL 1 secretion, which is also an early differentiation marker of osteoblasts, was monitored after 14 days of culture (Figures 6B and 7) [29]. Sr surface coated substrates displayed higher COLL 1 production. The majority of MC3T3-E1 cells on Sr immobilized substrates, in particular the cells on Ti-pTAN-1Stp-Sr, exhibited cuboidal morphology (Figure 6), which is a characteristic feature of differentiating cells [38]. On the other hand, the majority of cells cultured on Ti and Ti-pTAN displayed a fusiform appearance (Figure 6), which is a predominant feature of proliferating cells [38]. This further corroborates the osteogenic potential of Sr immobilized substrates. Furthermore, the Sr-based surface coatings also augmented in the osteocalcin and matrix mineralization by the osteoblasts (Figure 8).

Studies had shown that strontium supplementation in media inhibited osteoblast mineralization in vitro [39,40]. Contrastingly, another report established that Sr enhanced ALP, osteocalcin secretion and the mineralization of mouse bone marrow stromal cells [41], and human primary osteoblasts [42]. Sr supplemented in cell culture media as a $SrCl_2$ triggered osteogenic differentiation of human mesenchymal stem cells (MSCs) and stimulated in vivo bone deposition when delivered locally as

Sr-HA-collagen scaffold to treat rat calvarial defect via Wnt signal transduction [43]. In this study, Sr-immobilized substrates such as Ti-pTAN-1Stp-Sr and Ti-pTAN-2Stp-Sr augmented ALP activity, COLL 1 production, osteocalcin secretion and calcium deposition of MC3T3-E1 cells. However, Ti-pTAN-2Stp-Sr showed a higher surface density of Sr as compared to Ti-pTAN-1Stp-Sr, although the potential to augment osteogenesis was similar for both the substrates. This suggests that even if the two-step method of coating resulted in a higher efficiency of Sr binding properties, a low Sr density on the surface, such as 0.32 ± 0.02 $\mu g/cm^2$, is sufficient in promoting osteoblast differentiation.

Osteoclasts are important cells for bone remodeling [44]; hence, the regulation of their activity on implant surfaces is crucial for osseointegration. The effects of osteoclasts on Sr-modified surfaces were also investigated. As a RANKL-sensitive pre-osteoclast population, RAW 264.7 can be easily procured and is easy to handle as compared to primary cultures, which makes is an attractive in vitro model for osteoclast research [45]. Osteoclastic differentiation of RAW 264.7 cells on Ti, Ti-pTAN, Ti-pTAN-1Stp-Sr and Ti-pTAN-2Stp-Sr were studied by estimating TRAP activity. TRAP activity was used as a marker for osteoclast differentiation and resorption [46–48]. pTAN coated substrates such as Ti-pTAN, Ti-pTAN-1Stp-Sr and Ti-pTAN-2Stp-Sr decreased TRAP activity and total DNA (Figure 9). However, on day 5, the decrease in TRAP production was greatest in Ti-pTAN-2Stp-Sr. Likewise, immunofluorescence labelling of intracellular TRAP production was also lowest on the Ti-pTAN-2Stp-Sr substrate (Figure 10d3). Strontium ranelate was shown to inhibit RANKL mediated osteoclastogenesis of RAW 264.7 and human peripheral blood monocyte cells [49]. As shown by Chung et al., Ti modified with Sr substituted HA by micro-arc treatment reduced osteoclast differentiation of RAW 264.7 cells [50]. Equally, this study demonstrated that Ti-pTAN-2Stp-Sr reduced osteoclast activity as compared to Ti and Ti-pTAN. Florencio-Silva et al. reported that osteoclast precursors fuse, polarize, and rearrange actin cytoskeleton to form actin rings and a sealing zone [51]. The formation of multinucleated cells was also reduced on all the pTAN coated substrates such as Ti-pTAN, Ti-pTAN-1Stp-Sr and Ti-pTAN-2Stp-Sr (Figure 10). However, the inhibitory effects of Ti-pTAN-2Stp-Sr were superior to those of Ti, Ti-pTAN and Ti-pTAN-1Stp-Sr substrates.

5. Conclusions

In this study, a pTAN coating was used to immobilize Sr on a Ti surface. Although Ti-pTAN-2Stp-Sr showed a higher surface density of Sr as compared to Ti-pTAN-1Stp-Sr, the ability to augment osteogenesis was similar for both the substrates. Osteoclast differentiation studies revealed that the Ti-pTAN-2Stp-Sr substrates decreased the TRAP activity of osteoclast significantly as compared to Ti, Ti-pTAN and Ti-pTAN-1Stp-Sr. The pTAN coating also had an influence on the negative regulation of osteoclast differentiation, as reported [52]. Henceforth, this economical and simple coating of pTAN with inherent anti-osteoclastogenic properties could be exploited to immobilize osteogenic metal ions such as Sr^{2+} on surfaces. The two-step method was more efficient to immobilize Sr, as well as to modulate osteoblast and osteoclast cells.

It is hoped that these findings will be beneficial to deliver Sr in order to tailor cellular responses at bone-implant interfaces, especially for osteoporotic bone. The imbalance in osteoclasts and osteoblasts activity in patients with bone metabolic disease such as osteoporosis can be an obstacle to bone implant success [53–55]. Coatings that actively aim at correcting this imbalance are required to recover bone turnover and improve the osseointegration of bone implants [56,57]. We believe that these surface coatings will be beneficial to regulate osteoclast and osteoblast development at implant surfaces. Nonetheless, we acknowledge that the in vitro results presented here are conducted in controlled laboratory settings and could not be extrapolated to in vivo or clinical conditions. The coupling of bone formation and bone resorption in a basic multicellular unit of a healthy bone is a multifaceted process which involves the interplay of various regulatory molecules and bone cells [58,59]. Further studies are needed to understand the viability of coatings and the potential to regulate the molecular and cellular interplay in in vivo settings.

Author Contributions: Conceptualization, C.S., Z.S. and W.W.; Data curation, C.S.; Formal analysis, C.S., Z.S. and C.H.K.; Funding acquisition, W.W.; Investigation, C.S.; Methodology, C.S., S.W.C. and D.W.; Project administration, W.W.; Resources, C.S., Z.S., S.W.C., D.W. and W.W.; Supervision, Z.S. and W.W.; Validation, C.S., Z.S., C.H.K. and S.W.C.; Visualization, C.S. and C.H.K.; Writing–original draft, C.S.; Writing–review & editing, C.S., Z.S., C.H.K., D.W. and W.W.

Funding: The study was supported by NUSMed Post-Doctoral Fellowship (NO. NUHSRO/2017/078/PDF/08).

Conflicts of Interest: The authors declare no conflict of interest.

References

1. Marcus, R. Post-menopausal osteoporosis. *Best Pract. Res. Clin. Obstet. Gynaecol.* **2002**, *16*, 309–327. [CrossRef] [PubMed]

2. Sessa, G.; Evola, F.R.; Costarella, L. Osteosynthesis systems in fragility fracture. *Aging Clin. Exp. Res.* **2011**, *23*, 69–70. [PubMed]

3. Junker, R.; Dimakis, A.; Thoneick, M.; Jansen, J.A. Effects of implant surface coatings and composition on bone integration: A systematic review. *Clin. Oral Implant. Res.* **2009**, *20* (Suppl. 4), 185–206. [CrossRef] [PubMed]

4. Palmquist, A.; Omar, O.M.; Esposito, M.; Lausmaa, J.; Thomsen, P. Titanium oral implants: Surface characteristics, interface biology and clinical outcome. *J. R. Soc. Interface* **2010**, *7* (Suppl. 5), S515–S527. [CrossRef] [PubMed]

5. O'Brien, E.M.; Risser, G.E.; Spiller, K.L. Sequential drug delivery to modulate macrophage behavior and enhance implant integration. *Adv. Drug Deliv. Rev.* **2019**. [CrossRef] [PubMed]

6. Moroni, A.; Faldini, C.; Marchetti, S.; Manca, M.; Consoli, V.; Giannini, S. Improvement of the bone-pin interface strength in osteoporotic bone with use of hydroxyapatite-coated tapered external-fixation pins. *J. Bone J. Surg Am.* **2001**, *83*, 717–721. [CrossRef] [PubMed]

7. Marchisio, M.; Carmine, M.D.; Pagone, R.; Piattelli, A.; Miscia, S. Implant surface roughness influences osteoclast proliferation and differentiation. *J. Biomed. Mater. Res. B Appl. Biomater.* **2005**, *75*, 251–256. [CrossRef]

8. Schwartz, Z.; Lohmann, C.; Oefinger, J.; Bonewald, L.; Dean, D.; Boyan, B. Implant surface characteristics modulate differentiation behavior of cells in the osteoblastic lineage. *Adv. Dent. Res.* **1999**, *13*, 38–48. [CrossRef]

9. Klymov, A.; Prodanov, L.; Lamers, E.; Jansen, J.A.; Walboomers, X.F. Understanding the role of nano-topography on the surface of a bone-implant. *Biomater. Sci.* **2013**, *1*, 135–151. [CrossRef]

10. Poh, C.K.; Shi, Z.; Lim, T.Y.; Neoh, K.G.; Wang, W. The effect of VEGF functionalization of titanium on endothelial cells in vitro. *Biomaterials* **2010**, *31*, 1578–1585. [CrossRef]

11. Tan, H.C.; Poh, C.K.; Cai, Y.; Soe, M.T.; Wang, W. Covalently grafted BMP-7 peptide to reduce macrophage/monocyte activity: An in vitro study on cobalt chromium alloy. *Biotechnol. Bioeng.* **2013**, *110*, 969–979. [CrossRef] [PubMed]

12. Sachse, A.; Wagner, A.; Keller, M.; Wagner, O.; Wetzel, W.D.; Layher, F.; Venbrocks, R.A.; Hortschansky, P.; Pietraszczyk, M.; Wiederanders, B.; et al. Osteointegration of hydroxyapatite-titanium implants coated with nonglycosylated recombinant human bone morphogenetic protein-2 (BMP-2) in aged sheep. *Bone* **2005**, *37*, 699–710. [CrossRef] [PubMed]

13. Gao, Y.; Zou, S.; Liu, X.; Bao, C.; Hu, J. The effect of surface immobilized bisphosphonates on the fixation of hydroxyapatite-coated titanium implants in ovariectomized rats. *Biomaterials* **2009**, *30*, 1790–1796. [CrossRef] [PubMed]

14. Lenart, B.A.; Lorich, D.G.; Lane, J.M. Atypical Fractures of the Femoral Diaphysis in Postmenopausal Women Taking Alendronate. *N. Engl. J. Med.* **2008**, *358*, 1304–1306. [CrossRef] [PubMed]

15. Reginster, J.-Y.; Kaufman, J.-M.; Goemaere, S.; Devogelaer, J.P.; Benhamou, C.L.; Felsenberg, D.; Diaz-Curiel, M.; Brandi, M.-L.; Badurski, J.; Wark, J.; et al. Maintenance of antifracture efficacy over 10 years with strontium ranelate in postmenopausal osteoporosis. *Osteoporos. Int.* **2012**, *23*, 1115–1122. [CrossRef] [PubMed]

16. Cianferotti, L.; D'Asta, F.; Brandi, M.L. A review on strontium ranelate long-term antifracture efficacy in the treatment of postmenopausal osteoporosis. *Ther. Adv. Musculoskelet. Dis.* **2013**, *5*, 127–139. [CrossRef] [PubMed]

17. Bain, S.D.; Jerome, C.; Shen, V.; Dupin-Roger, I.; Ammann, P. Strontium ranelate improves bone strength in ovariectomized rat by positively influencing bone resistance determinants. *Osteoporos. Int.* **2009**, *20*, 1417–1428. [CrossRef] [PubMed]

18. Bonnelye, E.; Chabadel, A.; Saltel, F.; Jurdic, P. Dual effect of strontium ranelate: Stimulation of osteoblast differentiation and inhibition of osteoclast formation and resorption in vitro. *Bone* **2008**, *42*, 129–138. [CrossRef] [PubMed]

19. Andersen, O.Z.; Offermanns, V.; Sillassen, M.; Almtoft, K.P.; Andersen, I.H.; Sørensen, S.; Jeppesen, C.S.; Kraft, D.C.E.; Bøttiger, J.; Rasse, M.; et al. Accelerated bone ingrowth by local delivery of strontium from surface functionalized titanium implants. *Biomaterials* **2013**, *34*, 5883–5890. [CrossRef]

20. Tao, Z.S.; Bai, B.L.; He, X.W.; Liu, W.; Li, H.; Zhou, Q.; Sun, T.; Huang, Z.L.; Tu, K.K.; Lv, Y.X.; et al. A comparative study of strontium-substituted hydroxyapatite coating on implant's osseointegration for osteopenic rats. *Med. Biol. Eng. Comput.* **2016**, *54*, 1959–1968. [CrossRef]

21. Dang, Y.; Zhang, L.; Song, W.; Chang, B.; Han, T.; Zhang, Y.; Zhao, L. In vivo osseointegration of Ti implants with a strontium-containing nanotubular coating. *Int. J. Nanomed.* **2016**, *11*, 1003–1011.

22. Tao, Z.S.; Zhou, W.S.; He, X.W.; Liu, W.; Bai, B.L.; Zhou, Q.; Huang, Z.L.; Tu, K.K.; Li, H.; Sun, T.; et al. A comparative study of zinc, magnesium, strontium-incorporated hydroxyapatite-coated titanium implants for osseointegration of osteopenic rats. *Mater. Sci. Eng. C Mater. Biol. Appl.* **2016**, *62*, 226–232. [CrossRef] [PubMed]

23. Sileika, T.S.; Barrett, D.G.; Zhang, R.; Lau, K.H.; Messersmith, P.B. Colorless multifunctional coatings inspired by polyphenols found in tea, chocolate, and wine. *Angew. Chem.* **2013**, *52*, 10766–10770. [CrossRef] [PubMed]

24. Kim, S.; Pasc, A. Advances in Multifunctional Surface Coating Using Metal-Phenolic Networks. *Bull. Korean Chem. Soc.* **2017**, *38*, 519–520. [CrossRef]

25. Geissler, S.; Barrantes, A.; Tengvall, P.; Messersmith, P.B.; Tiainen, H. Deposition Kinetics of Bioinspired Phenolic Coatings on Titanium Surfaces. *Langmuir* **2016**, *32*, 8050–8060. [CrossRef] [PubMed]

26. Ejima, H.; Richardson, J.J.; Liang, K.; Best, J.P.; van Koeverden, M.P.; Such, G.K.; Cui, J.; Caruso, F. One-step assembly of coordination complexes for versatile film and particle engineering. *Science* **2013**, *341*, 154–157. [CrossRef]

27. Robinson, K.S.; Sherwood, P. X-ray photoelectron spectroscopic studies of the surface of sputter ion plated films. *Surf. Interface Anal.* **1984**, *6*, 261–266. [CrossRef]

28. Pan, L.; Wang, H.; Wu, C.; Liao, C.; Li, L. Tannic-Acid-Coated Polypropylene Membrane as a Separator for Lithium-Ion Batteries. *ACS Appl. Mater. Interfaces* **2015**, *7*, 16003–16010. [CrossRef] [PubMed]

29. Huang, W.; Yang, S.; Shao, J.; Li, Y.-P. Signaling and transcriptional regulation in osteoblast commitment and differentiation. *Front. Biosci.* **2007**, *12*, 3068–3092. [CrossRef]

30. Jones, G.L.; Motta, A.; Marshall, M.J.; El Haj, A.J.; Cartmell, S.H. Osteoblast: Osteoclast co-cultures on silk fibroin, chitosan and PLLA films. *Biomaterials* **2009**, *30*, 5376–5384. [CrossRef]

31. Rose, F.R.; Cyster, L.A.; Grant, D.M.; Scotchford, C.A.; Howdle, S.M.; Shakesheff, K.M. In vitro assessment of cell penetration into porous hydroxyapatite scaffolds with a central aligned channel. *Biomaterials* **2004**, *25*, 5507–5514. [CrossRef]

32. Surleva, A.; Atanasova, P.; Kolusheva, T.; Costadinnova, L. Study of the complex equilibrium between titanium (IV) and tannic acid. *J. Chem. Technol. Metall.* **2014**, *49*, 594–600.

33. Caverzasio, J. Strontium ranelate promotes osteoblastic cell replication through at least two different mechanisms. *Bone* **2008**, *42*, 1131–1136. [CrossRef]

34. Almeida, M.M.; Nani, E.P.; Teixeira, L.N.; Peruzzo, D.C.; Joly, J.C.; Napimoga, M.H.; Martinez, E.F. Strontium ranelate increases osteoblast activity. *Tissue Cell* **2016**, *48*, 183–188. [CrossRef]

35. Quarles, L.D.; Yohay, D.A.; Lever, L.W.; Caton, R.; Wenstrup, R.J. Distinct proliferative and differentiated stages of murine MC3T3-E1 cells in culture: An in vitro model of osteoblast development. *J. Bone Miner. Res. Off. J. Am. Soc. Bone Mineral. Res.* **1992**, *7*, 683–692. [CrossRef]

36. Gu, Y.X.; Du, J.; Si, M.S.; Mo, J.J.; Qiao, S.C.; Lai, H.C. The roles of PI3K/Akt signaling pathway in regulating MC3T3-E1 preosteoblast proliferation and differentiation on SLA and SLActive titanium surfaces. *J. Biomed. Mater. Res. Part A* **2013**, *101*, 748–754. [CrossRef]

37. Lian, J.B.; Stein, G.S. Concepts of osteoblast growth and differentiation: Basis for modulation of bone cell development and tissue formation. *Crit. Rev. Oral Biol. Med.* **1992**, *3*, 269–305. [CrossRef]

38. Hong, D.; Chen, H.-X.; Yu, H.-Q.; Liang, Y.; Wang, C.; Lian, Q.-Q.; Deng, H.-T.; Ge, R.-S. Morphological and proteomic analysis of early stage of osteoblast differentiation in osteoblastic progenitor cells. *Exp. Cell Res.* **2010**, *316*, 2291–2300. [CrossRef]

39. Verberckmoes, S.C.; De Broe, M.E.; D'Haese, P.C. Dose-dependent effects of strontium on osteoblast function and mineralization. *Kidney Int.* **2003**, *64*, 534–543. [CrossRef]

40. Wornham, D.P.; Hajjawi, M.O.; Orriss, I.R.; Arnett, T.R. Strontium potently inhibits mineralisation in bone-forming primary rat osteoblast cultures and reduces numbers of osteoclasts in mouse marrow cultures. *Osteoporos. Int.* **2014**, *25*, 2477–2484. [CrossRef]

41. Choudhary, S.; Halbout, P.; Alander, C.; Raisz, L.; Pilbeam, C. Strontium ranelate promotes osteoblastic differentiation and mineralization of murine bone marrow stromal cells: Involvement of prostaglandins. *J. Bone Mineral. Res.* **2007**, *22*, 1002–1010. [CrossRef]

42. Atkins, G.; Welldon, K.; Halbout, P.; Findlay, D. Strontium ranelate treatment of human primary osteoblasts promotes an osteocyte-like phenotype while eliciting an osteoprotegerin response. *Osteoporos. Int.* **2009**, *20*, 653–664. [CrossRef]

43. Yang, F.; Yang, D.; Tu, J.; Zheng, Q.; Cai, L.; Wang, L. Strontium enhances osteogenic differentiation of mesenchymal stem cells and in vivo bone formation by activating Wnt/catenin signaling. *Stem Cells* **2011**, *29*, 981–991. [CrossRef]

44. Sims, N.A.; Gooi, J.H. Bone remodeling: Multiple cellular interactions required for coupling of bone formation and resorption. *Semin. Cell Dev. Biol.* **2008**, *19*, 444–451. [CrossRef]

45. Collin-Osdoby, P.; Osdoby, P. RANKL-mediated osteoclast formation from murine RAW 264.7 cells. *Methods Mol. Biol.* **2012**, *816*, 187–202.

46. Kirstein, B.; Chambers, T.J.; Fuller, K. Secretion of tartrate-resistant acid phosphatase by osteoclasts correlates with resorptive behavior. *J. Cell. Biochem.* **2006**, *98*, 1085–1094. [CrossRef]

47. Halleen, J.M.; Alatalo, S.L.; Suominen, H.; Cheng, S.; Janckila, A.J.; Väänänen, H.K. Tartrate-resistant acid phosphatase 5b: A novel serum marker of bone resorption. *J. Bone Mineral. Res.* **2000**, *15*, 1337–1345. [CrossRef]

48. Lv, Y.; Wang, G.; Xu, W.; Tao, P.; Lv, X.; Wang, Y. Tartrate-resistant acid phosphatase 5b is a marker of osteoclast number and volume in RAW 264.7 cells treated with receptor-activated nuclear κB ligand. *Exp. Ther. Med.* **2015**, *9*, 143–146. [CrossRef]

49. Caudrillier, A.; Hurtel-Lemaire, A.-S.; Wattel, A.; Cournarie, F.; Godin, C.; Petit, L.; Petit, J.-P.; Terwilliger, E.; Kamel, S.; Brown, E.M. Strontium ranelate decreases receptor activator of nuclear factor-κB ligand-induced osteoclastic differentiation in vitro: Involvement of the calcium-sensing receptor. *Mol. Pharmacol.* **2010**, *78*, 569–576. [CrossRef]

50. Chung, C.-J.; Long, H.-Y. Systematic strontium substitution in hydroxyapatite coatings on titanium via micro-arc treatment and their osteoblast/osteoclast responses. *Acta Biomater.* **2011**, *7*, 4081–4087. [CrossRef]

51. Florencio-Silva, R.; Sasso, G.R.; Sasso-Cerri, E.; Simoes, M.J.; Cerri, P.S. Biology of Bone Tissue: Structure, Function, and Factors That Influence Bone Cells. *BioMed Res. Int.* **2015**, *2015*, 421746. [CrossRef]

52. Steffi, C.; Shi, Z.; Kong Chee, H.; Wang, W. Bioinspired polydopamine and polyphenol tannic acid functionalized titanium suppress osteoclast differentiation: A facile and efficient strategy to regulate osteoclast activity at bone–implant interface. *J. R. Soc. Interface* **2019**, *16*, 20180799. [CrossRef]

53. Schulze, M.; Gehweiler, D.; Riesenbeck, O.; Wähnert, D.; Raschke, M.J.; Hartensuer, R.; Vordemvenne, T. Biomechanical characteristics of pedicle screws in osteoporotic vertebrae—Comparing a new cadaver corpectomy model and pure pull-out testing. *J. Orthop. Res.* **2017**, *35*, 167–174. [CrossRef]

54. Apostu, D.; Lucaciu, O.; Berce, C.; Lucaciu, D.; Cosma, D. Current methods of preventing aseptic loosening and improving osseointegration of titanium implants in cementless total hip arthroplasty: A review. *J. Int. Med. Res.* **2017**, *46*, 2104–2119. [CrossRef]

55. Weiser, L.; Huber, G.; Sellenschloh, K.; Viezens, L.; Püschel, K.; Morlock, M.M.; Lehmann, W. Insufficient stability of pedicle screws in osteoporotic vertebrae: Biomechanical correlation of bone mineral density and pedicle screw fixation strength. *Eur. Spine J.* **2017**, *26*, 2891–2897. [CrossRef]

56. Civantos, A.; Martínez-Campos, E.; Ramos, V.; Elvira, C.; Gallardo, A.; Abarrategi, A. Titanium Coatings and Surface Modifications: Toward Clinically Useful Bioactive Implants. *ACS Biomater. Sci. Eng.* **2017**, *3*, 1245–1261. [CrossRef]

Polymers **2019**, *11*, 1256

57. Zhao, H.; Huang, Y.; Zhang, W.; Guo, Q.; Cui, W.; Sun, Z.; Eglin, D.; Liu, L.; Pan, G.; Shi, Q. Mussel-Inspired Peptide Coatings on Titanium Implant to Improve Osseointegration in Osteoporotic Condition. *ACS Biomater. Sci. Eng.* **2018**, *4*, 2505–2515. [CrossRef]

58. Sims, N.A.; Martin, T.J. Coupling the activities of bone formation and resorption: A multitude of signals within the basic multicellular unit. *Bonekey Rep.* **2014**, *3*, 481. [CrossRef]

59. Kohli, N.; Ho, S.; Brown, S.J.; Sawadkar, P.; Sharma, V.; Snow, M.; García-Gareta, E. Bone remodelling in vitro: Where are we headed? A review on the current understanding of physiological bone remodelling and inflammation and the strategies for testing biomaterials in vitro. *Bone* **2018**, *110*, 38–46. [CrossRef]

 *polymers*

Article

Basic Properties of a New Polymer Gel for 3D-Dosimetry at High Dose-Rates Typical for FFF Irradiation Based on Dithiothreitol and Methacrylic Acid (MAGADIT): Sensitivity, Range, Reproducibility, Accuracy, Dose Rate Effect and Impact of Oxygen Scavenger

Muzafar Khan [1,2,3], Gerd Heilemann [4], Wolfgang Lechner [4], Dietmar Georg [4] and Andreas Georg Berg [1,2,*]

1   Center for Medical Physics and Biomedical Engineering, Medical University of Vienna, Waehringer Guertel 18-23, A-1090 Vienna, Austria; mohmand169@yahoo.com
2   High-Field MR-Center (MRCE), Medical University of Vienna; Lazarettg 14, A-1090 Vienna, Austria
3   Nuclear Medicine, Oncology and Radiotherapy Institute (NORI), G-8/3 Islamabad 44000, Pakistan
4   Department of Radiation Oncology, Medical University of Vienna/AKH Vienna, Waehringer Guertel 18-20, 1090 Vienna, Austria; gerd.heilemann@meduniwien.ac.at (G.H.); wolfgang.lechner@akhwien.at (W.L.); dietmar.georg@meduniwien.ac.at (D.G.)
*   Correspondence: andreas.berg@meduniwien.ac.at

Received: 29 August 2019; Accepted: 10 October 2019; Published: 19 October 2019

Abstract: The photon induced radical-initiated polymerization in polymer gels can be used for high-resolution tissue equivalent dosimeters in quality control of radiation therapy. The dose (D) distribution in radiation therapy can be measured as a change of the physical measurement parameter T2 using T2-weighted magnetic resonance imaging. The detection by T2 is relying on the local change of the molecular mobility due to local polymerization initiated by radicals generated by the ionizing radiation. The dosimetric signals R2 = 1/T2 of many of the current polymer gels are dose-rate dependent, which reduces the reliability of the gel for clinical use. A novel gel dosimeter, based on methacrylic acid, gelatin and the newly added dithiothreitol (MAGADIT) as an oxygen-scavenger was analyzed for basic properties, such as sensitivity, reproducibility, accuracy and dose-rate dependence. Dithiothreitol features no toxic classification with a difference to THPC and offers a stronger negative redox-potential than ascorbic acid. Polymer gels with three different concentration levels of dithiothreitol were irradiated with a preclinical research X-ray unit and MR-scanned (T2) for quantitative dosimetry after calibration. The polymer gel with the lowest concentration of the oxygen scavenger was about factor 3 more sensitive to dose as compared to the gel with the highest concentration. The dose sensitivity ($\alpha = \Delta R2/\Delta D$) of MAGADIT gels was significantly dependent on the applied dose rate $\dot{D}$ ($\approx$48% reduction between $\dot{D}$ = 0.6 Gy/min and $\dot{D}$ = 4 Gy/min). However, this undesirable dose-rate effect reduced between 4–8 Gy/min ($\approx$23%) and almost disappeared in the high dose-rate range ($8 \leq \dot{D} \leq 12$ Gy/min) used in flattening-filter-free (FFF) irradiations. The dose response varied for different samples within one manufacturing batch within 3%–6% (reproducibility). The accuracy ranged between 3.5% and 7.9%. The impact of the dose rate on the spatial integrity is demonstrated in the example of a linear accelerator (LINAC) small sized 5×10 mm^2 10 MV photon field. For MAGADIT the maximum shift in the flanks in this field is limited to about 0.8 mm at a FFF dose rate of 15 Gy/min. Dose rate sensitive polymer gels likely perform better at high dose rates; MAGADIT exhibits a slightly improved performance compared to the reference normoxic polymer gel methacrylic and ascorbic acid in gelatin initiated by copper (MAGIC) using ascorbic acid.

Keywords: responsive gels in biomedical and diagnostic applications; polymer; gel; precision; radiation therapy; dosimetry; 3D; flattening filter free; FFF; oxygen scavenger; dose rate; magnetic resonance

---

1. Introduction

Modern advanced radiation therapy allows for a highly conformal application of high radiation doses to a well-defined target volume while keeping the dose to the surrounding tissue at a relatively low level. Technological advances over the past decades include intensity-modulated radiation therapy (IMRT) and volumetric modulated arc therapy (VMAT), which allow to deliver the complex dose distributions in three-dimensions (3D) [1–3] with a nominal spatial accuracy of a few mm adapted to the planning target volume (PTV); the PTV is usually defined as the area in the human body, to which a prescribed dose is to be applied to, related, e.g., to a tumor region sparing nearby dose sensitive healthy tissue. Absolute values of dose and even the applied dose rate might vary from step to step [4–7]. For extended quality control, especially in the process of introducing new technologies and treatment techniques, it is necessary to validate the calculated dose distributions experimentally. This is typically done on the basis of simple cylindrical or rectangular phantoms with insert openings for point-wise detection (0-dimensional) of dose (e.g., ionization chambers). Alternatively, 2-dimensional (2-D) dose distributions can be obtained with dosimetric films or detector arrays. Even 3-dimensional (3-D) dose distributions can be evaluated with a 3-D dosimeter, positioned in cavities in phantoms simulating the human body, which can measure a spatially different dose in all of the three dimensions at the same time.

Ideally, a 3-D-dose detector should provide the ability to measure the 3-D dose distribution quantitatively at high spatial resolution Furthermore, it should feature dose linearity and tissue equivalence. The dose signal should be energy and dose rate independent. Both of these features might be varied during delivery with modern linear accelerators (LINAC). Most of the proposed tissue equivalent 3-D integrative dosimeters are based mainly on three types of detection mechanisms: (1) Radiochromic dosimeters such as plastic dosimeters, e.g., PRESAGE® [8] gel, micelle dosimeters [9,10] and tetrazolium hydrogels rely on color changes of a dye after irradiation. Tetrazolium gellan gum gel dosimeters [11] and the newly proposed flexible polymer/silicone based dosimeters also rely on color changes during irradiation [12]. (2) "Fricke" types of gels are sensitive to the oxidation of ferrous to ferric ions during irradiation [13,14]. The change of oxidation status can be detected by optical methods in radiochromic gels or magnetic resonance imaging (MRI) using R1 = 1/T1 or R2 = 1/T2 contrast. (3) Radiation induced polymerization (the corresponding dosimeters are usually specified as "polymer gels" and can be read out by optical methods [15] using an increased scattering of light). Their dose response can also be visualized by computed tomography (CT) [16], ultrasonography [17] and magnetic resonance imaging (MRI).

MRI has been mostly used as a reading method of dosimetric response using relative changes in the MRI contrast parameter T2, i.e., the transverse relaxation time. The transverse relaxation rate R2 = 1/T2 might exhibit an increase by several 100% after irradiation and thus serves very well as a sensitive measurement parameter of the applied dose [18]. Radiation therapy planning is increasingly performed on the basis of MRI due to the high soft tissue contrast with comparison to CT. Thus, MRI-scanners also become increasingly available in radiation therapy sites at hospitals. We focused subsequently on this sensitive MR-based polymer gel dosimetry (MRPD).

1.1. Principles of MR-Based Polymer Gel Dosimetry: An Introduction

Polymer gels used for the dosimetric imaging of ionizing radiation [17] mainly consist of radiation sensitive chemicals and a mobility reducing agent, e.g., gelatin or agarose and water. The radiation sensitive agent polymerizes in several steps from dimers to multimers even building up networks upon irradiation according to a dose response function, i.e., the change of a measurement parameter with dose, which is specific for the individual chemical composition. The polymerization degree can be measured and visualized using T2-weighted magnetic resonance imaging (MRI) [18]. MRI has been mostly applied [19–23] as the reading method of the dosimetric response using relative changes in the MRI contrast parameter T2 (transverse relaxation time). The measured reciprocal R2 = 1/T2 features in most of the gel dosimeters in use a linear dependence on the applied dose D within a certain dose range typically used in radiation therapy (e.g., 0–20 Gy):

$$R2 = R20 + \alpha \, D \tag{1}$$

R20 indicates the transverse relaxation rate of the polymer gel at dose D = 0 Gy. In the following, the measured relaxation rate R2 as a function of the dose is called the dose response ($R2_{(D)}$). The polymer gel response might be characterized by the dose sensitivity defined as the slope α in the linear increase of the relaxation rate R2 with dose:

$$\alpha = \Delta R2/\Delta D \tag{2}$$

The chemicals mostly used for radiation detection by polymerization are based on acrylic monomers, e.g., (a) methacrylic acid (MA) in the gel types methacrylic and ascorbic acid in gelatin initiated by copper (MAGIC) [22] and MAGAT (Methacrylic acid gel and (hydroxymethyl)phosphonium chloride (THPC)) [17]; MA is also used within this manuscript in methacrylic acid, gelatin and the newly added dithiothreitol (MAGADIT); (b) acryl-amide and the co-monomer *N,N*-methylene-bis-acrylamide (BIS), applied, e.g., in the polymer gels: Bis acryl amide nitrogen gel (BANG) [16], polyacrylamide gel (PAG) [21] and PAGAT (PAG and THPC) [17]); (c) *N*-isopropylacrylamide (NIPAM) [24] replacing the extremely toxic acrylamide in PAGAT gels and (d) *N*-vinyl pyrrolidone (NVP) [25], e.g., in the polymer gel dosimeters VIPAR (VIPAR N-vinylpyrrolidone argon) [25], N-vinylpyrrolidone with computed tomography (VIC) [26] and N-vinylpyrrolidone and THPC with inorganic add-on (iVIPET) [27].

Before ionizing radiation the monomer–gelatin–water mixture is usually heated (e.g., 50 °C) above the gelification temperature of the gelatin such that the polymer gel becomes liquid (polymer gel solution) and it may be poured into arbitrarily shaped containers. After cooling below gelification temperature (e.g., about 26 °C) the polymer gel becomes semi-rigid for preserving the multimer or polymer network location and consequently dosimetric spatial information after being irradiated. The gelatin or agarose ingredient is mainly used for the immobilization of the initially produced multimer/polymer after irradiation.

The oxygen scavenging ingredient is added to the polymer gel, usually in millimolarconcentrations, to remove chemically active, dissolved oxygen, which suppresses the radical initiated polymerization. The mostly used oxygen scavenging chemicals are ascorbic acid with initiation by copper sulfate as proposed in the pioneering article by Fong et al. [22]. This type of gel was called MAGIC (methacrylic and ascorbic acid in gelatin initiated by copper). THPC represents a very effective oxygen scavenger and is therefore often used in different types of polymer gels [17].

Polymer gels feature mainly the subsequent advantages with comparison to standard one-dimensional dosimeters, e.g., ionization chambers:

(1) The dose distribution can be read out by three-dimensional or multi slice 2D- imaging techniques within one scanning procedure (3D-dosimetric imaging) and a 3D dose representation might be calculated quickly.

(2) The polymer gels are tissue equivalent and consequently, do not disturb the fluency and the spatial distribution of ionizing radiation. This is relevant especially with reference to a possible multi array 3D-matrix of single point detectors, e.g., ionization chambers, which might absorb parts of the radiation due to, e.g., metal housing. Such a 3D-arrangement might result in a reduction of the measured dose in lower layers due to absorption in the upper layers when being irradiated from the top.

(3) Polymer gels can be designed easily by pouring them in liquid status using arbitrarily shaped containers to cover and simulate different shapes of organs or tissue.

(4) The polymer agglomerates and network structure initiated by the irradiation dose feature only small size ($\cong \mu$m [18]) and thus high resolution comparable to film dosimetry might be obtained in principle. However, the spatial resolution is ultimately limited by the voxel size of the imaging modality, typically about 1 mm^3.

These advantages are especially important for high resolution dosimetric imaging quality control i.e., checking the spatial accuracy of dose delivery in irradiation fields with strong dose gradients on tumor patients. These strong dose fall-offs are especially necessary when there are areas that have to be avoided by irradiation namely risk organs, e.g., the optical nerve in the brain close to tumor areas, which demand the maximum dose.

In quantitative dosimetric imaging usually the dose response function for a certain manufacturing batch of polymer gels is obtained (Equation (1)) for a subset of calibrations samples. Alternatively a reference irradiation with known dose distribution with minimum two measurements of R2i at two different dose levels Di can be evaluated. Thus the two unknown parameters characterizing the dose response: the polymer gel relaxation rate without dose (R20) and the sensitivity α can be evaluated. For relative dosimetry also a region outside the irradiated region (D = 0 Gy) can be evaluated for obtaining R20 without dose (R20).

The difference ΔR2 = R2 − R20 of the locally measured R2 to the background (R20) is according to Equation (1) proportional to the applied dose and thus can be used for quantifying the relative changes in the measured dose distribution (relative dosimetry).

1.2. Actual Status in Polymer Gel Dosimetry with Respect to Dose Rate and Motivation

In the last few years, there has been a growing interest in magnetic resonance imaging based polymer gel dosimetry (MRPD) due to its tissue equivalence and potential to determine quantitatively complex 3D dose distributions with high spatial resolution [19–23,28–33].

Since the proposal and manufacturing of the first polymer gels, there is a continuous process to check individual preparation and evaluation parameters for achieving precision and accuracy for different radiation properties over various experiments [16,34]. Good reproducibility of gel dosimeter needs strict control of ingredient purity, manufacturing, storage and measurement parameters [34,35]. Distinctive factors like gel cooling, gel structure, MRI related artifacts and dose-rate affect the accuracy of the polymer gel [20,34–36]. Other sources of deviations in dose reading are represented by temperature variations of the gel dosimeter during scanning and irradiation [15,17,18,37–39].

Unfortunately, the dosimetric signals of many of the existing polymer gels are dose-rate dependent. This dose-rate impact—among other factors—reduces in many cases the dosimetric accuracy below the clinical standard [40].

It has been shown, that the acrylamide based PAGAT (polyacryl-amide gel and THPC) type gel exhibited only a small dose rate dependence of the sensitivity α (dose rate effect). Consequently, a PAGAT dosimeter passed the quality criterion checks for clinical use [41]. However, the used acrylamide is extremely toxic [24] and features mutagenic and teratogenic risks. Polymer gels using *N*-vinylpyrrolidone (NVP) [25] as a monomer were reported to exhibit a low dose rate dependence [42] but featured significantly reduced sensitivity [15]. Recently modifications of these NVP gels using specific inorganic salts have been proposed ("iVIPET"-gels) for increased sensitivity [27]. These are reported to exhibit higher sensitivities in comparison to the VIPET dosimeter ($\alpha_{iVIPET}/\alpha_{VIPET} \approx 3$) up to $\alpha_{iVIPET} = 0.26$ Gy^{-1} s^{-1}, which is still significantly less than some of the MA based polymer gels (e.g., $\alpha_{MAGAT} \cong 2$ Gy^{-1} s^{-1}) [39]. These modified NVP polymer gels (iVIPET) are very promising also for reduced dose rate dependence. However, NVP is categorized (level 2) as likely carcinogenic in humans in addition to inhalation, skin and eye harms and the very effective but toxic THPC is used as an oxygen scavenger.

In parallel, several dosimetric scanning methods using radiation induced polymerization have been commissioned [43–46]. In particular, little research has been conducted, to our knowledge, to show quantitative proposals for the improvements of the dose rate dependence of the most sensitive polymer gels based on MA [17,39]. The dose rate might vary for different points in a treatment volume even in single beam irradiation by an order of magnitude [20,35] due to the lateral dose distribution created by collimation. In the case of intensity modulated radiation therapy (IMRT) and volumetric modulated arc therapy (VMAT), characterized by varying collimation beam areas during irradiation, the lateral dose rate variation of a single beam might even be changed afterwards by the time course of beam directions. Even the applied dose rates might vary from step to step. However, the dose-rate dependence has not been documented for many gel dosimeter systems [17]. Previous research showed that the sensitivity of polymer gels was enhanced for a lower dose rate, e.g., 0.6 Gy/min with comparison to the higher dose rates ($\approx$2–5 Gy/min) as applied often in standard clinical radiation therapy on a linear accelerator (LINAC) [20,47]. For instance, methacrylic acid gel and THPC (MAGAT) gels, which were found to be most sensitive to irradiation, showed maximum dose rate dependence [39]. The first type of a normoxic polymer gel, which was manufactured using ascorbic acid as an oxygen scavenger (MAGIC), had also been shown to feature this dose-rate effect [17,47]. Adding oxygen scavengers to polymer gels avoids the application of nitrogen or other passive gases as argon [25,42] to blow out the oxygen, which suppresses the radical initiated polymerization in polymer gels [17]. The most efficient scavenger (THPC) in MAGAT-type gels features severe toxicity and danger for acid burns. We suspected a possible impact of the oxygen scavenger concentration on the dose-rate effect. We also assumed dithiothreitol (dithio, IUPAC name: 1,4-bis (sulfanyl) butane-2.3-diol) due to its low redox potential (-0.33 V) [48] in comparison to ascorbic acid ($E_h = -0.081$ V) [49] and usage in biochemistry [50] for preventing oxidation of thiol groups, being a candidate for a new oxygen scavenger in polymer gel dosimetry with less impact on radical initiated polymerization. High concentrations of oxygen scavengers are capable of reducing the dose-rate effect, but also reduce the sensitivity of the polymer gel to dose [47]. Dithio features the advantage of non-toxic nature. After having checked the principle effectiveness of this new oxygen scavenger with reference to radiation initiated polymerization in MA based polymer gels, we report here about the basic properties of this new polymer gel type (MAGADIT) with three different concentration levels of dithio (2, 10 and 50 mmol/kg).

First of all, the impact of the oxygen scavenger concentration on sensitivity α is reported in Section 3.1. The dose-rate dependence of the sensitivity was investigated for all of the different MAGADIT gels, particularly in the range of clinically applied high dose rates as, e.g., in FFF irradiations (5–12 Gy/min), but also the low dose rate range down to $\dot{D}_{min} = 0.6$ Gy/min. Absolute and sensitivity normalized dose rate dependence $\Delta\alpha/\dot{D}$ were evaluated comparing the performance of the three different polymer gel implementations (Section 3.2). In Section 3.3 the reproducibility (σ_r) i.e., the relative standard deviation of the dose response, measured as relaxation rate $R2_{(D)}$ for a set of different polymer gel dosimeters and relative accuracy a_r defined as normalized deviation between applied and measured dose are presented for all of the different dithio concentration levels. Finally, the impact of potential dose rate effects, i.e., the influence of varying dose rates across the delivered dose distribution on the geometric accuracy is shown for a small 5×10 mm^2 10 MV photon field on a clinically used LINAC. Two protocols at different dose rates applied in the center are compared: a) at standard IMRT dose rate of $\dot{D} = 4.2$ Gy/min and b) that of a FFF high dose rate ($\dot{D} = 15$ Gy/min). The results on geometric accuracy, evaluated as deviation of the width of the measured dose profile from the true one, are compared for the FFF and standard dose rate protocol in Section 3.4.

2. Materials and Methods

2.1. Gel Manufacturing

The polymer gel was prepared at normal atmospheric conditions in a chemical laboratory with standard equipment [47]. The MAGADIT preparation involved methacrylic acid, porcine gelatin,

(300 Bloom, SIGMA-ALDRICH, Vienna, Austria) and dithiothreitol (Lactan Chemikalien und Laborger., Graz, Austria) as an oxygen scavenger. Dithiothreitol (dithio) is also known as Cleland's reagent and used in biochemistry as a reducing and protective reagent for SH groups [51]. Table 1 indicates the gel composition for three different concentrations of dithio: DIT1, DIT2 and DIT3.

Table 1. MAGADIT polymer gel composition with three different oxygen scavenger concentrations. The relative weight of the chemicals is indicated (*w/w*) unless otherwise mentioned.

Ingredients	DIT 1	DIT 2	DIT 3
Distilled water	82%	82%	82%
Gelatin	10%	10%	10%
Methacrylic acid	8%	8%	8%
Dithio	2 mmol/kg	10 mmol/kg	50 mmol/kg

The gelatin was added to distilled water at around 50 °C within a procedure of steady magnetic stirring and moderate mixing. The dissolving process required approximately 50 minutes until a transparent solution was obtained. Then, MA was added to the solution accompanied by continuous mixing for 5 minutes. Lastly, dithio was added to the gel solution followed by thorough mixing using an automated high-speed propeller machine. High speed mixing was necessary for direct chemical contact between the scavenger molecules and oxygen. After mixing the entire mixture including the monomer MA, gelatin, dithio and water was still in the liquid state due to the temperature above gelification (T $\approx$ 26 °C). However, the entire solution turned into a cloudy froth-type shape. The foamy polymer gel then was moved to a water bath at about 54 °C to guarantee for the settling of the foam until a clear solution was achieved. Polyethylene terephthalate (PET) containers of about 34 mL volume (wall thickness $\approx$ 1 mm) were filled with the liquid gel solution. An additional protection for oxygen penetration at cap was provided using a layer of oxygen dense Saran^{TM} film (Merkur-market, Vienna, Austria) at the top opening side. All the filled containers were placed at room temperature in the dark to cool down in a position of topside upward to collect left oxygen bubbles at the top side far distant from the irradiation target region at the bottom. After cooling down to room temperature, the gel containers were positioned in a separate big glass container filled with an oxygen scavenger solution (2–5 mmol/L ascorbic acid and 100 µmol/L copper-sulfate) to avoid oxygen penetration in addition.

We add some information on our practical experience and aspects related to manufacturing and storage of polymer gels in general and in specific for the MAGADIT species:

(a) Retaining oxygen scavenging is crucial for a stable dose response. Especially small containers have to be oxygen sealed. Standard plastic material like polyethylene at mm thickness is not sufficient for avoiding oxygen penetration through the plastic wall container material. Otherwise, the in-diffusing oxygen saturates after some time the oxygen scavenging capacity in any scavenger add-on. The material best suited as the container material with photon absorption characteristics close to tissue—in our experience—is represented by a copolymer of acrylonitrile and methyl methacrylate (BAREX^R). However, this material is hardly available on the market anymore. Therefore other protection modalities, e.g., thick container walls or additional closed glass containers, being removed just before irradiation, are proposed to be used. For these reasons we stored the MAGADIT polymer gel containers about 12–24 hours after manufacturing in an oxygen impermeable glass container with a metal cap filled with an aqueous solution of an oxygen scavenger (e.g., 2–5 mmol/L ascorbic acid and 100 µmol/L copper-sulfate). Such a procedure is especially recommended if the normoxic dosimeter container is made of plastics and features small wall thickness.

(b) The time and temperature course between preparation, irradiation and evaluation might in principle have an impact on the absolute values of the dose response [17,18,33,52]. We, therefore, advise (see also [20]) for accurate absolute dosimetry to prepare, irradiate, store (incl. time after irradiation) and evaluate the calibration gels in the same way as the 3D-dosimeters for quantitative evaluation. Thus eventual changes in dose response due to co-factors are minimized.

(c) From our practical experience, the polymerization in the clear MAGADIT polymer gels can be visualized at medium dose levels within linear range as turbid areas within 1 hour after irradiation without visual changes after about 3 hours. After measuring (we advise, e.g., up to 5 days after irradiation) and storing in the fridge these polymer gels have not changed their visual appearance (turbid area size) even after 3 months. We, therefore, think that these types of polymer gels (MAGADIT) still can be quantitatively accurately evaluated after this time (month), if the calibration and evaluation polymer gels are treated and stored in the same way. For relative dosimetry they might be used even after longer adequate storage times in a fridge at 4–8 °C. Actually the FFF photon field was applied to the MAGADIT D3 dosimeter about 12 months after preparation. The stability of the linear dose response might be quantitatively proved in further investigations related to the practical aspects of applications (see also Section 4.2. *Limitations of the Study and Possible Future Investigations,* in the chapter discussion).

2.2. Irradiation and MR-Dosimetric Evaluation

Two different types of irradiation protocols were performed: (a) A low energy 200 kV protocol on a preclinical irradiation unit for the investigation of the basic properties of the MAGADIT dosimeter and best control of the dose rate and (b) a high energy (10 MV), close to clinical IMRT irradiation protocol with two different dose rates applying an (a): standard dose rate (FF) ($\cong$4.2 Gy/min) and (b): a high dose rate, FFF-typical protocol ($\cong$15 Gy/min). A 5×10 mm^2 small photon field on a clinically used LINAC was utilized for demonstrating the impact of the LINAC accelerator dose rate, applied to the center region, on the measured lateral spatial dose distribution determined by the collimation set up in the specific irradiation set-up. These different protocols are described subsequently:

2.2.1. Low Energy 200 kV-Protocol (Yxlon) for Basic Properties of MAGADIT

A preclinical research, X-ray machine (YXLON International GmbH, Hamburg, Germany, with X-ray tube Y.TU/320-D03, E = 200 kV, filtered with 3 mm Al and 3 mm Be) was utilized to irradiate MAGADIT dosimeters [53]. This machine offered an improved availability and simple adjustment of dose rate at low drop-out probability (interlocking) and high dose levels within short administration time. LINACs for clinical therapy at high photon energy might experience short deficiencies and drop-out in continuous dose delivery, which results in prolonged dose delivery time. Thus the dose rate might be effectively reduced and not constant in time. No significant difference in the dose response of polymer gels is expected between low and high energy (>1 MeV) LINAC photons [54]. All gel containers were kept in the X-ray machine room for at least 1 hour before irradiation in order to achieve nearly uniform temperature for radiation. The first irradiations (see Figure 1) were performed for calibration and checking the sensitivity and linear dose range at $\dot{D}$ = 2 Gy/min. A second gel set was used for measuring the dose-rate impact on the different types of gel: DIT1, DIT2 and DIT3. The last set served for checking the reproducibility and accuracy.

For the systematic dose measurements, the gel containers were placed in a Perspex®holder with different holes in a position such that the bottom parts of the containers were directed towards the radiation source (Figure 2). After the irradiation, all the containers were moved and retained for 36 hours in the MR room for temperature equilibration and subsequent MR scanning.

Calibration

Gel containers for calibrations were irradiated at YXLON with 200 kV x-rays. A set of three different types of gels (DIT1, DIT2 and DIT3) was irradiated one day after manufacturing. For calibration, the irradiations of the gel dosimeters were performed such that the bottom surfaces of the gel containers were positioned perpendicular to the radiation beam (Figure 2). All the applied dose levels and dose rates were cross-checked with a calibrated ionization chamber (type 31013, PTW-Freiburg, Freiburg, Germany) with a sensitive volume of 0.3 cm^3. For accurate dose measurement, the ionization chamber was placed in the same gel container type filled with water at the same reference level (at 1 cm depth from bottom side), as used for the gel evaluation.

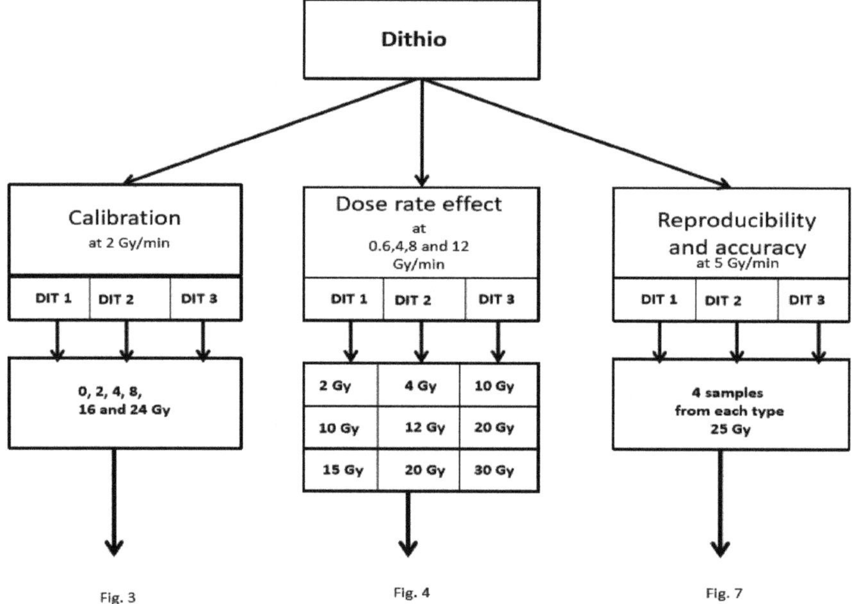

Figure 1. Irradiation scheme procedure for the measurements of the basic dosimetric characteristics of the three types of MAGADIT polymer gels: dose response, calibration, reproducibility and accuracy. In addition, the dependence of the dose response on the dose rate was evaluated.

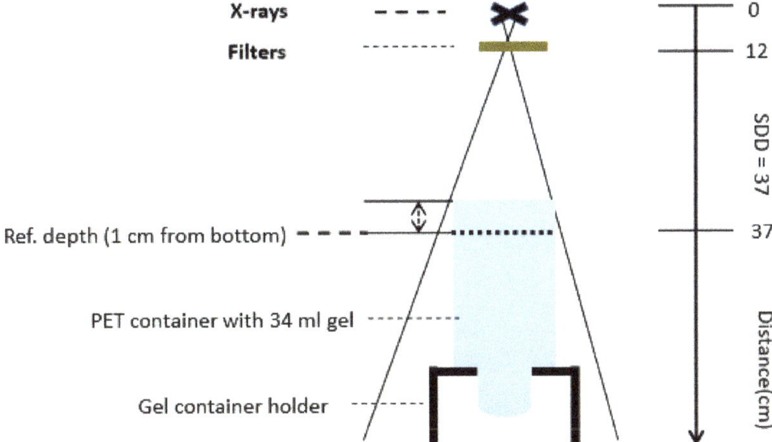

Figure 2. Gel irradiation set-up (non-scaled) indicating the position of the gel container with reference to the radiation source without collimation. The source to detector distance (SDD) was adjusted to 37 cm for the calibration set-up. For investigations of the dose-rate effect the SDD and tube current was adjusted differently (SDD = 23, 25, 27 and 65 cm). The reference position at 1 cm depth distant from the bottom side in the container is indicated.

The various absolute dose levels of 2, 4, 8, 16 and 24 Gy ($\dot{D}$ = 2 Gy/min) were delivered to the gel samples for obtaining the calibration curves for each type of dosimeter. The polymer gels were positioned at a source to surface (gel container bottom side) distance of about 36 cm. A current of 10.4 mA resulted in a dose rate of 2 Gy/min at the prefixed distance.

Reproducibility and Accuracy

We defined reproducibility here as the variation of outcome (dose response: relaxation rate $R2 = 1/T2$) for different samples of the same type of gel, manufactured in the same batch within one MRI measurement sequence. For the quality check on reproducibility, four samples of each type of gel (DIT1, DIT2 and DIT3) of the same batch were irradiated at a dose level of 25 Gy at 5 Gy/min (Figure 1 right). The set of gels used for calibration was not used for the reproducibility test. Reproducibility (σ_r) was quantitatively evaluated using the standard deviation in the relaxation rate (σ_{R2}) and the average value (μ_{R2}) of the measured relaxation rate, obtained from an ROI analysis (about 60–100 pixel) for four different samples (Equation (3)):

$$\sigma_r = \frac{\sigma_{R2}}{\mu_{R2}} \tag{3}$$

For accuracy, four samples from the manufacturing set for precision were used, but the quantitative dose levels were calculated using the calibration curve obtained from the fifth polymer gel out of the same manufacturing batch, irradiated at 5 Gy/min and a reference sample without dose (D = 0 Gy). The calibration curve was therefore obtained for the same batch of samples but an independent calibration sample set. We thus used only two reference points for calibration similar to an evaluation with subtracted background R2–R20 imaging, with R20 obtained from regions outside the irradiated area at D = 0 Gy. This procedure is often performed in simple dosimetric imaging with a limited number of calibration reference data.

In addition we aimed to demonstrate also the consequences of evaluating dose levels from calibration data with different dose rates (here 2 Gy/min instead of correct 5 Gy/min). Therefore, we also used calibration data from gel set 1 (Figure 1) obtained at $\dot{D}=2$ Gy/min and evaluated the corresponding dose for this calibration data. The corresponding differences in evaluated dose D from those actually applied are listed in Table 3. The relative accuracy (a_r) was defined as the difference (ΔD) of the actual dose (D_a), measured by the ionization chamber in the same reference position as the gel in the container and the calculated dose (D_m) from the calibration curve, divided by the actual dose.

$$a_r = \frac{D_m - D_{ar}}{D_a} \tag{4}$$

The results for reproducibility and accuracy are shown in Table 3 and Figure 6.

Dose Rate Dependence

Polymer gels from the same manufacturing batch as used for calibration and reproducibility served for the investigations on the dose-rate impact. Four different dose rates ($\dot{D}$) ranging from $\dot{D}_{min} = 0.6$ Gy/min to $\dot{D}_{max} = 12$ Gy/min with three different dose levels were applied. The dose rates were adjusted either by varying the current of the x-ray tube or by changing the source detector distance (SDD). Varying both at the same time allowed for the broad range of dose rates. Three different (low, medium and high) dose levels and a sample with zero dose at all were investigated for each dosimeter type assuming linear dose range and sensitivity to a maximum dose of 15, 20 and 30 Gy respectively (Figure 4). The absolute reference dose was delivered at 1 cm distance from the bottom of the gel container.

MRI Measurements on the Basic Properties

All the gel containers for Yxlon irradiation were scanned at a minimum of 36 hours after the irradiation using a high-field 7 T MR scanner (Siemens, Magnetom 7T, Erlangen, Germany) installed at the High field MR Center Vienna. The small gel containers were kept in the scanning room at least two hours before the measurements to reduce any variations on T2 values during measurements due to temperature imbalance. For the purpose of comparability, the reference (non-irradiated) polymer gels ($D = 0$ Gy) were controlled throughout the process in the same way as the irradiated ones. Using a head coil, parameter selective T2 mapping was performed on all gels. T2-weighted imaging was applied with a multiple spin echo sequence of 15 echoes, echo time spacing TE = 10 ms (10, ..., 150 ms) for accurate measurement of fast decaying magnetization in the high dose range, repetition time TR = 8.4 s, field of view (FOV) = 100 mm, matrix size (Mtx) = 128 × 128, slice thickness (slth) = 5 mm, number of slices (nrsl) = 5. The analysis for quantitative T2 evaluation was performed using a mono-exponential fitting tool, which was developed for ImageJ (National Institute of Health, USA) by Karl Schmidt (Harvard, USA). In order to get more accurate T2-results, the first echo was removed before processing as proposed in the standard for highly accurate T2-measurements [47,55].

2.2.2. Dosimetry of a Small Sized 5×10 mm^2 FFF Field of a LINAC Used for Clinical Radiation therapy

Using a linear accelerator (LINAC, Elekta Instrument AB Stockholm, Sweden) for patient therapy a flattening filter free (FFF) photon (10 MV) radiation field of small asymmetric size (5 mm × 10 mm) was applied to (a) polymer gel D3 and (b) Gafchromic®EBT3 film used as reference for high resolution dosimetry (SSD = 90 cm and ref. depth = 24 mm). Two different dose rates were applied in order to simulate and demonstrate the impact of applying very high dose rates as typical for FFF irradiation vs. standard high dose rate on the geometric dosimetric accuracy:

(a) High dose rate ($\dot{D} = 15.3$ Gy/min) protocol as used for clinical FFF-irradiations and;

(b) Lower dose rate ($\dot{D} = 4.2$ Gy/min) protocol with the same geometric configuration.

Polymer gels of type D3 in the small containers (outer diameter ≈ 30 mm) were left from the same batch as for the investigations on the basic properties and irradiated. They did not exhibit any color change or polymerization after being stored in a fridge in an oxygen-sealed water container for about 12 months. After thermal equilibration two of these were exposed to the small sized $5 \times 10\ \mathrm{mm}^2$ radiation field. A high resolution MR protocol, slightly different from the investigations on the basic properties, was used in order to investigate the dose profiles of the small radiation field for the two different dose rates (FOV: 30 mm Mtx: 192×192; TR = 3960 ms; TE = 15, 30, ..., 300 ms and voxel size: $0.156 \times 0.156 \times 1\ \mathrm{mm}^3$). The first echo was removed in the evaluation of T2-relaxation maps for avoiding non-exponential decay artifacts due to stimulated echoes [55]. The dose response was calibrated using linear interpolation and two reference regions of interest (ROIs) in the polymer gel of dose levels known from the treatment planning system (TPS). The TPS represents a software interface for calculation of an expected dose distribution from radiation parameters, its visualization and comparison of the planned dose distribution with reference to the object, e.g., patient in radiation therapy. These ROIs have been: a circular rim outside of the central $5 \times 10\ \mathrm{mm}^2$ LINAC photon field for reference dose (D1 = 0 Gy) and a small ROI in the center of the field (D2 = 60 Gy). Dose images were obtained from R2 maps (images of the reciprocal T2) using this calibration data according to Equation (1) [44] assuming a linear dose response (Figures 3 and 4) between these reference dose levels (D1 = 0 Gy and D2 = 60 Gy).

Film dosimetric evaluations were performed with radiochromic films (Gafchromic®EBT3). They were based on the evaluation of the optical density using all of the three channels and six calibration data points up to D_{max} = 6 Gy [56]. For the evaluation of the impact of dose rate on the geometry of the profile, MR and film dosimetric profiles were normalized (relative dosimetry) to the maximum dose evaluated as the mean value in the center of the $5 \times 10\ \mathrm{mm}^2$ radiation field such, that the dose profiles from film and polymer gel can be compared.

3. Results

3.1. The Impact of the Oxygen Scavenger Concentration on Polymer Gel Sensitivity

Gels from all three types were irradiated for calibration with 200 kV of x-rays up to 24 Gy dose at a dose rate of 2 Gy/min. Thereafter, MRI was applied to these irradiated dosimeters samples together with non-irradiated polymer gels as a reference. The gel with the lowest oxygen scavenger concentration (DIT1) featured a relaxation rate R2 = 10 s^{-1} at the highest dose level applied (24 Gy), representing high sensitivity (Figure 3).

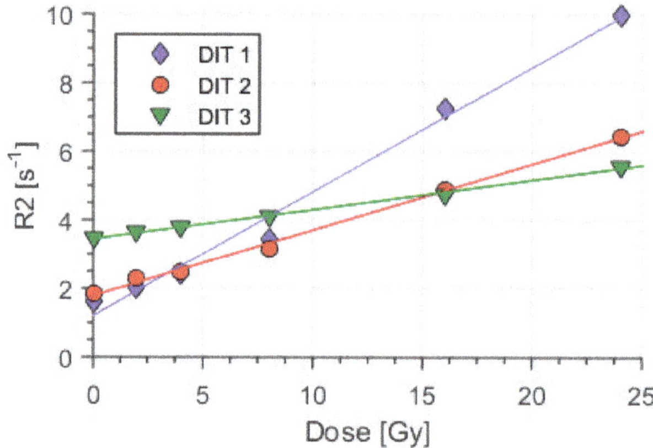

Figure 3. Calibration gel (Gel_ref, $\dot{D}_{\text{stand}}$=2 Gy/min) dose response with three concentrations of dithio oxygen scavenger. DIT1 featured minimum (c_d = 2 mmol/kg) and DIT3 (c_d = 50 mmol/kg) maximum concentration. A reduction in dose response R2 can be seen from low scavenger concentration (DIT1) to the highest scavenger concentration (DIT3) for dose levels D > 7.5 Gy. This drop was highest for the high dose level (D = 24 Gy). Note the high offset R2 values below 7.5 Gy for DIT3. Error bars were acquired as the standard deviation of R2 in the selected regions of interest (ROI). These were generally smaller than the symbol for measurement data.

Further enhancing the concentration levels of oxygen scavenger resulted in lowered relaxation rates in MRI for DIT2: $R2_{\text{DIT2}}$ = 6.5 s^{-1} and for DIT3: $R2_{\text{DIT3}}$ = 5.6 s^{-1} (D = 24 Gy). The highest concentration of oxygen scavenger decreased the dose sensitivity measured as the slope of the relaxation rate R2 with dose ($\alpha = \Delta R2/\Delta D$; Figure 3). In general, increasing the concentration of oxygen scavengers decreased the sensitivity of the polymer gels (Figure 3).

3.2. Dose Rate Dependence of the Dose Response in Dithio Gel Samples

The dose-rate effect was quantitatively evaluated using the sensitivity of the polymer gel measured as the slope ($\alpha = \Delta R2/\Delta D$) in dependence of the dose rate varying the dithio concentration as a control parameter. At first, the impact of the dose rate on the absolute R2 response at certain discrete dose levels was evaluated (Figure 4). The reference gel with the lowest concentration, DIT1, showed a significant decrease in the dose response from low to high dose rates at D = 15 Gy: $R2_{0.6\text{Gy/min}}$ = 9.8 s^{-1} and $R2_{12\text{Gy/min}}$ = 4.9 s^{-1}. This reduction corresponds to about 50% relative change.

We further increased the scavenger concentration (c_d = 10 mmol/kg) in DIT2 as compared to DIT1 (c_d = 2 mmol/kg). All of the gel containers of DIT2 were irradiated with the same dose rates as in DIT1, but low, medium and high dose ranges were set differently due to the lower sensitivity of DIT2 (Figure 3). The results of the MRI showed, that the R2 values were reduced for all dose rates as compared to DIT1 (Figure 4b). Concerning the dose-rate impact, at 20 Gy, R2 dropped from about 8.5 to 4.2 s^{-1} while changing the dose rate from 0.6 to 12 Gy/min. The relative dose-response reduction with increasing dose rate reported here amounted to about 63.6%. The dose-rate effect measured as a change of sensitivity (slope) with a difference in dose rate reduced approaching the higher dose rates, typical for flattening filter free irradiation (Figure 5 left). DIT2 resulted almost in a similar percentage drop in the sensitivity, as in DIT1 in the high dose rate region.

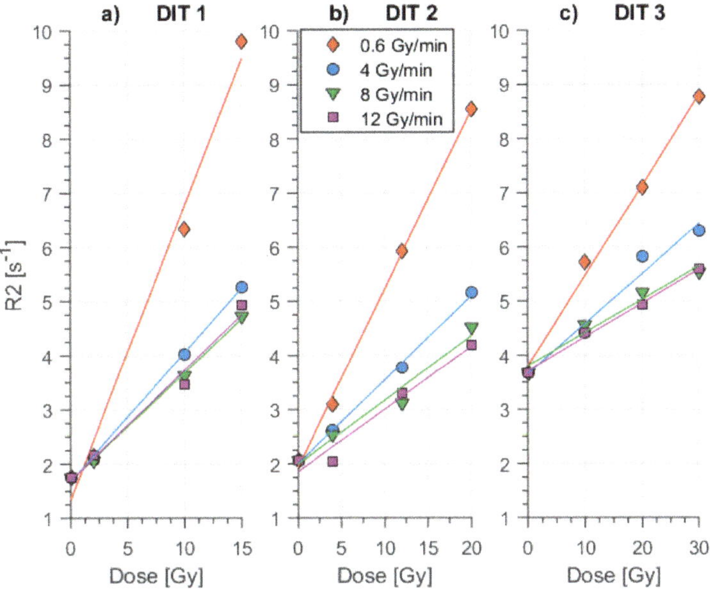

Figure 4. Dithio polymer gel dose response with three concentrations of oxygen scavenger for various dose rates ranging up to $\dot{D}_{max} = 12$ Gy/min. Error bars were acquired from the standard deviation of R2 in ROI. Commonly these were smaller than the measurement data symbols. Results of linear regression analysis are plotted. **a (left)**: DIT1 dose response for the lowest concentration of oxygen scavenger. **b (middle)**: Dose rate dependence of DIT2. **c (right)**: DIT3 dose response for highest concentration (50 mmol/kg). The relaxation rate R2 at the high dose level (e.g., D = 15 Gy) was strongly reduced for all types of gel with higher dose rates. The R2 offset value ($R2_{D = 0Gy}$) was quite prominent in DIT3 for all dose rates. Please note, that the variation in slope with high dose rates $\dot{D} \geq 8$ Gy/min almost disappeared for all types.

Finally, we investigated DIT3 for the dose rate-effect using the same dose rates as DIT1 and DIT2. Due to the low sensitivity of DIT3 (Figure 3), the low, medium and high-dose levels were chosen higher than for DIT1 and DIT2. The MRI results showed that the relaxation rate R2 was reduced by about 64% while changing dose rates from 0.6 Gy/min to 12 Gy/min at 30 Gy (Figure 4c). Here, the drop in sensitivity was nearly the same as in the case of DIT1 and DIT2. Please note that the changes in the sensitivities reduced at higher dose rates $\dot{D} \geq 4$ Gy/min and finally disappeared within errors at a dose rate: $\dot{D} > 7$ Gy/min.

The dose sensitivities of the different polymer gel types were evaluated as the slope from Figure 4 using linear regression analysis of the data. The results are summarized in Table 2 and visualized in Figure 5.

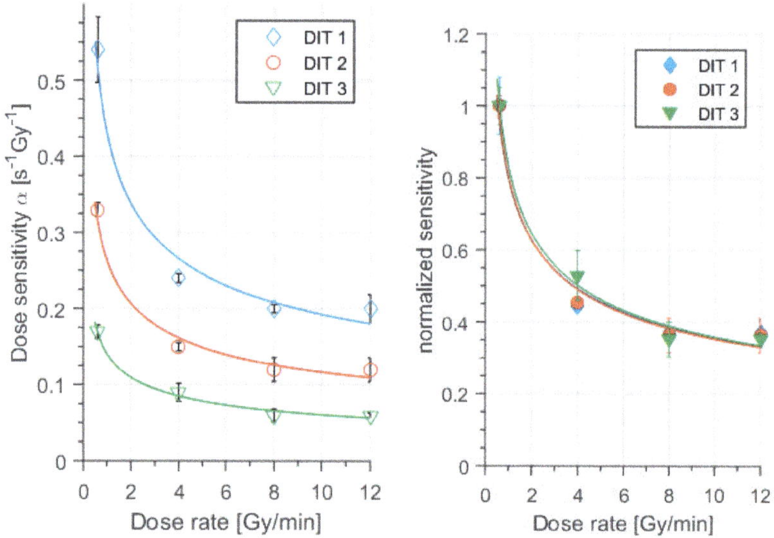

Figure 5. Absolute (**left**) and normalized (**right**) sensitivity versus dose rate for different dithio concentrations (DIT1, DIT2 and DIT3). Note the substantial variation of sensitivity with dose rate, particularly in the region between 0.6 Gy/min to 4 Gy/min, for all polymer gels, but almost vanishing for dose rates $\dot{D} > 4$ Gy/min. A significant reduction in dose rate sensitivity at higher dose rates was observed. The regression analysis result was added using a power law fitting curve.

Table 2. Dose sensitivity variations with dose rates for different gels. The relative change in sensitivity apparently disappeared (within errors) for the high dose rate range between $\dot{D}$ = 8 Gy/min and $\dot{D}$ = 12 Gy/min for all types of MAGADIT.

Dose Rate (Gy/min)	Dose Sensitivity α (s^{-1} Gy^{-1})		
	DIT1	DIT2	DIT3
0.6	0.54 ± 0.04	0.33 ± 0.01	0.17 ± 0.01
4	0.24 ± 0.01	0.15 ± 0.01	0.09 ± 0.01
8	0.20 ± 0.01	0.12 ± 0.02	0.06 ± 0.01
12	0.20 ± 0.02	0.12 ± 0.02	0.06 ± 0.00

For quantitative dose calculations and relative errors due to the dose rate effect the relative change of sensitivities with dose rate normalized to the specific sensitivity of each gel is relevant. Therefore we normalized the dose sensitivities for each type of gel to the dose sensitivity at the lowest dose rate ($\dot{D}$ = 0.6 Gy/min): $n\alpha = \alpha/\alpha_{0.6\,\text{Gy/min}}$ (Figure 5 right). The reduction of sensitivity with dose rate appeared to be almost identical for all of the different types of polymer gel (Figure 5 right).

3.3. Reproducibility and Accuracy

Table 3 lists the reproducibility and accuracy results of different gel types.

Using Equations (3) and (4), the calculated reproducibility for DIT1, DIT2 and DIT3 were equal to 3.3%, 6% and 3.6%, respectively.

Table 3. Summary of the evaluations on reproducibility σ_r and accuracy a_r, defined as the relative deviation of measured dose values from the actually applied dose. Reproducibility and accuracy were evaluated for the highest dose level (D = 25 Gy). Note the high errors in the accuracy ($a_r \cong 16\%$–20%), if the calibration is performed at a dose rate ($\dot{D}_{cal}$ = 2 Gy/min), different from the accuracy irradiations ($\dot{D}_{dos}$ = 5 Gy/min).

Gel Type	Relative Accuracy a_r (%)		Reproducibility σ_r (%) at 5 Gy/min	Average R2 of 4 Samples (s^{-1}) and Standard Deviation
	5 Gy/min	2 Gy/min		
DIT1	3.5	20.2	3.3	8.5 ± 0.28
DIT2	7.4	23.6	6	5.5 ± 0.33
DIT3	7.9	16.0	3.6	5.3 ± 0.19

The evaluation of accuracy resulted in significantly improved values, when calibration and accuracy irradiations were performed at the same dose rates (3.5%, 7.4% and 7.9% respectively). We found significantly worse values in the accuracy, when calibration and accuracy experiments were practiced at different dose rates (2 and 5 Gy/min) for DIT1, DIT2 and DIT3 (20.2%, 23.6% and 16.0% respectively).

An overview of the results for the different types of polymer gels (DIT1, DIT2 and DIT3) is shown in Figure 6.

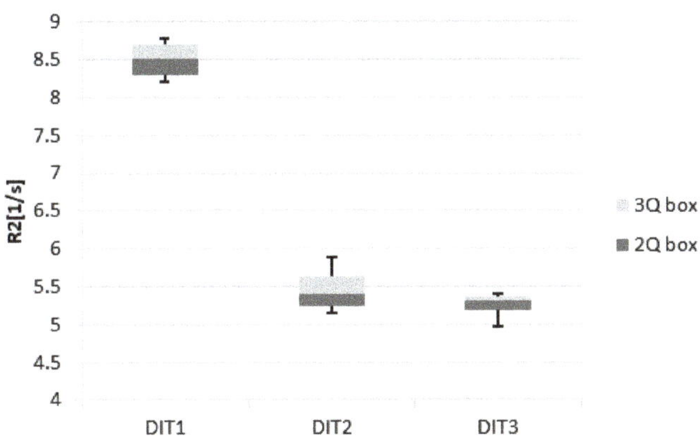

Figure 6. Box plots of the dose response R2 for the evaluation of reproducibility of the three different types of gel dosimeters. For each MAGADIT type, four samples were irradiated to 25 Gy at 5 Gy/min. Quartiles, median levels and whiskers for minimum and maximum values are also indicated.

3.4. Dosimetry of a Small Sized 5 × 10 mm^2 FFF Field of a LINAC used for Clinical Radiation Therapy

Two dose profiles were obtained for:

(a) The FFF type irradiation protocol at a dose rate of 15.3 Gy/min (FFF) and;
(b) A standard high dose rate (4 Gy/min) protocol (FF).

These were compared to the results of the film dosimetry assumed to represent the best reference for high resolution dosimetry after normalization to the dose levels in the same region of interest in the center of the radiation field. The impact of the dose rate on the dosimetric geometric accuracy, as evaluated by the dose profiles along the 10 mm axis, is shown in Figure 7.

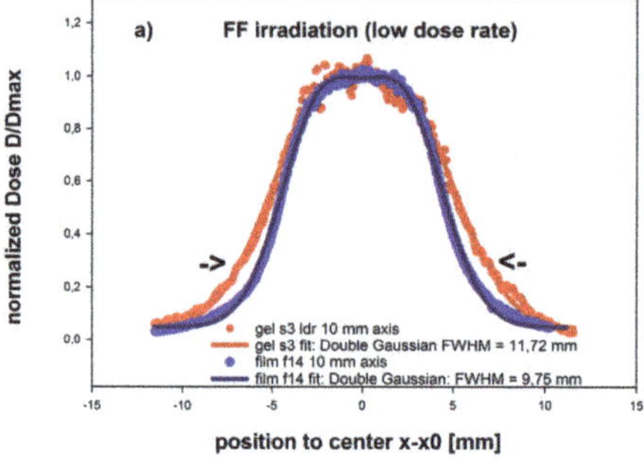

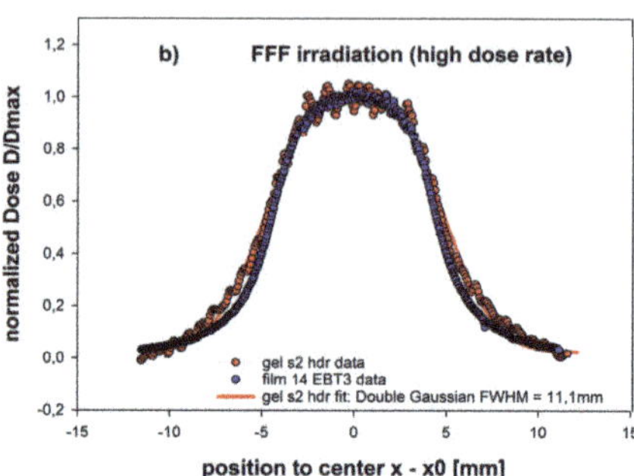

Figure 7. Lateral dose profiles along the 10 mm axis of a small sized 10 MV linear accelerator (LINAC) photon field for the MAGADIT polymer gel DIT3, applied at a dose rate of (**a**) FF: 4.2 Gy/min (center of field) and (**b**) FFF: 15.3 Gy/min (center of field). Dose profiles from the film as references are indicated in blue color. Note the small but significant geometric deviations in the profiles between MAGADIT and film in the flanks of the dose profiles at about 25% isodose levels (see arrows) for the relatively low incident dose rate in (a) FF and the significantly reduced deviations in (b) FFF for the high dose rate. The width of the measured dose profiles is evaluated as the full width at half maximum height (FWHM) and indicated in legend for the film and MAGADIT polymer gels separately.

The dose and dose rates along the lateral profile varied proportionally and significantly in the same way due to the same geometric application. Thereby consequences of a reduced relative dose rate dependence at the higher dose rate range as typical for FFF applications were demonstrated. In both cases, the dose rate effect on the spatial accuracy of the results could be detected. For the low dose rate of 4.2 Gy/min applied in the center, the increased polymer gel sensitivity at 25% isodose line, corresponding to the reduced rate ($\dot{D}$ = 1.05 Gy/min), results in an increased relative dose response, which appeared geometrically in a widening of the profile and a lateral shift of the flank (see arrows in Figure 7) at that position by about 1.5 mm (15% of 10 mm radiation field). The full width at half maximum height (FWHM) obtained from the overall profile regression analysis was increased to 11.72 mm instead of 9.75 mm as obtained from the film measurements, assumed as the "gold standard" for high resolution dosimetry.

When applying the high dose rates typical for FFF at 15.3 Gy/min this shift in the flanks (relevant, e.g., for quality control using gamma criterion) was significantly reduced to about 0.8 mm.

4. Discussion

4.1. MAGADIT in Comparison to Existing Methacrylic Acid Based Polymer Gels (MAG)

This study reported on the main fundamental characteristics of a methacrylic acid based polymer gel (MA) using a new oxygen scavenger (dithio). The novel MAGADIT gels showed promising results and an improved performance regarding the dose rate dependence in the high dose rate range, a well-known issue observed in MAGAT and MAGIC type of polymer gels. This dose-rate effect was one of the key factors in limiting the dosimetric accuracy compared to other dosimetric systems, especially in complex dose distributions with steep dose gradients. Most of the investigations on the dose-rate effect and the impact of oxygen scavenger concentration on sensitivity have been performed on the very sensitive MAGAT-type of gels using THPC, an oxygen scavenger of toxic nature [20,35,39,52,57,58].

Dithio features a redox potential E_h = −0.33 V [48] and thus represents a stronger reduction agent than ascorbic acid (E_h = −0.081 V) [46]. It is used as Cleland's reagent in biochemistry due to its oxygen protective behavior for avoiding disulfide bonds in cysteine based proteins [51]. Though it features some hazard warnings (H302, H315, H319 and H335 according to EC regulation (No 1272/2008) it is classified as non-toxic in contrast to the very effective—but skin contaminative—oxygen scavenger THPC, used in MAGAT gels. However, we would like to indicate here, that also the monomer (MA) used in the MAGADIT gels at 8% concentration, is classified as toxic in contact with skin and demands for corresponding additional protective, storage and administrative efforts. However, MA does not feature mutagenic (H340) and teratogenic risks (H361f) compared to the toxic acrylamide as used in the relatively dose rate tolerant PAGAT gel. Non-toxicity and reduced harm in manufacturing are subject to continuous improvements in polymer gel manufacturing [24]. Recently Abtahi [59] proposed to replace the extremely toxic acrylamide by the less harmful 2-acrylamido-2-methyl-propane sulfonic acid (AMPS); the sensitivity of this new polymer gel was reported to be an order of magnitude less than that for MAGAT (methacrylic acid gel and THPC) gels using methacrylic acid (MA). Future variations might use these other monomers along with dithio.

The basic properties of the novel MAGADIT have been analyzed with respect to: sensitivity, linearity, precision and accuracy as well as dependence on dose rate. To evaluate the best performance, especially regarding dose rate independency, all tests were performed with three different levels of dithio concentration.

The linear change of the dose response (Figures 3 and 4) with dose demonstrated the efficient oxygen scavenging in MA based polymer gels even at low concentrations (2 mmol/kg) of Cleland's reagent combined with the high sensitivity of $\Delta R2/\Delta D = 0.54$ Gy^{-1} s^{-1} ($\dot{D} = 0.6$ Gy/min). However, the sensitivity reduced (Figures 3–5) with higher dithio concentrations, similar to the reports for other oxygen scavengers [39,47] down to about $\Delta R2/\Delta D = 0.17$ /Gy^{-1} s^{-1} ($\dot{D} = 0.6$ Gy/min) at highest dithio concentration (50 mmol/kg). This impact of oxygen scavenger concentration on sensitivity is generally observed. We attributed this observation to a capacity of strong reducing agents, as oxygen scavengers usually are, for chemical interaction with molecules carrying unpaired electrons by electron release. The number of polymerization initiating radicals was reduced within this model. Even higher sensitivities might be obtained for even less concentrations of dithio.

In contrast to MAGIC type polymer gels using ascorbic acid [27] an increase in relaxation rate offset (at dose D = 0 Gy) with oxygen scavenger concentration was observed (Figure 3) from $R20_{Dit1} \cong 1.7$ s^{-1} up to $R20_{Dit3} \cong 3.5$ s^{-1} for the highest dithio concentration (c $_{DIT3}$ = 50 mmol/kg). We attributed this increase in the R20 response to a direct impact of the non-oxidized dithio molecule on the relaxation rate R2 of the water molecules in their vicinity. This might be mediated by a dissociation of the protons connected to the end sulfur atoms of the thiol-group in the dithio molecule in aqueous solution.

Regarding the dose rate dependency of MAGDIT-gels we evaluated a broad dose rate range starting at the low dose rate (0.6 Gy/min), which might be used in brachytherapy, and ranging up to the high dose rates (12 Gy/min) applied in modern radiotherapy and flattening filter free (FFF) applications [17,47,60–62]. The MAGADIT sample with the lowest oxygen scavenger concentration (DIT1) yielded the highest sensitivity and thus showed the largest absolute decrease in dose sensitivity with dose rate. Sensitivity was reduced from $\alpha = 0.54$ s^{-1} Gy^{-1} for 0.6 Gy/min to $\alpha = 0.2$ s^{-1} Gy^{-1} for 12 Gy/min, resulting in a difference of $\Delta\alpha_{DIT1} = 0.34$ s^{-1} Gy^{-1}, thus demonstrating also a dose-rate effect for dithio type polymer gels. For the highest concentration of oxygen scavenger (DIT3) the absolute difference in sensitivity between the lowest and highest dose rate was $\Delta\alpha_{DIT3} = 0.11$ s^{-1} Gy^{-1} (Figure 5; Table 2). Thus the absolute values in dose rate sensitivity changed by a factor of three ($\Delta\alpha_{DIT1}/\Delta\alpha_{DIT3} \approx 3$).

However, as a downside, the sensitivity was also reduced with increasing oxygen scavenger concentration. When normalizing the dose rate dependence to the sensitivity of the lowest dose rate for each type of the MAGADIT gels (Figure 5 right) the following observations could be made:

(1). There was a strong dose rate-effect for dose rates in between $\dot{D} = 0.6$ Gy/min and about $\dot{D} = 8$ Gy/min for all investigated dithio concentrations.

(2). This dose rate dependence of sensitivity almost disappeared within the accuracy of the results in the high dose rate range $\dot{D} \geq 8$ Gy/min (Figure 5).

(3).   The change of normalized dose sensitivities with dose rate appeared to be independent of the oxygen scavenger concentration for MAGADIT dosimeters. This suggests a more universal correlation of the impact of oxygen scavengers (or presumably other additives) on dose rate and sensitivity evaluated as the slope of the dose response.

We would like to note here, that—in a previous investigation on the influence of ascorbic acid on the dose rate dependence [47]—we found a similar behavior up to about $\dot{D} = 5$ Gy/min. Even beyond that a dose rate dependence was documented, with a drop in sensitivity of up to 20% between 5 Gy/min and 12 G/min for all of the different ascorbic acid concentrations [47]. For MAGADIT we did not observe any change of sensitivity beyond the critical dose rate $\dot{D}_{crit} \cong 8$ Gy/min within measurement accuracy (Table 2, Figure 5). Therefore this critical dose rate $\dot{D}_{crit}$ might depend on the specific oxygen scavenger and its effectiveness on radical scavenging. In a THPC based MA gel (MAGAT) a decrease by about 34% in the dose response was reported, if the dose rate increased from 0.3 Gy/min to 4 Gy/min [39]. Data provided in the same source [39,52] already indicates that a reduction in dose rate dependency might be observed also for MAGAT gels, supported by measurement at the highest investigated dose rate of 4–5 Gy/min. We more systematically investigated this indication by extending the dose rate range to the high dose rates typical of FFF irradiation schemes in clinics and the impact of the oxygen scavenger dithio on this dose rate dependence (4 Gy/min << 12 Gy/min). Moreover, we demonstrated the consequences for this reduced dose rate dependence at FFF typical dose rates for a 10×5 mm^2 photon field (10 MV) applied on a clinically used LINAC by quantitative comparisons to calibrated reference film dosimetry.

It has been reported also, that the dose rate impacts R2 response at higher dose regions more effectively than at low measurement ranges [52]. Our results (Figures 3 and 4) showed a similar behavior for low dose rates and high dose regions. In the case of radiation fields using collimation the dose rate impact definitely changes the measured dose distribution. This is due to the fact that in collimation treatment techniques, the dose rate is highly non-uniform across the treated volume [17,47].

Polyacrylamide-type gels had indicated less dose rate dependence, but feature toxic behavior [20]. Similarly, NIPAM gels showed negligible variation in dose response due to the dose-rate effect [63] but also exhibited a significantly reduced sensitivity in comparison to methacrylic acid based polymer gels [17]. Different methacrylic acid (MA) based gels containing THPC as an oxygen scavenger had been distinctly impacted by the dose rate [39,52,58]. In general, polymer gels with higher sensitivity showed a linear dose-response up to lower-level dose [64]. Highly sensitive polymer gels appeared to be more sensitive to the dose-rate effect, e.g., MAGAT type of gel dosimeters. We also observed a higher linear dose range for the less sensitive type of dithio polymer gels (Figure 4).

For MAGADIT the dose-rate effect was more pronounced for the more sensitive type of dithio dosimeters at lower concentration with regard to absolute value changes in sensitivity (Figure 5). This might be explained by the non-linear dependence of radical recombination and multimer termination on dose, effective at high dose rates: Radical generation and concentration are high at high dose rates. If the number of radical carrying molecules created in a gel becomes high in the local vicinity of radical generation, the probability, that the radicals react with each other and terminate polymerization reactions is significantly increased [20,58]. Therefore, the radical recombination rate increases non-linearly with dose rate and the dose sensitivity evaluated as a change of R2 with dose is reduced [35].

In this study, we exemplified the importance of calibrating the polymer gel at the same dose rate range as the target measurement, especially below the critical threshold $\dot{D}_{crit} \cong 8$ Gy/min. When calibrated at $\dot{D} = 2$ Gy/min the accuracy for polymer gels irradiated at the same dose with an increased dose rate $\dot{D} = 5$ Gy/min dropped significantly, with deviations of up to 20% for all types of polymer gels. This example demonstrates already the amount of error when using non-adapted dose rate calibration. Standard accuracy varied around 7% for dosimeters originating from the same batch and using calibration data at the same dose rate (Table 3). Consequently, calibration is strongly recommended to be performed at the same dose rates or dose fraction delivery as for the actual final gel dosimetric evaluation below $\dot{D}$ = 8 Gy/min.

The applicability of the MAGADIT gels for clinical small fields (5×10 mm^2) has been demonstrated at high dose rates of 15 Gy/min in an FFF beam. Inaccurate measurements of dose distributions could be reduced to a maximum shift of the dose side flanks in the dose profile to about 0.8 mm compared to radiochromic film as a gold standard. Lower dose rates applied to the center result in significantly larger inaccuracies in spatial dosimetric measurements, especially in the low dose region outside the center of the field, where even lower dose rates are present. We therefore strongly recommend that the novel MAGADIT polymer gels are used for any complex irradiations at the highest available dose rate spectrum. Within the range of the indicated dose rates ($\dot{D}$ > 8 Gy/min) MAGADIT polymer gels are well suited for FFF irradiations on LINACS and also pre-clinical investigations on animals using low energy, e.g., 50–200 kV radiation machine with high dose rates. Dithio polymer gels perform even slightly better than MAGIC type of gels with this respect.

4.2. Limitations of the Study and Possible Future Investigations

There are several aspects of polymer gel dosimeters in general, which could not be covered by this first report on the basic dosimetric properties using the new sulfur based oxygen scavenger dithiothreitol due to the limited scope. We will discuss shortly these aspects in the following in order to offer additional information on possible dosimetric performance limitations for clinical applications and possible future scientific work.

4.2.1. Edge Enhancement Effect

The loss of spatial integrity and overshoot of dose response near sharp dose gradients (edge enhancement effect) is related to the diffusivity of monomers as a consequence of the inhomogeneous monomer consumption close to dose edges [17,57,65,66]. This edge enhancement effect has been mainly investigated on acrylamide based BANG polymer gels [18,65,66] and methacrylic acid based polymer gels [39]. Most of the other polymer gels have not been explored systematically. For any polymer gel type, the relevance of gelatin concentration can be assumed due to its impact on the diffusivity of the monomers. Edge enhancement is mainly present at low concentrations of gelatin at about 2%–6% [66] as a consequence of the higher monomer diffusivity at low gelatin concentration. No edge enhancement is observed for MAGAS (methacrylic acid gel and ascorbic acid) polymer gels at 8% gelatin concentration [57]. For MAGAT polymer gels at 8% gelatin concentration a significant overshoot of the dose response has only been observed at dose levels in the nonlinear dose response range close to saturation end especially prominent 6 days after irradiation [39]. For the reason of a reduced edge enhancement we therefore used in our experiments the even higher concentration of 10% gelatin with high bloom strength (300) for reduced mobility and diffusivity of the polymer agglomerates in MAGADIT gels. The irradiation was applied in the linear range and the high resolution MR reading was performed within 3 days after irradiation. Due to the same diffusivity limiting ingredient in the polymer recipe i.e., gelatin it could, therefore, be expected that the edge enhancement effect in MAGADIT gels did not differ significantly from the previously reported ones at high gelatin concentration showing no appearance of edge enhancement. We, therefore, did not expect prominent edge enhancement effects for the proposed MAGADIT compositions. Actually, therefore, the results on the dosimetric profiles for the small field of 5×10 mm^2 with dose gradients of 6 Gy/mm (Figure 7) did not show an overshoot of dose response at the dose edge.

4.2.2. Temperature Dependence

The polymer gel dose response for all polymer gels evaluated as relaxivity R2 is dependent on the temperature during manufacturing, irradiation, storage and MR-reading and also depends on the specific composition, mainly of the monomer and gelatin type, but also on their concentration in quantitative numbers, which has been documented in several publications, e.g., in [17,18,33,39].

This observation represents a consequence of the chemical processes during polymerization involved, as these are sensitive to temperature. The relaxivity R2 measured by MRI during the reading of the polymer gels depends on the mobility of (mainly water) molecules according to basic physical principles (BPP-theory [67]). As temperature represents the main physical parameter on the mobility of biomolecules and water, the impact of temperature on T2-measurement and consequently dose response in polymer gels is likely and has been reported already in the first proposals and investigations on polymer gels, e.g., [18,33,39]. It is therefore only rarely reported for each new proposal of a polymer gel composition.

We also did not investigate this temperature dependence, as the main mechanisms and dependencies for the most important acrylic acid, methacrylic acid and acrylamide/bisacrylamide gel systems have already been explored [17,18,39] and no fundamental differences can be expected for temperatures distant from sol–gel transition for exchanging the low concentration ascorbic acid by dithio. In general, identical experimental protocols including temperature time courses and MR scanning temperature are strongly recommended for the measurement of calibration gels and final evaluation for dosimetric accuracy in absolute 3D-dosimetry based on polymer gels. We would like to indicate that a wide range of temperature during measurement of the gel response might be used for relative dosimetry as long as the dose response is linearly dependent on dose, the temperature distribution is homogeneous in the polymer gel and the gelatin remains in a rigid state below getting liquid.

4.2.3. Stability of the Dose Response with Time after Manufacturing

The absolute values in dose sensitivity might vary with time after manufacturing and irradiation and time interval after irradiation and MR scanning due to the time dependent behavior of the chemical processes involved and additional gelatinization taking place (cf., e.g., [39,52]. However, this might not be relevant for relative dosimetry as long as the polymer gel dosimeter remains homogeneous and the response of the polymer is linearly dependent on dose for relative dosimetric purposes such that known or measured applied dose levels in single points can be used for linear interpolation.

We did not investigate that aspect systematically but here report about some aspects and qualitative experience on the investigated MAGADIT and the previously investigated MAGIC-type of polymer gels:

(a) Chemical Stability

We did observe for MAGIC type gels (2 mmol/kg oxygen scavenger ascorbic acid concentration) a change from transparent to a red brown color, even when being stored in the fridge in oxygen dense containers after several weeks. This change in color indicates chemical processes ongoing in the chemical dosimeter and might be related to the degradation of gelatin [68]. The relaxivity at zero dose (R20 offset) also slightly increased with time for the used high gelatin concentrations (8%–10%). We attributed this change in relaxivity to an ongoing gelatinization process even at low temperatures. Such an impact has already been assumed and reported earlier [17,69].

We did not observe a corresponding change in transparency or color for the MAGADIT gels even 12 months after manufacturing with storage in the fridge at 4–8 °C. The dithio polymer gels still showed a response R2 linearly dependent on dose. The results for the dose profiles at a center dose rate of 15.3 Gy/min are shown in Figure 7 and did exhibit agreement with the film dosimetric reference data within 0. 8 mm distance.

(b) Oxygen Penetration

As indicated in the chapter materials and methods, polymer gel containers often are made of plastics, as this material behaves similarly to water with regard to absorption properties for different radiation qualities. However, the available container materials for polymer gels are often restricted with regard to oxygen impermeability. We used PET containers with a wall thickness of about 1 mm, instead of the previously used BAREX^R material, which is not available any more on the market, but actually had to realize that even after few weeks after manufacturing the dose response was suppressed, especially at the edges close to the container wall. The scavenger add-on in the polymer gel is capable of trapping some amount of the penetrating oxygen, but when its capacity is exhausted the dose response at the corresponding locations especially at the rim to the container material will be suppressed. We think that oxygen transport from outside passing the low-permeable PET barrier is responsible for that observation. We recommend storing normoxic polymer gel containers, made of plastics after manufacturing in an oxygen impermeable outer container filled with an aqueous solution of an oxygen scavenger, e.g., 2–10 mmol/L ascorbic acid (cheap and non-toxic).

5. Conclusions

Polymer gels with the novel oxygen scavenger dithiothreitol (MAGADIT) were investigated at varying concentrations for their basic dosimetric properties: linear dose range, sensitivity, precision, accuracy and the dose rate dependence of their dose sensitivity. The dithio polymer gels have demonstrated a linear dose-response up to a dose range of a minimum of 24 Gy. The dosimeter with the lowest investigated scavenger concentration (2 mmol/kg) featured the highest sensitivity (0.54 Gy^{-1} s^{-1}) at $\dot{D} = 0.6$ Gy/min. This sensitivity was a little lower than reported for the MAGIC-type of polymer gels. All types of investigated polymer gels featured a significant dose-rate effect in the dose rate range between 0.6 Gy/min to about 8 Gy/min. The dose rate dependence decreased for higher dose rates and apparently vanished for $\dot{D} \geq 8$ Gy/min within measurement errors for all types of dithio polymer gels. The absolute sensitivity change was distinctly reduced for higher dithio concentrations on the cost of absolute sensitivity. The absolute accuracy in gel dosimetry might be strongly degraded in the low dose rate range, if the dose rates for the calibration differ from the dose rates of the actual measurement.

An application of MAGADIT for small field measurements in a clinical 10MV FFF beam demonstrated its suitability as a multi-dimensional dose detector providing good agreement with reference to radiochromic film measurements. However, the MAGADIT gels, similarly to other sensitive MA based gels, are not suited for the low dose rate range for complex radiation fields with locally strongly varying dose rates.

Methacrylic acid based polymer gels using the new oxygen scavenger dithio (MAGADIT) feature advantages over standard type of polymer gels like MAGIC or MAGAT type of dosimeters, i.e., fewer hazards in comparison to MAGAT and almost disappearing dose rate effect at high dose rates $\dot{D} \geq 8$ Gy/min in comparison to MAGIC. This high dose rate region is generally utilized as a part of contemporary clinical treatments with FFF irradiations. The qualitative results presented in this study should be applicable to other types of polymer gels as well.

The source files of the investigations referring to the tables and figures i.e., excel evaluation data incl. graphs, MRI T2-images of the polymer gels can be downloaded from a server at the Medical University of Vienna after contact to the author per e-mail.

Author Contributions: Conceptualization: (a) MR-based polymer gel dosimetry (MRPD) on MAGADIT: A.G.B.; (b) FFF-LINAC 5-10mm irradiation scheme: A.G.B., W.L., G.H., D.G. Methodology: (a) MRPD: A.G.B.; film dosimetry: W.L., G.H. Sample preparation: M.K. Evaluation, formal analysis and investigation: (a) polymer gel basic properties: M.K.; (b) film dosimetry: G.H., W.L.; (c) MRPD on FFF-LINAC 5-10mm irradiation scheme and comparison to film dosimetric data: A.G.B. Visualization: M.K, G.H., W.L., A.G.B. Data curation: M.K, G.H, W.L, A.G.B. Writing—original draft preparation: M.K. Writing—review and editing: M.K., A.G.B., G.H., W.L., D.G. Supervision: A.G.B. Project administration: A.G.B. Funding acquisition: A.G.B., M.K. Parts of the content of the manuscript are contained in the PhD-dissertation of Muzafar Khan at the Medical University of Vienna [70].

Funding: This work was funded by "Hochschuljubiläumsfonds" (HSJ) grant number H-1208/2003 of the City of Vienna/Austria. The position of M. Khan is financed by a stipend from the higher education commission (HEC) of Pakistan. Materials were partly funded by intramural support of the Medical University of Vienna.

Polymers **2019**, *11*, 1717

Acknowledgments: Thanks are due to the administrative support of the Austrian agency for international mobility and cooperation in education, science, and research (OEAD). An introduction to the operation of the Yxlon-apparatus was kindly offered by Peter Kuess (Department of Radiation Oncology, Medical University of Vienna). We acknowledge the hint by Arash Mirzahosseini, Semmelweis University Budapest, who indicated to us the usage of dithio as oxygen scavenging agent in Biochemistry.

Conflicts of Interest: The authors declare no conflict of interest. The funders had no role in the design of the study, in the collection, analyses or interpretation of data, in the writing of the manuscript or in the decision to publish the results.

MESH Terms: Gels (D20.280.320), Diagnostic Imaging (E01.370.350), Polymers (D05.750), Radiometry/methods (E05.799), Radiotherapy (Q000532)

Abbreviations

AMPS	2-acrylamido-2-methyl-propane sulfonic acid
BIS	*N,N*-methylene-bis-acrylamide
α	Dose sensitivity ($\Delta R2/\Delta D$)
$\Delta\alpha/\Delta\dot{D}$	Absolute dose rate dependence ($\Delta R2/\Delta D)/\Delta\dot{D}$ of the dose sensitivity
BANG	Bis acryl amide nitrogen gel
c	Concentration of the oxygen scavenger
CT	Computer tomography
D	Dose
$\dot{D}$	Dose rate: ($\dot{D} = \Delta D/\Delta t$)
Dithio	Dithiothreitol (IUPAC name: 1,4-bis(sulfanyl)butane-2.3-diol)
DIT1	Polymer gel with dithiothreitol at lowest concentration (c_{dit1} = 2 mmol/kg)
DIT2	Polymer gel with dithiothreitol at medium concentration (c_{dit2} = 10 mmol/kg)
DIT3	Polymer gel with dithiothreitol at highest concentration (c_{dit3} = 50 mmol/kg)
EBT3	Classification specification of a commercial radiochromic film
FFF	Flattening filter free
FWHM	Full width at half maximum
IMRT	Intensity modulated radiation therapy
iVIPET	Vinylpyrrolidone and THPC with inorganic add-on
LINAC	Linear accelerator
MA	Methacrylic acid (IUPAC name: 2-methylprop-2-enoic acid)
MAGIC	Methacrylic and ascorbic acid in gelatin initiated by copper
MAGAS	Methacrylic acid gel and ascorbic acid
MAGAT	Methacrylic acid gel and THPC
MRI	Magnetic resonance imaging
MRPD	Magnetic resonance imaging based polymer gel dosimetry
NIPAM	*N*-isopropylacrylamide
NVP	*N*-vinylpyrrolidone
PAG	Polyacrylamide gel
PAGAT	Polyacrylamide gel and THPC
PTV	Planning target volume
ROI	Region of interest, frame in the image to be investigated
R2	Transverse relaxation rate (R2 = 1/T2)
R20	Transverse relaxation rate of the polymer gel at Dose D = 0 Gy
T2	Transverse relaxation time (spin-spin relaxation time)
SSD	Source to surface distance
THPC	Tetrakis(hydroxymethyl)phosphonium chloride
TPS	Treatment planning system
VIC	Vinylpyrrolidone with computed tomography
VIPAR	VIPAR N-vinylpyrrolidone argon
VIPET	Vinylpyrrolidone and THPC
VMAT	Volumetric modulated arc therapy

References

1. Lee, N.; Xia, P.; Quivey, J.M.; Sultanem, K.; Poon, I.; Akazawa, C.; Akawzawa, P.; Weinberg, V.; Fu, K.K. Intensity-modulated radiotherapy in the treatment of nasopharyngeal carcinoma: An update of the UCSF experience. *Int. J. Radiat. Oncol. Biol. Phys.* **2002**, *53*, 12–22. [CrossRef]

2. Purdy, J.A. Dose to normal tissues outside the radiation therapy patient's treated volume: A review of different radiation therapy techniques. *Health Phys.* **2008**, *95*, 666–676. [CrossRef] [PubMed]

3. Nguyen, N.P.; Ceizyk, M.; Vos, P.; Vinh-Hung, V.; Davis, R.; Desai, A.; Abraham, D.; Krafft, S.P.; Jang, S.; Watchman, C.J.; et al. Effectiveness of image-guided radiotherapy for laryngeal sparing in head and neck cancer. *Oral Oncol.* **2010**, *46*, 283–286. [CrossRef]

4. Zelefsky, M.J.; Fuks, Z.; Happersett, L.; Lee, H.J.; Ling, C.C.; Burman, C.M.; Hunt, M.; Wolfe, T.; Venkatraman, E.S.; Jackson, A.; et al. Clinical experience with intensity modulated radiation therapy (IMRT) in prostate cancer. *Radiother. Oncol.* **2000**, *55*, 241–249. [CrossRef]

5. Teh, B.S.; Woo, S.Y.; Butler, E.B. Intensity modulated radiation therapy (IMRT): A new promising technology in radiation oncology. *Oncologist* **1999**, *4*, 433–442. [PubMed]

6. Cheung, K. Intensity modulated radiotherapy: Advantages, limitations and future developments. *Biomed. Imaging Interv. J.* **2006**, *2*, e19. [CrossRef] [PubMed]

7. Bortfeld, T. IMRT: A review and preview. *Phys. Med. Biol.* **2006**, *51*, R363–R379. [CrossRef] [PubMed]

8. Adamovics, J.; Maryanski, M.J. Characterisation of PRESAGE: A new 3-D radiochromic solid polymer dosemeter for ionising radiation. *Radiat. Prot. Dosim.* **2006**, *120*, 107–112. [CrossRef] [PubMed]

9. Jordan, K.; Avvakumov, N. Radiochromic leuco dye micelle hydrogels: I. Initial Investigation. *Phys. Med. Biol.* **2009**, *54*, 6773. [CrossRef] [PubMed]

10. Babic, S.; Battista, J.; Jordan, K. Radiochromic leuco dye micelle hydrogels: II. Low diffusion rate leuco crystal violet gel. *Phys. Med. Biol.* **2009**, *54*, 6791–6808. [CrossRef]

11. Penev, K.I.; Wang, M.; Mequanint, K. Tetrazolium salt monomers for gel dosimetry I: Principles. *J. Phys. Conf. Ser.* **2017**, *847*. [CrossRef]

12. De Deene, Y.; Skyt, P.S.; Hil, R.; Booth, J.T. FlexyDos3D: A deformable anthropomorphic 3D radiation dosimeter: Radiation properties. *Phys. Med. Biol.* **2015**, *60*, 1543. [CrossRef] [PubMed]

13. Schreiner, L.J. Review of Fricke gel dosimeters. *J. Phys. Conf. Ser.* **2004**, *3*, 9–21. [CrossRef]

14. d'Errico, F.; Lazzeri, L.; Dondi, D.; Mariani, M.; Marrale, M.; Souza, S.O.; Gambarini, G. Novel GTA-PVA Fricke gels for three-dimensional dose mapping in radiotherapy. *Radiat. Meas.* **2017**, *106*, 612–617. [CrossRef]

15. Gore, J.C.; Ranade, M.; Maryansky, M.J.; Schulz, R.J. Radiation dose distributions in three dimensions from tomographic optical density scanning of polymer gels: I. Development of an optical scanner. *Phys. Med. Biol.* **1996**, *41*, 2695–2704. [CrossRef] [PubMed]

16. Hill, B.; Venning, A.J.; Baldock, C. The dose response of normoxic polymer gel dosimeters measured using X-ray CT. *Br. J. Radiol.* **2005**, *78*, 623–630. [CrossRef]

17. Baldock, C.; De Deene, Y.; Doran, S.; Ibbott, G.; Jirasek, A.; Lepage, M.; McAuley, K.B.; Oldham, M.; Schreiner, L.J. Polymer gel dosimetry. *Phys. Med. Biol.* **2010**, *55*, R1–R63. [CrossRef]

18. Maryanski, M.J.; Schulz, R.J.; Ibbott, G.S.; Gatenby, J.C.; Xie, J.; Horton, D.; Gore, J.C. Magnetic resonance imaging of radiation dose distributions using a polymer-gel dosimeter. *Phys. Med. Biol.* **1994**, *39*, 1437. [CrossRef]

19. Gore, J.C.; Kang, Y.S. Measurement of radiation dose distributions by nuclear magnetic resonance (NMR) imaging. *Phys. Med. Biol.* **1984**, *29*, 1189. [CrossRef]

20. De Deene, Y.; De Wagter, C.; Van Duyse, B.; Derycke, S.; De Neve, W.; Achten, E. Three-dimensional dosimetry using polymer gel and magnetic resonance imaging applied to the verification of conformal radiation therapy in head-and-neck cancer. *Radiother. Oncol.* **1998**, *48*, 283–291. [CrossRef]

21. Baldock, C.; Burford, R.P.; Billingham, N.; Wagner, G.S.; Patval, S.; Badawi, R.D.; Keevil, S.F. Experimental procedure for the manufacture and calibration of polyacrylamide gel (PAG) for magnetic resonance imaging (MRI) radiation dosimetry. *Phys. Med. Biol.* **1998**, *43*, 695–702. [CrossRef]

22. Fong, M.P.; Derek, C.K.; Mark, D.D.; Gore, J.C. Polymer gels for magnetic resonance imaging of radiation dose distributions at normal room atmosphere. *Phys. Med. Biol.* **2001**, *46*, 3105. [CrossRef]

23. Berg, A.; Ertl, A.; Moser, E. High-resolution polymer gel dosimetry by parameter selective MR-microimaging on a whole body scanner at 3T. *Med. Phys.* **2001**, *28*, 833–843. [CrossRef] [PubMed]

24. Senden, R.J.; De Jean, P.; McAuley, K.B.; Schreiner, L.J. Polymer gel dosimeters with reduced toxicity: A preliminary investigation of the NMR and optical dose response using different monomers. *Phys. Med. Biol.* **2006**, *51*, 3301–3314. [CrossRef] [PubMed]

25. Pappas, E.; Maris, T.; Angelopoulos, A.; Paparigopoulou, M.; Sakelliou, L.; Sandilos, P.; Voyiatzi, S.; Vlachos, L. A new polymer gel for magnetic resonance imaging (MRI) radiation dosimetry. *Phys. Med. Biol.* **1999**, *44*, 2677. [CrossRef] [PubMed]

26. Kozicki, M.; Jaszczak, M.; Maras, P.; Dudek, M.; Cłapa, M. On the development of a VIPARnd radiotherapy 3D polymer gel dosimeter. *Phys. Med. Biol.* **2017**, *62*, 986–1008. [CrossRef] [PubMed]

27. Watanabe, Y.; Mizukami, S.; Eguchi, K.; Maeyama, T.; Hayashid, S.; Muraishi, H.; Terazaki, T.; Gomi, T. Dose distribution verification in high-dose-rate brachytherapy using a highly sensitive norrnoxic N-vinylpyrrolidone polymer gel dosimeter. *Phys. Med.* **2019**, *57*, 72–79. [CrossRef] [PubMed]

28. Berg, A.; Pernkopf, M.; Waldhäusl, C.; Schmidt, W.; Moser, E. High resolution MR based polymer dosimetry versus film densitometry: A systematic study based on the modulation transfer function approach. *Phys. Med. Biol.* **2004**, *49*, 4087–4108. [CrossRef]

29. De Deene, Y.D. Essential characteristics of polymer gel dosimeters. *J. Phys. Conf. Ser.* **2004**, *3*, 34–57. [CrossRef]

30. Bayreder, C.; Schon, R.; Wieland, M.; Georg, D.; Moser, E.; Berg, A. The spatial resolution in dosimetry with normoxic polymer-gels investigated with the dose modulation transfer approach. *Med. Phys.* **2008**, *35*, 1756–1769. [CrossRef]

31. Olding, T.; Holmes, O.; Dejean, P.; McAuley, K.B.; Nkongchu, K.; Santyr, G.; Schreiner, L.J. Small field dose delivery evaluations using cone beam optical computed tomography-based polymer gel dosimetry. *Med. Phys.* **2011**, *36*, 3–14. [CrossRef] [PubMed]

32. Razak, N.; Rahman, A.; Kandaiya, S.; Mustafa, I.; Yahaya, N.; Mahmoud, A.; Maizan, R. Accuracy and Precision of Magat Gel As a Dosimeter. *Mater. Sci. Res. India* **2015**, *12*, 1–7. [CrossRef]

33. Watanabe, Y.; Warmington, L.; Gopishankar, N. Three-dimensional radiation dosimetry using polymer gel and solid radiochromic polymer: From basics to clinical applications. *World J. Radiol.* **2017**, *9*, 112–125. [CrossRef] [PubMed]

34. Baldock, C. Historical overview of the development of gel dosimetry: Another personal perspective. *J. Phys. Conf. Ser.* **2009**, *164*. [CrossRef]

35. Jirasek, A.; McAuley, K.B.; Lepage, M. How does the chemistry of polymer gel dosimeters affect their performance? *J. Phys. Conf. Ser.* **2009**, *164*. [CrossRef]

36. De Deene, Y.; Reynaert, N.; De Wagter, C. On the accuracy of monomer/polymer gel dosimetry in the proximity of a high-dose-rate 192Ir source. *Phys. Med. Biol.* **2001**, *46*, 2801–2825. [CrossRef]

37. Spevacek, V.; Novotny, J., Jr.; Dvorak, P.; Novotny, J.; Vymazal, J.; Cechak, T. Temperature dependence of polymer-gel dosimeter nuclear magnetic resonance response. *Med. Phys.* **2001**, *28*, 2370–2378. [CrossRef]

38. Berg, A.; Bayreder, C.; Georg, D.; Bankamp, A.; Wolber, G. Aspects of radiation beam quality and their effect on the dose response of polymer gels: Photons, electrons and fast neutrons. *J. Phys. Conf. Ser.* **2009**, *164*. [CrossRef]

39. De Deene, Y.; Vergote, K.; Claeys, C.; De Wagter, C. The fundamental radiation properties of normoxic polymer gel dosimeters: A comparison between a methacrylic acid based gel and acrylamide based gels. *Phys. Med. Biol.* **2006**, *51*, 653–673. [CrossRef]

40. Crescenti, R.A.; Scheib, S.G.; Schneider, U.; Gianolini, S. Introducing gel dosimetry in a clinical environment: Customization of polymer gel composition and magnetic resonance imaging parameters used for 3D dose verifications in radiosurgery and intensity modulated radiotherapy. *Med. Phys.* **2007**, *34*, 1286–1297. [CrossRef]

41. Vandecasteele, J.; De Deene, Y. On the validity of 3D polymer gel dosimetry: I. reproducibility study. *Phys. Med. Biol.* **2013**, *58*, 19–42. [CrossRef] [PubMed]

42. Kipouros, P.; Pappas, E.; Baras, P.; Hatzipanayoti, D.; Karaiskos, P.; Sakelliou, L.; Sandilos, P.; Seimenis, I. Wide dynamic dose range of VIPAR polymer gel dosimetry. *Phys. Med. Biol.* **2001**, *46*, 2143–2159. [CrossRef] [PubMed]

43. Grebe, G.; Pfaender, M.; Roll, M.; Luedemann, L.; Wurm, R.E. Dynamic arc radiosurgery and radiotherapy: Commissioning and verification of dose distributions. *Int. J. Radiat. Oncol. Biol. Phys.* **2001**, *49*, 1451–1460. [CrossRef]

44. Thomas, A.; Newton, J.; Adamovics, J.; Oldham, M. Commissioning and benchmarking a 3D dosimetry system for clinical use. *Med. Phys.* **2011**, *38*, 4846–4857. [CrossRef] [PubMed]

45. Newton, J.; Oldham, M.; Thomas, A.; Li, Y.; Adamovics, J.; Kirsch, D.G.; Das, S. Commissioning a small-field biological irradiator using point, 2D, and 3D dosimetry techniques. *Med. Phys.* **2011**, *38*, 6754–6762. [CrossRef] [PubMed]

46. Johnston, H.; Hilts, M.; Jirasek, A. SU-E-T-70: Commissioning a Multislice CT Scanner for X-ray CT Polymer Gel Dosimetry. *Med. Phys.* **2014**, *41*, 238. [CrossRef]

47. Khan, M.; Heilemann, G.; Kuess, P.; Georg, D.; Berg, A. The impact of the oxygen scavenger on the dose-rate dependence and dose sensitivity of MAGIC type polymer gels. *Phys. Med. Biol.* **2018**, *63*, 06NT1. [CrossRef]

48. O'Neil, M.J.; Heckelman, P.E.; Koch, C.B.; Roman, K.J. The Merck Index: An Encyclopedia of Chemicals, Drugs, and Biologicals. *J. Am. Chem. Soc.* **2007**, *129*, 2197.

49. Fruton, J.S. Oxidation-reduction potentials of ascorbic acid. *J. Biol. Chem.* **1934**, *105*, 79–85.

50. Katsumi, H.; Nishikawa, M.; Nishiyama, K.; Hirosaki, R.; Nagamine, N.; Okamoto, H.; Mizugichi, H.; Kusamori, K.; Yasui, H.; Yamashita, F.; et al. Development of PEGylated serum albumin with multiple reduced thiols as a long-circulating scavenger of reactive oxygen species for the treatment of fulminant hepatic failure in mice. *Free Radic. Biol. Med.* **2014**, *69*, 318–323. [CrossRef]

51. Cleland, W.W. Dithiothreitol, A New Protective Reagent for SH Groups. *Biochemistry* **1964**, *3*, 480–482. [CrossRef] [PubMed]

52. Bayreder, C.; Georg, D.; Moser, E.; Berg, A. Basic investigations on the performance of a normoxic polymer gel with tetrakis-hydroxy-methyl-phosphonium chloride as an oxygen scavenger: Reproducibility, accuracy, stability, and dose rate dependence. *Med. Phys.* **2006**, *33*, 2506–2518. [CrossRef] [PubMed]

53. Kuess, P.; Bozsaky, E.; Hopfgartner, J.; Seifritz, G.; Dorr, W.; Georg, D. Dosimetric challenges of small animal irradiation with a commercial X-ray unit. *Z. Med. Phys.* **2014**, *24*, 363–372. [CrossRef] [PubMed]

54. Farajollahi, A.R.; Bonnett, D.E.; Ratcliffe, A.J.; Aukett, R.J.; Mills, J.A. An investigation into the use of polymer gel dosimetry in low dose rate brachytherapy. *Br. J. Radiol.* **1999**, *72*, 1085–1092. [CrossRef]

55. Milford, D.; Rosbach, N.; Bendszus, M.; Heiland, S. Mono-Exponential Fitting in T2-Relaxometry: Relevance of Offset and First Echo. *PLoS ONE* **2015**, *10*, e0145255. [CrossRef]

56. Dreindl, R.; Georg, D.; Stock, M. Radiochromic film dosimetry: Considerations on precision and accuracy for EBT2 and EBT3 type films. *Z. Med. Phys.* **2014**, *24*, 153–163. [CrossRef]

57. Deene, Y.D.; Hurley, C.; Venning, A.; Vergote, K.; Mather, M.; Healy, B.J.; Baldock, C. A basic study of some normoxic polymer gel dosimeters. *Phys. Med. Biol.* **2002**, *47*, 3441–3463. [CrossRef]

58. Karlsson, A.; Gustavsson, H.; Mansson, S.; McAuley, K.B.; Back, S.A.J. Dose integration characteristics in normoxic polymer gel dosimetry investigated using sequential beam irradiation. *Phys. Med. Biol.* **2007**, *52*, 4697–4706. [CrossRef]

59. Abtahi, S.M. Characteristics of a novel polymer gel dosimeter formula for MRI scanning: Dosimetry, toxicity and temporal stability of response. *Phys. Med.* **2016**, *32*, 1156–1161. [CrossRef]

60. Georg, D.; Knoos, T.; McClean, B. Current status and future perspective of flattening filter free photon beams. *Med. Phys.* **2011**, *38*, 1280–1293. [CrossRef]

61. Zehtabian, M.; Faghihi, R.; Zahmatkesh, M.H.; Meigooni, A.S.; Mosleh-Shirazi, M.A.; Mehdizadeh, S.; Sina, S.; Bagheri, S. Investigation of the dose rate dependency of the PAGAT gel dosimeter at low dose rates. *Radiat. Meas.* **2012**, *47*, 139–144. [CrossRef]

62. Prendergast, B.M.; Fiveash, J.B.; Popple, R.A.; Clark, G.M.; Thomas, E.M.; Minnich, D.J.; Jacob, R.; Spencer, S.A.; Bonner, J.A.; Dobelbower, M.C. Flattening filter-free linac improves treatment delivery efficiency in stereotactic body radiation therapy. *J. Appl. Clin. Med. Phys.* **2013**, *14*, 64–71. [CrossRef]

63. Jirasek, A.; Johnston, H.; Hilts, M. Dose rate properties of NIPAM-based x-ray CT polymer gel dosimeters. *Phys. Med. Biol.* **2015**, *60*, 4399–4411. [CrossRef] [PubMed]

64. Massilon, J.L.; Minniti, R.; Soares, C.G.; Maryanski, M.J.; Robertson, S. Characteristics of a new polymer gel for high-dose gradient dosimetry using a micro optical CT scanner. *Appl. Radiat. Isotopes* **2009**, *68*, 144–154. [CrossRef] [PubMed]

65. Fuxman, A.M.; McAuley, K.B.; Schreiner, L.J. Modeling of polyacrylamide gel dosimeters with spatially non-uniform radiation dose distributions. *Chem. Eng. Sci.* **2005**, *60*, 1277–1293. [CrossRef]

66. Vergote, K.; De Deene, Y.; Vanden Bussche, E.; De Wagter, C. On the relation between the spatial dose integrity and the temporal instability of polymer gel dosimeters. *Phys. Med. Biol.* **2004**, *49*, 4507–4522. [CrossRef]

67. Bloembergen, N.; Purcell, E.M.; Pound, R.V. Relaxation Effects in Nuclear Magnetic Resonance Absorption. *Phys. Rev.* **1948**, *7*, 679–746. [CrossRef]
68. Jaszczak, M.; Kolesińska, B.; Wach, R.; Maras, P.; Dudek, M.; Kozicki, M. Examination of THPC as an oxygen scavenger impacting VIC dosimeter thermal stability and comparison of NVP-containing polymer gel dosimeters. *Phys. Med. Biol.* **2019**, *64*, 035019. [CrossRef]
69. De Deene, Y.; Hanselaer, P.; De Wagter, C.; Achten, E.; De Neve, W. An investigation of the chemical stability of a monomer/polymer gel dosimeter. *Phys. Med. Biol.* **2000**, *45*, 859–878. [CrossRef]
70. Khan, M. Magnetic Resonance Imaging Based Polymer Gel Dosimetry for Radiation Therapy: Basic Properties of New Normoxic Polymer Gels for High Dose Rates. Ph.D. Thesis, Medical University of Vienna, Vienna, Austria, May 2018.

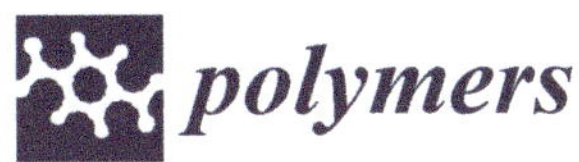

Article

FluoroTome 1: An Apparatus for Tomographic Imaging of Radio-Fluorogenic (RFG) Gels

John M. Warman [1,*], Matthijs P. de Haas [1], Leonard H. Luthjens [1], Tiantian Yao [1], Julia Navarro-Campos [1], Sölen Yuksel [1], Jan Aarts [2], Simon Thiele [2], Jacco Houter [2] and Wilco in het Zandt [2]

[1] Department of Radiation Science and Technology, Faculty of Applied Sciences, Delft University of Technology, Mekelweg 15, 2629 JB Delft, The Netherlands; m.p.dehaas@tudelft.nl (M.P.d.H.); l.h.luthjens@tudelft.nl (L.H.L.); tiant.yao@gmail.com (T.Y.); julianavarro.c@gmail.com (J.N.-C.); S.Yuksel@tudelft.nl (S.Y.)
[2] 4PICO B.V., Jan Tinbergenstraat 4B, 5491 DC Sint-Oedenrode, The Netherlands; Jan@4pico.nl (J.A.); Simon@4pico.nl (S.T.); Jacco@4pico.nl (J.H.); Wilco@4pico.nl (W.i.h.Z.)
* Correspondence: J.M.Warman@TUDelft.nl

Received: 17 September 2019; Accepted: 22 October 2019; Published: 23 October 2019

Abstract: Radio-fluorogenic (RFG) gels become permanently fluorescent when exposed to high-energy radiation with the intensity of the emission proportional to the local dose of radiation absorbed. An apparatus is described, FluoroTome 1, that is capable of taking a series of tomographic images (thin slices) of the fluorescence of such an irradiated RFG gel on-site and within minutes of radiation exposure. These images can then be compiled to construct a 3D movie of the dose distribution within the gel. The historical development via a laboratory-bench prototype to a readily transportable, user-friendly apparatus is described. Instrumental details and performance tests are presented.

Keywords: fluorescent gels; radio-fluorogenic (RFG) gel; tomographic fluorescence imaging; polymer-gel radiation dosimetry; 3D radiation dosimetry

---

1. Introduction

Several bulk media have been proposed for the creation of three-dimensional images of the energy deposited in materials by complex fields of high-energy photon or particle radiation [1]. The need for such media has been driven by the increasing sophistication and subtlety of dose delivery in radiotherapy clinics and the requirement for quality control of treatment protocols and equipment [2–4]. The use of such media as phantoms for the training of clinical personnel is an additional, important role they may play.

The media proposed are based on physico-chemical effects induced at a molecular level that change the local optical, dielectric or nuclear spin properties of the medium. The changes are spatially "fixed" in a quasi-rigid-gel or polymer matrix. The apparatus and data handling required to derive 3D dose mapping using such systems has been covered in the literature [1,5]. None of the methods reviewed are capable of rapid, on-site data analysis and no single method has received universal acceptance in the clinic. Multiple-arrays of solid-state devices or multi-layers of radiochromic films have also been proposed but are incapable of providing the sub-millimeter spatial resolution desired.

Here we describe a readily transportable, user-friendly apparatus that has been developed to allow on-site analysis of radio-fluorogenic (RFG) gel media within minutes of radiation exposure. The use of RFG gels to monitor dose deposition by a variety of radiation sources, from a brachytherapy radioisotope to an 80 MeV proton beam, has been recently reviewed [6]. In these early measurements, images of the *bulk* medium were made. An advance has been the incorporation of slits, forming a thin sheet of UV excitation light, that allowed the apparatus to make tomographic (multiple slice) images of the radiation-induced fluorescence within the gel. These slices can then be compiled to form 3D images that can be displayed in video mode [7,8]. This report describes the design, construction and testing of an apparatus based on this laboratory prototype by a firm that specializes in devices based on UV solid state sources.

2. Historical Background

The first application of a bulk radio-fluorogenic gel to monitor a non-homogeneous radiation field was reported in 2011 by Warman et al. [9]. Application to a variety of radiation sources, including x-ray beams, proton beams and radio-isotopic seeds followed [6]. These measurements of bulk fluorescence were carried out initially with an extremely simple set-up consisting of two Mercury discharge lamps and a standard Ricoh "Caplio RX" digital camera, as illustrated in Figure 14 of reference 6. The mercury lamps were eventually replaced by LED lamps from 4PICO.BV. These had a 16 cm, linear array of UV light emitting diodes with a linear collimating Fresnel lens. Using these lamps pseudo-3D images of bulk $40 \times 40 \times 40$ mm^3 gel samples irradiated with multiple proton beams were made, as shown in Figure 2 of reference [10].

The development to the prototype of the present apparatus required the introduction of plates on both sides of the sample that had slits which restricted the UV light to a thin sheet passing through the sample. A translatable stage was incorporated to transport the sample through the UV sheet and multiple images were taken of the fluorescence as a function of position. This procedure has been fully described and illustrated in reference 7. A photograph of the bench-top prototype is shown in Figure 1.

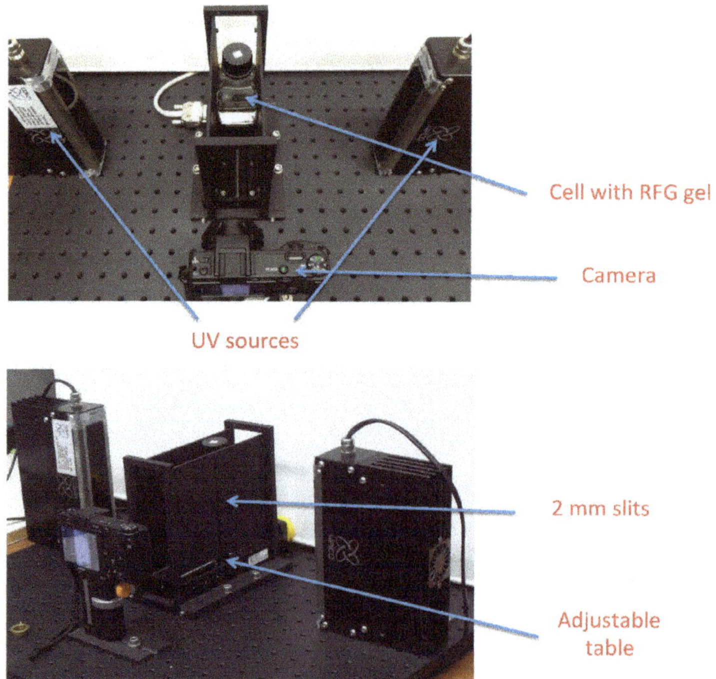

Figure 1. The bench-top prototype to FluoroTome 1.

The operation of the prototype in making tomographic, "sliced" images of the fluorescence of a gel sample is illustrated schematically in Figure 2. The application to crossed-beam irradiation of an RFG gel [7] and to irradiation of an eye phantom with a 3 mm X-ray beam [8] have been reported.

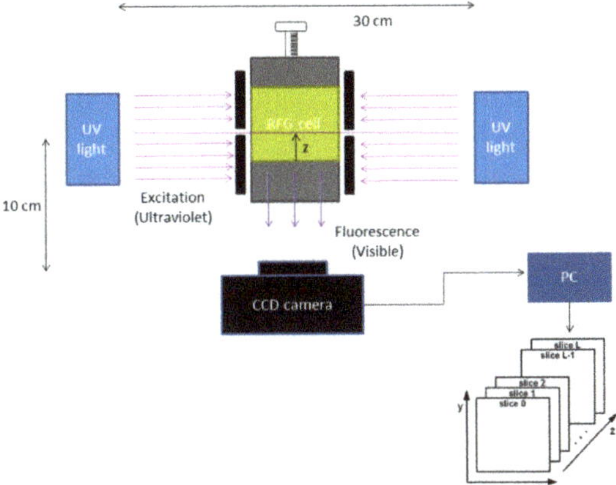

Figure 2. A schematic of the prototype shown in Figure 1 illustrating the mode of operation.

What is hidden in Figure 1 is the spaghetti-like complexity of power supplies and cable connections that made transportation to and operation at distant venues difficult. The prototype also required a dark room in the vicinity of the measurements for fast read-out. Clearly a compacter, transportable and user-friendly apparatus was required.

3. The FluoroTome 1 Apparatus

FluoroTome 1, shown in Figure 3, was designed and constructed by 4PICO BV (Sint Oedenrode, The Netherlands) in consultation with members of the Fluorodose project at the Reactor Institute of the Technical University of Delft, which provided the funding.

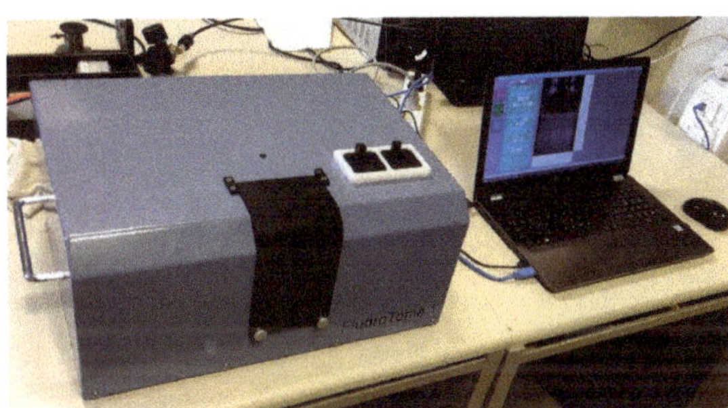

Figure 3. The complete FluoroTome 1 apparatus including laptop with software for camera and stage control and data management.

The apparatus was delivered on September 10th 2018 and further development of the software continued into January 2019.

Detailed schematics of the external and internal aspects of the apparatus are shown in Figure 4a,b with downloadable versions given in the supplemental information. A schematic of the basic layout is shown in Figure 5. This is somewhat different to that of the prototype shown in Figure 2, with the major difference being the placement of the camera on the opposite side of the sample to the observer with a 45-degree mirror acting to turn the fluorescent image of the sample through 90 degrees. This results in compactness and allows for a longer, 300 mm, image distance and a greater depth of field. As can be seen in the figure, because of the mirror, the left side of an image as taken by the camera is also the left side as viewed by the operator from the front.

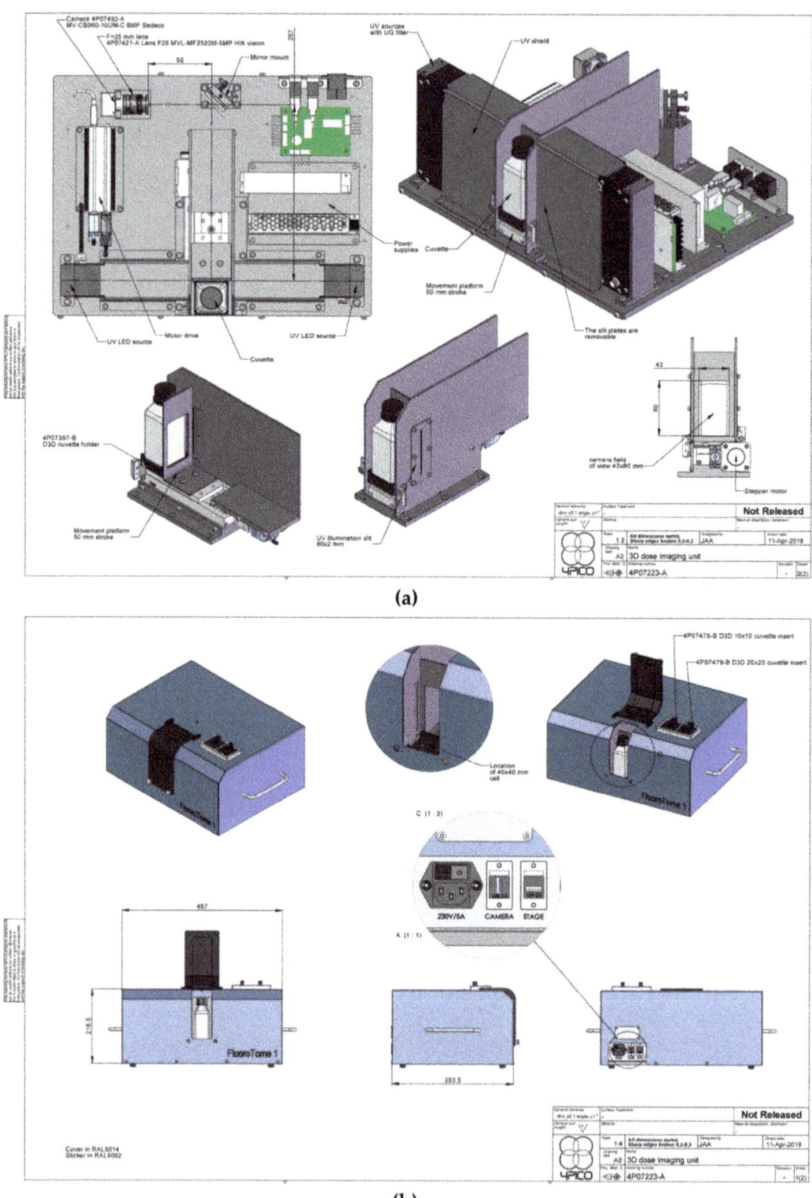

Figure 4. (**a**) (**upper**) and (**b**) (**lower**) are schematic drawings of the internal and external details of the apparatus respectively. See supplemental information for downloadable .jpg files.

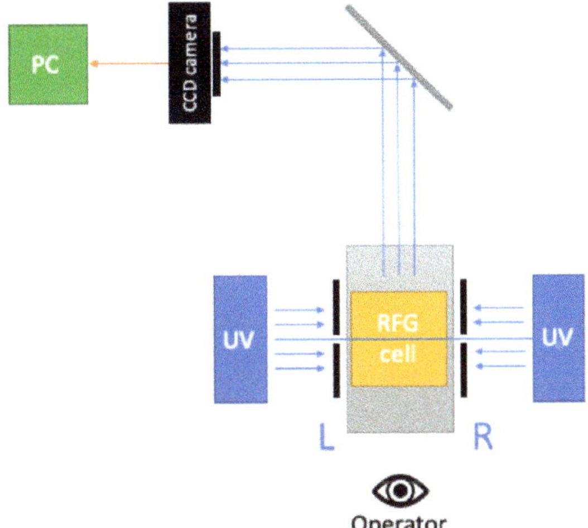

Figure 5. A schematic of the basic design of FluoroTome 1 with LED-array UV lamps, a CCD camera, and a translation stage in grey.

The basic functioning of the apparatus is as follows: with the light-tight flap open a 40 mm square borosilicate glass cell containing a fluorescent gel or solution is positioned in a holder on the translation stage. For the smaller (20 or 10 mm) sized cells the adapters shown adjacent to the flap in Figure 3 are used. The flap is closed and the translation stage with cell is moved to a given position with respect to the UV slits using the laptop software. The maximum travel of the stage is 50 mm; ±25 mm about the central position. An image of the fluorescence is taken and the raw file downloaded via a USB3 cable to the laptop where it is converted to 16-bit TIFF or 8-bit JPEG files. The camera exposure time can be varied by the user but the camera gain and color balance are fixed. Multiple images can be made in scanning mode with the width of the scan about the central position and the total number of images chosen. The images can then be individually analyzed using ImageJ or other software applications for compiling and 3D reconstruction of multiple tomographic images. A user manual has been compiled and is included in the supplemental information.

The main apparatus can be lifted and carried by a moderately fit person, see Figure 6. For transport over substantial distances a more stable and reliable form of carriage is advised in view of the somewhat critical optical alignments involved. A small bench space with two 220 V outlets is sufficient for operation. Darkroom facilities are not necessary. This makes it possible to carry out on-site measurements and analysis of the radiation-induced fluorescence in RFG gels within minutes of their radiation exposure.

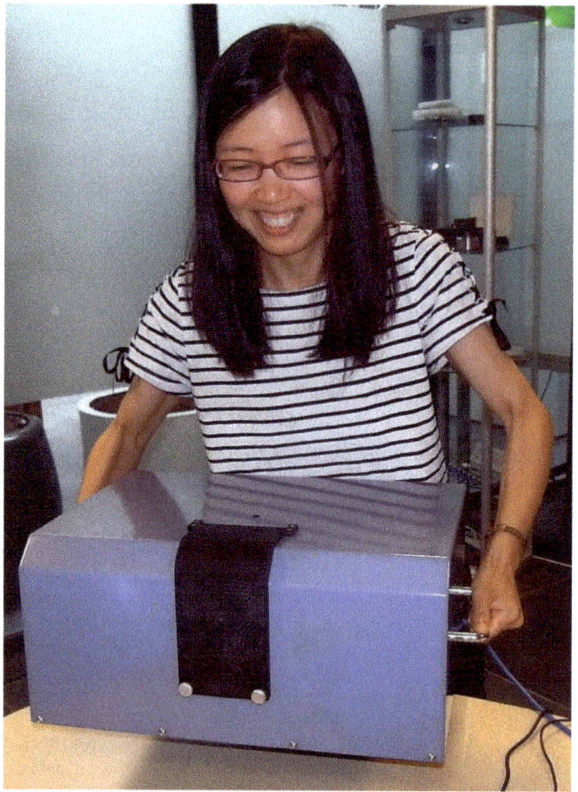

Figure 6. Portability test of FluoroTome 1.

4. Test Procedures

Prior and subsequent to delivery tests have been carried out on the various components, the operational functions and the data acquisition and handling operations of the complete apparatus. These tests are described in this section. Use of the apparatus for preliminary measurements on an irradiated gel sample are the subject of Section 5.

4.1. UV Excitation

The collimated, linear-array UV-LED lamps used were designed and constructed by 4PICO. They consist of a 16 cm long, vertical linear-array of 22 LEDs, covered by a rectangular Fresnel lens creating a 20 mm wide collimated beam. The individual LEDs were provided by High Power Lighting Corporation, New Taipei City, Taiwan (model HPL-H44DV1C0, High efficiency 3W UV LED). Passage through the slits results in a uniform sheet of UV light 2 mm thick, 60 mm high and 60 mm wide in the cell containment region. The output spectrum of the UV lamps is Gaussian with a wavelength maximum at 384 nm and a half-width (FWHM) of 10 nm. A 1 mm thick UG-1 polished optical filter (Phillips Safety Products, Middlesex, NJ, USA) attenuated any visible-light components.

The intensity of the beam transmitted by the slits was measured as a function of height above the stage using a photodiode detector and the values are listed in Table 1. The left and right side slit intensities differ by less than 2% and the sum is constant within ±2% over the distance from 30 to 90 mm. At 20 mm, the smallest height possible using the photodiode, the intensity is 10% lower than this average. This indicated that raising the base level of the cell mounts by 20 mm would result in better uniformity of illumination, as has actually been found.

Table 1. The UV intensity, measured using the photodiode, of light transmitted by the left and right side slits as a function of height above the cell holder.

Probe Height (mm)	Intensity (mW/cm^2)		
	Left	Right	Mean
20	10.2	9.6	9.9
30	11.2	10.5	10.9
40	11.0	10.4	10.7
50	11.2	10.9	11.1
60	11.5	11.0	11.3
70	11.1	11.0	11.1
80	11.1	11.0	11.1
90	10.4	11.2	10.8
100	3.0	6.5	4.8

A further test of beam uniformity was carried out using a very dilute (micromolar) solution in cyclohexane of the fluorescence standard diphenyl-anthracene (DPA). This has a fluorescence maximum at approximately 400 nm (close to that of the fluorescent version of the RFG gel) with a high extinction coefficient of 14,000 L/mol.cm and a quantum yield of fluorescence close to unity. The fluorescence of such a solution, by nature of its low-viscosity, is directly proportional to the intensity of the incident excitation light. It gives therefore a measure of any spatial variation in the incident UV light intensity.

A fluorescence image of a 40 mm ID square cell containing 30 mm of a DPA solution is shown in Figure 7 together with vertical and horizontal scans of the blue pixel intensity using ImageJ. Aside from optical artifacts at the solution/air meniscus, and solution/glass interface in the former and the solution/glass/air interfaces in the latter, the intensities are seen to be fairly uniform. The horizontal intensities display an increase in the center of the cell by 5%. Since, if anything, a slight decrease would have been expected due to UV absorption, this indicates a non-uniformity on the detection side of the apparatus which should be corrected for. The levels at the left and right side of the scan are equal within 1%, in agreement with the close similarities of the photodiode measurements of the beam intensities at the left and right hand side slits given in Table 1.

The high-power LEDs used displayed an appreciable decrease in intensity with time after turning on due to heating. Using the photodiode detector this was found to be 13 % after 5 min. Due to this, the maximum "on time" of the lamps was automatically limited to 5 min. A test, using the fluorescent solution mentioned in the previous paragraph, was also carried out and is shown in Figure 8. The decrease in the fluorescence intensity (the blue-pixel grey-scale value, P_B) of 13% over 5 min matched that found for the photodiode output measurements. Due to these observations the lamps were held in a default, "standby" *off* mode when carrying out measurements.

When taking an image the lamps were triggered to be *on* for only a short period (seconds) before and after the camera exposure that was set to a maximum value of 2 s. A dummy test has been carried out using the DPA solution in which a scan involving taking 40 images was carried out without the stage moving an appreciable distance. The change in the intensity of the image with scan time is shown in Figure 9 to be less than 1% over the total scan. Figure 9 also shows the statistical error from sample to sample in the scan to be extremely small.

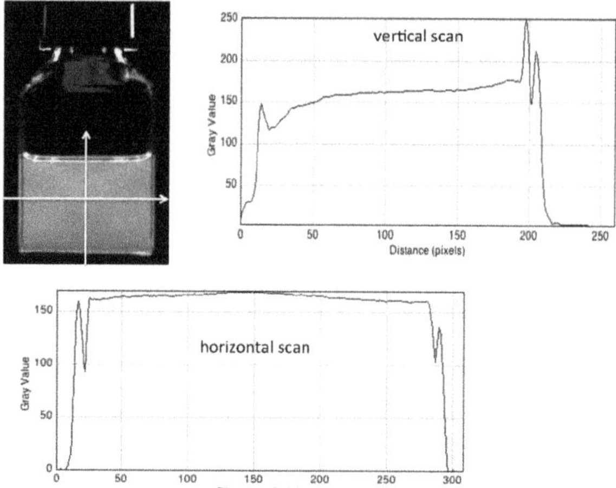

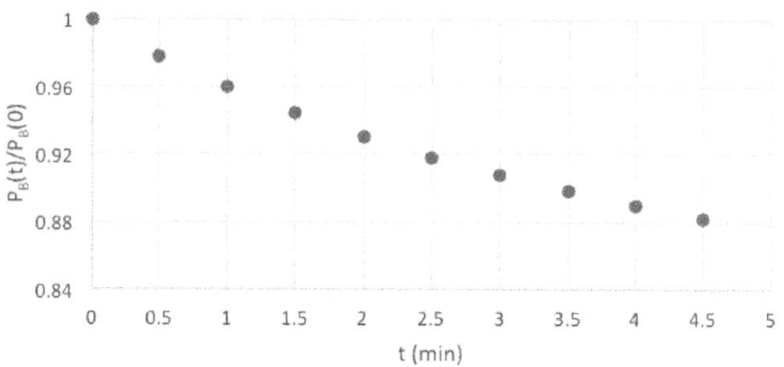

Figure 7. A fluorescence image of a 40 mm square cell containing 30 mm of a dilute liquid solution of diphenyl anthracene with vertical and horizontal intensity profiles made using ImageJ at 6.6 pixels per mm.

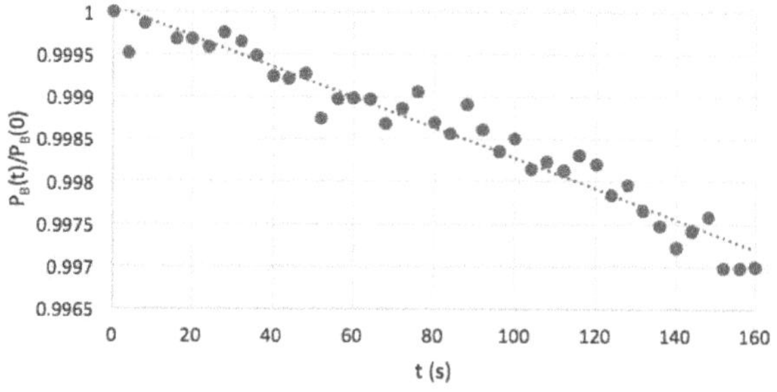

Figure 8. The decrease in fluorescence intensity of a DPA solution with time after turning on the UV lamps continuously.

Figure 9. The variation in intensity of images of a DPA solution during a dummy 40 image scan without appreciable movement of the stage using the default standby mode of operation with pulsed UV illumination.

4.2. Fluorescence Detection

Images were made of the cell, containing a fluorescent solution or gel, using an HIKVISION (Hangzhou, China) model MV-CB060-10UC camera with a 4P07421-A lens. An example of such an image of a fluorescent DPA solution is shown in Figure 7. When the stage is at its median position, the center of a cell is 300 mm from the camera. As can be seen in Figures 4b and 5, the image is taken in reflection via a 45 degree mirror which results in the view of the image by the camera being the same as that of the operator; i.e., the left side of the camera image (viewed from the rear of the cell) is also the left side as viewed from the front by the operator.

The raw images taken by the camera are transferred directly via a USB3 cable to the laptop where they are converted to 16-bit TIFF and 8-bit "JPEG" files. These files can be uploaded to the freely down-loadable software program ImageJ [11] which is used for data analysis including RGB color separation, determination of the average pixel grey levels for a specific rectangular area of an image (as in Figures 8 and 9) and making profile scans of pixel intensity (as in Figure 7).

In Figure 10 blue-pixel grey-scale values are plotted as a function of the camera exposure time for a fixed area of a fluorescent DPA solution. As can be seen, both the TIFF and "JPEG" files display a very good linear dependence on the integrated photon intensity up to saturation. In fact, the "JPEG" values are exactly 8 bits down on the TIFF values indicating that the more complex JPEG corrections, usual in commercial digital cameras, have not been applied. For most purposes therefore using the 8-bit JPEG files within the linear region is sufficient and has the advantage of being considerably smaller than the 16-bit TIFF version. This is a particularly important factor when making 3D reconstructions involving many files.

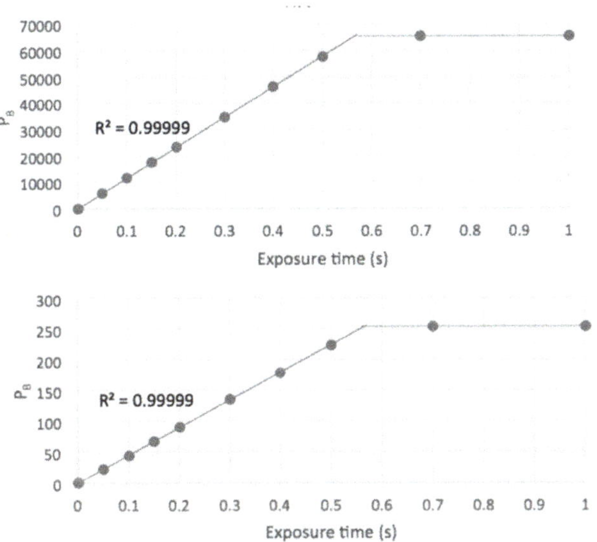

Figure 10. Blue-pixel grey-scale values of the 16-bit TIFF (upper) and 8-bit "JPEG" (lower) image files of an area of a fluorescent DPA solution as a function of the camera exposure time setting.

The camera outputs RGB files, and the color balance can be adjusted by varying the red, green and blue pixel levels individually between 0 and 4095, i.e., 12 bits. Since the fluorescence of present interest lies in the blue spectral region ($\pm$400 nm maximum), the sensitivities were set at red = 0, green = 0, and blue = 4095. This allows files in which the blue pixels in an RGB file may be saturated to be discarded. The overall gain factor could also be varied, between 1 and 20. The effect of varying the gain on the magnitude of the output for a constant light intensity is shown in Figure 11A. Changing the gain factor did not appear to influence the signal to noise ratio as shown in Figure 11B. It is important that any changes in the gain factor or the color balance are made only by acknowledged technicians and noted. These parameter changes should not be possible in the normal user interface.

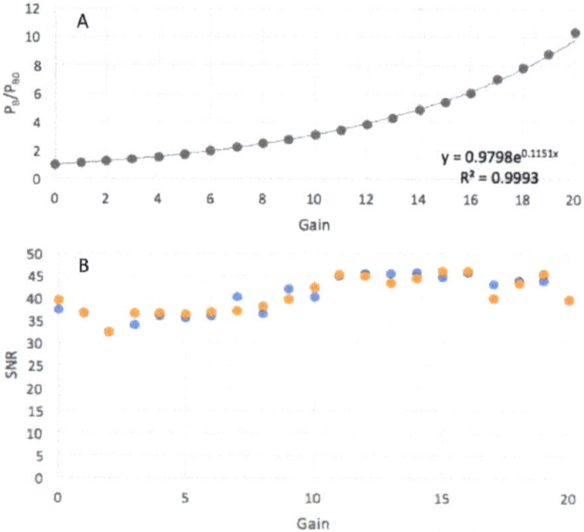

Figure 11. The dependence on the gain parameter of (**A**), the pixel level and (**B**), the signal-to-noise ratio with 16-bit images in blue and 8-bit images in orange.

A phenomenon that must be taken into account in the scanning measurements is that the length of the dielectric medium in the cell between the fluorescent sheet and the camera changes. This has the effect of changing the apparent dimensions of the object. This can be seen in Figure 12, where the apparent width of the cell measured in pixels is plotted against the length of the dielectric layer between the fluorescent layer and the camera as the scan proceeds. The width appears to decrease by approximately 10% over the 40 mm scan. This effect should be taken into account when compiling images for 3D representation.

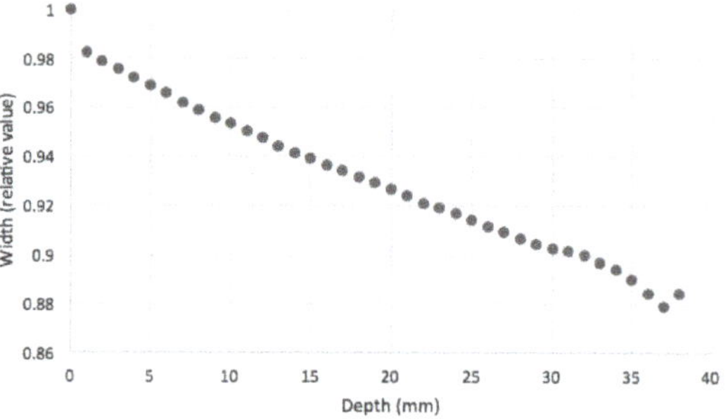

Figure 12. The apparent width of the cell as a function of the length of liquid dielectric between the UV sheet and the camera.

The magnitude of the average blue pixel level in a small, defined area of the bulk of the DPA solution as a function of the scan position is shown in Figure 13. Initial results indicated a non-uniform variation in the intensity (Figure 13 blue points). Masking of either the upper part of the slits or the cap and shoulder of the cell reduced the non-uniformity and the average pixel level by approximately 5%. The additional intensity is attributed to spurious reflections from the upper, empty regions of the cell as can be seen in Figure 14 (left). A black paper, non-fluorescent mask that covers the cap and shoulder of the cell is therefore advised. Using this the intensity is found to be constant within 3% over 35 mm for a 40 mm square cell.

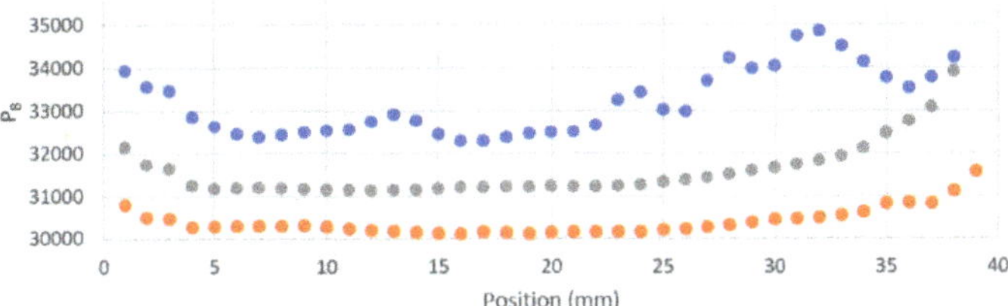

Figure 13. The average blue pixel level in a small area of the DPA solution as a function of the stage position. Upper, blue points: cell cap and slits uncovered; central, grey points: cell cap and shoulder masked; lower, orange points: top of slits masked.

Figure 14. Images of the cell containing the fluorescent DPA solution. Left; no cover. Right; black paper cover over cap and shoulder of the cell.

4.3. The Translation Stage

The translation stage used was a model 8MT175-50 from Standa Ltd. (Vilnius, Lithuania) with a maximum travel range of 50 mm and micrometer resolution. The scanning parameters were software controlled as described in the user manual (downloadable in the supplementary information). The position of the stage was defined with respect to the UV slits with position 25 corresponding to the center of the stage and the mounted cell. In positions 1 or 50 the stage is farthest from or nearest to the camera respectively. For scanning, a choice is made of the width of the scan about the central position, ΔW. As an example, a choice of 30 mm results in a scan from position 10 $(25 - \Delta W/2)$ to position 40 $(25 + \Delta W/2)$. The number of images to be taken during the scan, N, is then chosen with a choice of 31 resulting in images being taken every mm. A choice of 7 would result in images being taken every 5 mm. The step size between images is in general $\Delta W/(N - 1)$.

5. Preliminary, Irradiated RFG Gel Measurements

To test the overall functioning of the apparatus, we have carried out measurements on an RFG gel sample irradiated using our in-house x-ray source (YXLON model Y.MG325/4.5). The gel was contained in a 20 mm square cell and was 30 mm long. It was irradiated with 4, 3 mm square collimated 300 kVp x-ray beams separated vertically by approximately 7 mm center-to-center. The incident dose rates, calibrated using an ionization chamber, varied from 3.9 to 1.3 Gy/min. The exposures were adjusted to give the same (15.9 ± 0.3 Gy) accumulated dose. Images taken using FluoroTome 1 with the UV sheet at the center of the cell are shown in Figure 15 with the x-ray beam direction parallel or perpendicular to the plane of the UV sheet.

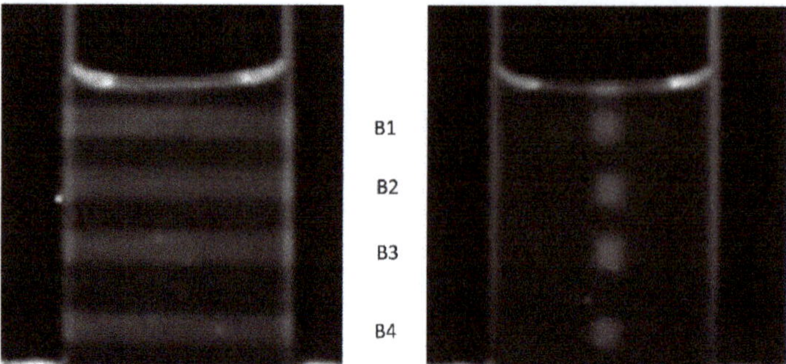

Figure 15. Images taken with Fluorotome 1 of a 20 × 20 mm square, ±30 mm long RFG gel that has been irradiated with four 3 mm square 300 kVp x-ray beams, denoted B1-B4 in the figure. All have total entrance doses of 15.9 ± 0.3 Gy with dose rates from 3.9 to 1.3 Gy/min (top to bottom). The plane of the UV-sheet was at the center of the cell (depth 10 mm) and parallel (left) or orthogonal (right) to the x-ray penetration direction.

Linear pixel profile scans, P_B versus distance, across the irradiated areas in Figure 15 (right) are shown in Figure 16, with and without background subtraction.

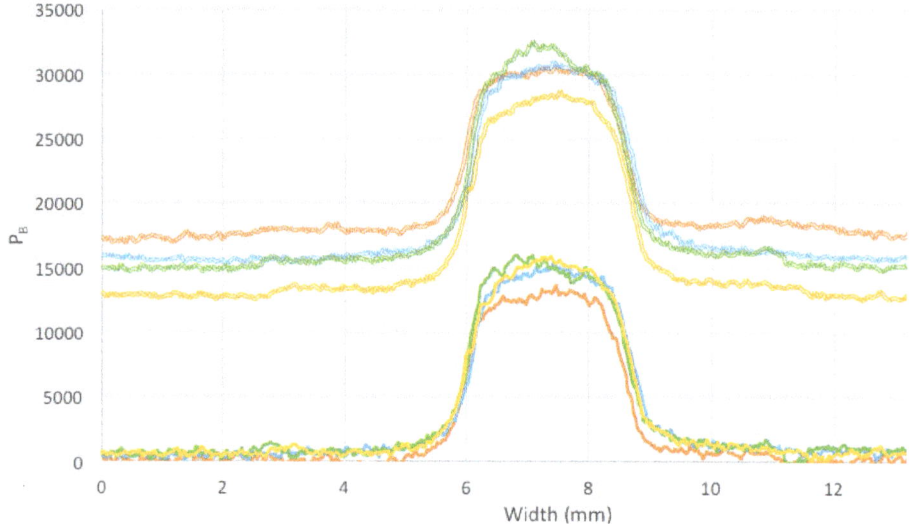

Figure 16. Linear pixel profile scans through the irradiated areas of image 15 (right) without, upper, and with, lower, baseline subtraction. The colour code is orange, B1; blue, B2; yellow, B3; green, B4. The total dose for all beams was 15.9 ± 0.3 Gy with dose rates 3.9, 2.7, 2.0, 1.3 Gy/min for B1, B2, B3, B4.

The following parameters derived from the profiles in Figure 16 are listed in Table 2: The maximum increase in intensity (pixel grey scale level, P_B) on irradiation measured at the center of the beam cross section, $\Delta I(10) = [P_{max} - P_{bkgd}](10)$; the FWHM of the beam image cross section, W; the 20–80% rise and fall of the beam image (the "penumbra"), δW.

Table 2. The parameters derived from the pixel profile scans in Figure 16, the symbols are defined in the text.

Beam	Dose Rate, D′ (Gy/min)	Dose, D (Gy)	ΔI(10)	ΔI(18)/ΔI(2)	W (mm)	δW (mm)
B1	3.9	15.6	12,800	0.834	2.62	0.41
B2	2.7	16.2	14,600	0.827	2.65	0.52
B3	2.0	16.0	15,000	0.841	2.64	0.50
B4	1.3	15.6	15,200	0.841	2.71	0.54
averages		15.9		0.836	2.66	0.49

The decrease in ΔI, by 16%, in going from a dose rate of 1.3 to 3.9 Gy/min is much greater than the 4% found between 1.3 and 2.7 Gy/min. We conclude that the dependence on dose rate is not strong but, because of the high background levels and signal noise in the present results, a more thorough investigation of this dependence should be carried out.

The FWHM of the fluorescence resulting from energy deposition by the 3.0 mm square collimated x-ray beams has a half-width of 2.7 mm with a 20–80% rise and fall of 0.49 mm. This 0.49 mm "penumbra" value is a measure of the overall sub-millimeter spatial resolution of detection including beam diffusion. The ultimate resolution possible is expected to be that corresponding to the interpixel distance, which is 0.15 mm/pixel for the present images.

In scanning the gel, images were taken every millimeter along the beam axis and the value of the radiation-induced fluorescence, ΔI(z), as a function of penetration depth z is plotted in Figure 17 for the four beams. Values for distances less than 2 mm from the cell walls have been omitted because of optical artifacts close to the interfaces. The ratio of the intensity at 18 mm to that at 2 mm is given in Table 2. This decrease with depth by a factor of 0.836 can be ascribed to attenuation of the x-ray beam over 1.6 cm in the medium i.e.,

$$\Delta I(18)/\Delta I(2) = \exp[-1.6\rho\mu] \tag{1}$$

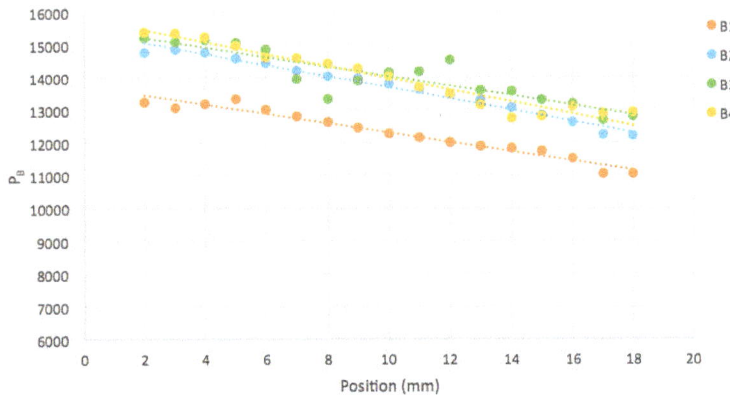

Figure 17. The radiation-induced fluorescence as a function of penetration depth along the x-ray beam for the four beams shown in Figure 15. The ratios of the value at 18 mm to that at 2 mm are listed in Table 2.

In (1), ρ is the density of the medium, 0.91 g/cm^3, and μ is the mass attenuation coefficient in g/cm^2. With ΔI(18)/ΔI(2) = 0.836, μ is 0.123 g/cm^2 for the 300 kVp x-rays used. Values of 0.154 g/cm^2 [12] and 0.163 g/cm^2 [8] for a similar tertiary-butyl acrylate gel have been determined previously for 200 kVp x-rays. The values found are close to those of 0.115, 0.138 and 0.164 g/cm^2 reported for 300, 200 and 100 keV x-rays for the chemically similar compound polymethylmethacrylate (PMMA) [13].

The procedure for forming 3D video images of the fluorescence within the irradiated gel from a series of tomographic images has been described in a previous publications [7,12]. A 3D video of the present irradiated gel, constructed from 20 images taken along the axis of the beams is given in the supporting information. In Figure 18, four still frames from the video are shown. The bright spots that appear outside of the irradiated beams arise from spurious fluorescent dust particles in the gel. They emphasize the importance of physical as well as chemical purity of the gels at all stages of their preparation.

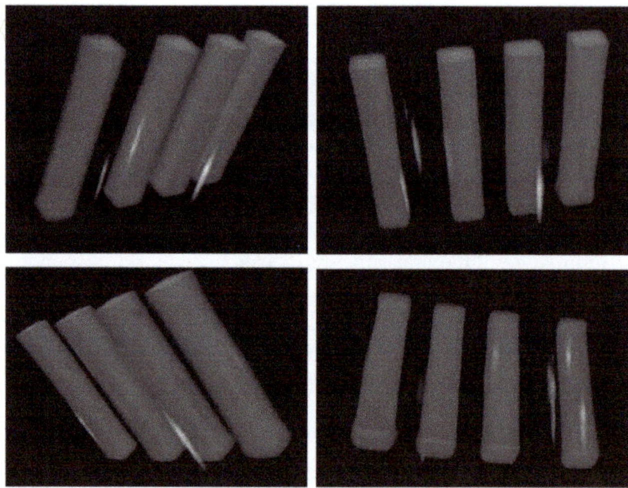

Figure 18. Still frames from the 3D video reconstructed from 20 tomographic images taken of the irradiated gel along the axis of the beams. The bright areas are related to spurious fluorescent dust particles in the gel.

6. Conclusions

FluoroTome 1 has been found to be capable of fulfilling the following product specifications:

(1)　producing a thin sheet of uniform ultraviolet light.
(2)　transporting a liquid or gel medium through this UV sheet using a remote control translation stage over a distance of at least 50 mm.
(3)　making multiple digital photographic images of the fluorescence of an irradiated medium produced by the UV excitation as the sheet is scanned through the medium.
(4)　having a light-tight encasement; no darkroom requirement.
(5)　being readily portable and transportable with no special local facilities required.
(6)　capable of preliminary data processing on-site, within minutes of radiation exposure of a gel medium.
(7)　having a user-friendly software interface.

It is intended to apply RFG gels and FluoroTome 1 to the study of energy deposition by proton (pencil) beams produced at the Holland Proton Therapy Centre which was recently constructed on a site adjacent to the Reactor Institute of the Delft University of Technology. In addition to fundamental studies of parameters affecting energy deposition, more general applications to dosimetric aspects of quality control and personnel training in radiotherapy are envisaged. Other formulations of radio-fluorogenic gels than that used by the present authors have been reported [14,15]. These should also be amenable to tomographic measurement using FluoroTome 1.

Supplementary Materials: Supplementary Materials: The following are available online at http://www.mdpi.com/2073-4360/11/11/1729/s1.

Author Contributions: Investigation, S.Y. and J.H.; methodology, J.M.W., M.P.d.H., L.H.L., T.Y. and J.A.; project administration, J.M.W., L.H.L., J.A. and W.i.h.Z.; resources, J.A.; software, S.T.; supervision, J.M.W. and J.A.; validation, J.N.-C., S.T. and J.H.; visualization, T.Y. and J.N.-C.; writing—original draft, J.M.W. and J.N.-C.

Funding: The research was partially funded by The Netherlands Organisation for Scientific Research (NWO); project "Protons4Vision" nr 14654. Julia Navarro-Campos was supported by the Erasmus program of the European Union.

Conflicts of Interest: The authors declare no conflict of interest.

References

1. Ibott, G. 9th International Conference on 3D Radiation Dosimetry. *J. Phys. Conf. Ser.* **2017**, *847*, 011001. [CrossRef]

2. Pasler, M.; Hernandez, V.; Jornet, N.; Clark, C.H. Novel methodologies for dosimetry audits: Adapting to advanced radiotherapy techniques. *Phys. Imaging Radiat. Oncol.* **2018**, *5*, 76–84. [CrossRef]

3. Watanabe, Y.; Warmington, L.; Gopishankar, N. Three-dimensional radiation dosimetry using polymer gel and solid radiochromic polymer: From basics to clinical applications. *World J. Radiol.* **2017**, *9*, 112–125. [CrossRef] [PubMed]

4. Kron, T.; Lehmann, J.; Greer, P. Dosimetry of ionising radiation in modern radiation oncology. *Phys. Med. Biol.* **2016**, *61*, R167–R205. [CrossRef] [PubMed]

5. Baldock, C.; De Deene, Y.; Doran, S.; Ibbott, G.; Jirasek, A.; Lepage, M.; McAuley, K.B.; Oldham, M.; Schreiner, L.J. Polymer gel dosimetry. *Phys. Med. Biol.* **2010**, *55*, R1–R63. [CrossRef] [PubMed]

6. Warman, J.M.; de Haas, M.P.; Luthjens, L.H.; Denkova, A.G.; Yao, T. A radio-fluorogenic polymer-gel makes fixed fluorescent images of complex radiation fields. *Polymers* **2018**, *10*, 685. [CrossRef] [PubMed]

7. Yao, T.; Gasparini, A.; De Haas, M.P.; Luthjens, L.H.; Denkova, A.G.; Warman, J.M. A tomographic UV-sheet scanning technique for producing 3D fluorescence images of x-ray beams in a radio-fluorogenic gel. *Biomed. Phys. Eng. Express* **2017**, *3*, 027004. [CrossRef]

8. Luthjens, L.H.; Yao, T.; Warman, J.M. A polymer-gel eye-phantom for 3D fluorescent imaging of millimetre radiation beams. *Polymers* **2018**, *10*, 1195. [CrossRef] [PubMed]

9. Warman, J.M.; Luthjens, L.H.; De Haas, M.P. High-energy radiation monitoring based on radio-fluorogenic co-polymerization. II: Fixed fluorescent images of collimated X-ray beams using an RFCP gel. *Phys. Med. Biol.* **2011**, *56*, 1487–1508. [CrossRef] [PubMed]

10. Warman, J.M.; De Haas, M.P.; Luthjens, L.H.; Denkova, A.G.; Kavatsyuk, O.; Van Goethem, M.-J.; Kiewiet, H.H.; Brandenburg, S. Fixed fluorescent images of an 80 MeV proton pencil beam. *Radiat. Phys. Chem.* **2013**, *85*, 179–181. [CrossRef]

11. Rasband, W.S. *1997–2009 ImageJ*; US National Institutes of Health: Bethesda, MD, USA, 2008. Available online: http://rsb.info.nih.gov/ij/ (accessed on 17 September 2019).

12. Yao, T. 3D Radiation Dosimetry Using a Radio-Fluorogenic Gel. Ph.D. Thesis, Technische Universiteit Delft, Delft, The Netherlands, 16 January 2017.

13. Hubbel, J.H.; Seltzer, S.M. *Tables of X-Ray Mass Attenuation Coefficients and Mass Energy-Absorption Coefficients from 1 keV to 20 MeV for Elements Z = 1 to 92 and 48 Additional Substances of Dosimetric Interest*; No. PB-95-220539/XAB; National Inst. of Standards: Gaithersburg, MD, USA, 2004.

14. Maeyama, T.; Shinnosuke, H. Nanoclay gel-based radio-fluorogenic gel dosimeters using various fluorescence probes. *Radiat. Phys. Chem.* **2018**, *151*, 42–46. [CrossRef]

15. Sandwall, P.A.; Bastow, B.P.; Spitz, H.B.; Elson, H.R.; Lamba, M.; Connick, W.B.; Fenichel, H. Radio-fluorogenic gel dosimetry with coumarin. *Bioengineering* **2018**, *5*, 53. [CrossRef] [PubMed]

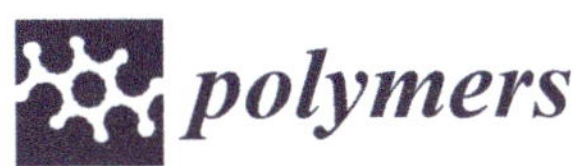

Article

Impact of Ultraviolet Radiation on the Aging Properties of SBS-Modified Asphalt Binders

Huanan Yu [1], Xianping Bai [2], Guoping Qian [1,*], Hui Wei [1], Xiangbing Gong [1], Jiao Jin [1] and Zhijie Li [3]

[1] National Engineering Laboratory for Highway Maintenance Technology, School of Traffic and Transportation Engineering, Changsha University of Science & Technology, Changsha 410114, China
[2] School of Traffic and Transportation Engineering, Changsha University of Science & Technology, Changsha 410114, China
[3] Guizhou Transportation Planning Survey & Design Academe Co., Ltd., Guizhou 550000, China
* Correspondence: guopingqian@sina.com

Received: 3 June 2019; Accepted: 20 June 2019; Published: 1 July 2019

Abstract: Styrene Butadiene Styrene (SBS) polymer-modified asphalt binders have become widely used in asphalt pavement because of their advantages in high- and low-temperature performance and fatigue resistance. Asphalt pavement is inevitably exposed to sunlight and ultraviolet (UV) radiation during its construction and service life. However, consideration of the aging effect of UV radiation is still limited in current pavement design and evaluation systems. In order to evaluate the impact of UV radiation on the aging properties of SBS-modified asphalt binders, UV aging tests were performed on Rolling Thin Film Oven Test (RTFOT)-aged samples with different UV radiation intensities and aging times. Sixteen different groups of tests were conducted to compare the rheological properties and functional group characteristics of SBS-modified asphalt binders. Dynamic Shear Rheometer (DSR), Bending Beam Rheometer (BBR), FTIR, and SEM tests were conducted to evaluate the aging mechanisms in various UV aging conditions. The results found that UV radiation seriously destroys the network structure formed by the cross-linking effect in SBS-modified asphalt binders, which aggravates the degradation of SBS and results in a great change of rheological properties after UV aging. The nature of SBS-modified asphalt binder aging resulted from the degradation of SBS and the changes of asphalt binder base composition, which lead to the transformation of colloidal structure and the deterioration of asphalt binder performance. The tests also found that continuous UV radiation can increase shrinkage stress in the asphalt binder surface and leads to surface cracking of the asphalt binder.

Keywords: SBS-modified asphalt binder; UV aging; rheological properties; functional group; cracking

1. Introduction

Asphalt pavements are widely used in the construction of roads because of their advantages of having a smooth surface, seamlessness, lower driving noise, easy maintenance, and recyclability [1,2]. Among all asphalt binder modification types, Styrene Butadiene Styrene (SBS)-modified asphalt binders are widely used in pavement areas because of their excellent performance under high and low temperature and their fatigue resistance, and they have been applied in a huge percentage of global asphalt pavement construction projects [3].

Normally, asphalt binder aging is characterized into short-term aging and long-term aging. The short-term aging typically occurs at the mixing and paving process, and the long-term aging occurs during the whole service life. Researches have pointed out that a wide range of factors impact the properties of the polymer/asphalt system [4,5]. As asphalt pavements are inevitably exposed to UV radiation, water, and thermal and other environments during their whole service life, the long-term aging is a combined impact of various factors. Consequently, different countries have developed several standardized accelerated aging methods in laboratory to simulate the in-field aging process. However, very little consideration has been given to the impact of UV radiation on aging.

Researchers have found that the impact of UV radiation has led to the aging of asphalt pavements, which results in the deterioration of road performance and the reduction of road service life [6]. Furthermore, UV radiation from sunlight has become more and more intense because of global warming and environmental changes, and thus the impact of UV aging on asphalt pavements is becoming more and more serious.

Some researchers pointed out that UV radiation might only impact the upper layers of the asphalt pavement surfacing, and the influence of UV radiation on asphalt aging has always been ignored in laboratory simulations of aging [7]. However, Durrieu [8] pointed out that the impact of UV radiation on the aging of asphalt pavements cannot be ignored, and the research found that the aging of asphalt samples under 10 hours of UV radiation in the lab was equal to the same level of Rolling Thin Film Oven Test (RTFOT) and Pressure Aging Vessel (PAV) aging, or aging equal to one year of service in field.

In order to quantify the rate and degree of UV aging of asphalt binders, Zheng et al. [9] established the nonlinear equation of the performance attenuation of asphalt binders after UV aging by using three indexes of penetration, viscosity, and ductility, respectively. Wang et al. [10] studied the effect of UV aging on the aggregated state of asphalt binders by various test methods, and found that UV aging reduced the rheological characteristics of asphalt binders, and that asphalt material is transformed from a viscous material to an elastic material under normal temperature. Xiao et al. [11] compared the impacts of long-term thermal and UV aging on foamed Warm Mix Asphalt (WMA) mixtures, and found that UV-aged mixtures had reduced the Indirect Tension Strength (ITS) values and increased dissipated energy compared to other asphalt mixtures. Kemp et al. [12] found that a 5–10 μm-thick layer of harder film forms above the asphalt binder under UV radiation. When the asphalt film thickness is greater than 200 μm, the change of the asphalt binder functional group is obviously decreased.

By means of Dynamic Shear Rheometer (DSR), FTIR, the Indirect Tension Strength (GPC), nuclear magnetic resonance hydrogen spectroscopy (1H-NMR), and other experiments, Ran [13] explored the changes of the macroscopic properties and microstructure of base asphalt binders and SBS-modified asphalt binders under the coupling of heat, light, and water, and, based on the physical and rheological properties, the nonlinear coupled aging rate prediction differential equation and the aromaticity microstructure aging kinetic equation were established. Wei et al. [14] evaluated the aging behavior of SBS-modified asphalt binders under various UV and water conditions, and found that water aggravates the UV aging of asphalt binder material, and the presence of acid or salt worsens UV aging. Zeng et al. [15] found that UV radiation can only transmit within 4.5 μm of the surface layer, but the UV aging depth can reach deeper and increases with the UV radiation time. The research pointed out that the different components of surface-aged asphalt binders diffuse to the lower parts of the asphalt binders, leading to the increasing of aging. Mouillet et al. [16] evaluated the impacts of UV radiation on elastomer-modified asphalt binders through FTIR and Scanning Electron Microscope (SEC) methods, and their results found that the elastomer architecture did not impact the degradation under UV radiation.

Hou et al. [17] pointed out that Fourier transform infrared spectroscopy (FTIR) is able to evaluate the aging characteristics and mechanisms of asphalt binder materials. Hu et al. [18] evaluated the impacts of different wavebands of UV radiation on the aging of asphalt binder materials, and found that UV aging has a great impact on the low-temperature performance of asphalt binders. Zeng et al. [19] pointed out that the effect of temperature on UV aging cannot be ignored, and suggested that the UV aging tests should be performed below 50 °C. Their research also pointed out that the coupling of high temperature and UV radiation accelerated the volatilization rate and oxidation rate of light components.

In recent years, scholars have made certain achievements in the study of asphalt binder UV aging. However, due to the variability of asphalt binder components and the complexity of the UV aging process, the understanding of the mechanism of asphalt binder UV aging is still not clear, and a standard asphalt binder UV aging evaluation system has not been established. In addition, the existing UV aging studies of asphalt binders are mainly focused on the base asphalt, while the studies on polymer-modified asphalt binders are relatively few in number. Therefore, in order to evaluate the macroscopic properties and the microstructure of SBS-modified asphalt binders, this paper conducted rheological tests and FTIR tests to reveal the aging characteristics and mechanisms of SBS-modified asphalt binders under different UV light intensities and time conditions.

2. Materials and Methods

2.1. Materials

SBS-modified asphalt binders supplied by Hua Te Asphalt Co., Ltd. (Xiamen, China) were used in this research, with an SBS concentration of 4%. The conventional tests were conducted following the procedures of "Standard Test Methods of Bitumen and Bituminous Mixtures for Highway Engineering" (JTG E20-2011) [20]. The test results and specification requirements are shown in Table 1, and it can be seen from that table that the asphalt binders satisfied the specification requirements.

Table 1. Physical properties of Styrene Butadiene Styrene (SBS)-modified asphalt binders.

Item		Value	Specification
Penetration (25 °C, 0.1 mm)		55	40–60
Ductility (5 °C, cm)		34	≥20
R&B temperature, softening point, (°C)		87	≥60
Viscosity (135 °C, Pa·s)		2.1	≤3
Flash point (COC, °C)		318	≥230
RTFOT	Mass loss (%)	0.06	≤±1.0
	Residual penetration ratio (%)	80	≥65
	Residual ductility ratio (%)	21.0	≥15

The flow chart of the experimental design is shown in Figure 1. Firstly, the virgin asphalt binder was short-time aged by RTFOT, and then UV aging was conducted on the short-time aged samples. Then, the Dynamic Shear Rheometer (DSR) and Bending Beam Rheometer (BBR) tests were conducted to evaluate the rheological properties of the aged samples, and Fourier Transform Infrared Spectroscopy (FTIR) and Scanning Electron Microscope (SEM) tests were conducted to evaluate the microstructure of the aged samples. Three duplicated samples were prepared for each test.

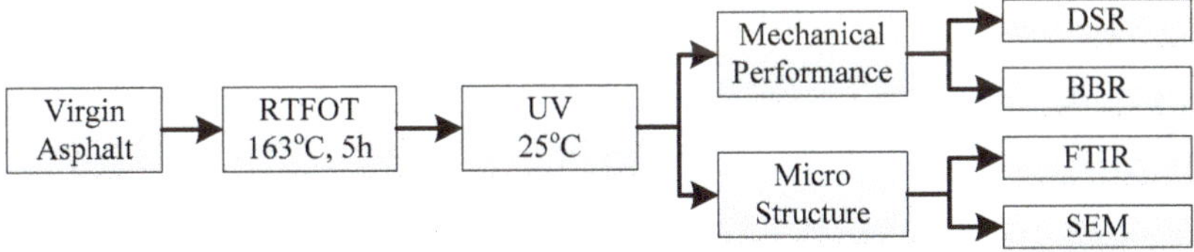

Figure 1. The flowchart of experimental design.

2.2. Aging Procedure

RTFOT short-term aging tests was conducted following the procedure of "Standard Test Methods of Bitumen and Bituminous Mixtures for Highway Engineering" (JTG E20-2011) [20]. UV aging tests were performed in an UV weathering chamber at a wavelength of ultraviolet long-wave (UVA) 365 nm. The steps for UV aging preparation are shown below:

Firstly, we poured around 15 ± 0.1 g of the RTFOT-aged SBS-modified asphalt binder into a 85 × 85 mm silicone paper carton, then placed it in an oven at 160 °C for 3 min, and then took it out to allow it cool down to room temperature. Then, a sample of asphalt binder with film thickness of 2 mm was prepared. Then, the samples were put into a UV radiation (UVA 365 nm; LED) chamber to conduct UV aging, and the UV radiation chamber was set at a temperature of 25 °C to eliminate the impact of temperature changes. The UV radiation tests were conducted at a range from 40 to 160 h, with an interval of 40 h. As the environmental UV radiation intensity between the Class-I and Class-II standards of China is close to 5 mW/cm^2, and for the consideration of accelerated aging, the UV intensity ranged from 5 to 20 mW/cm^2, with an interval of 5 mW/cm^2 [21]. In this research, the naming and UV radiation tests schematic are shown in Table 2.

Table 2. UV radiation tests schematic.

Naming (Radiation Intensity)	UV Radiation Time			
UV1 (UV = 5 mW/cm^2)	40 h	80 h	120 h	160 h
UV2 (UV = 10 mW/cm^2)	40 h	80 h	120 h	160 h
UV3 (UV = 15 mW/cm^2)	40 h	80 h	120 h	160 h
UV4 (UV = 20 mW/cm^2)	40 h	80 h	120 h	160 h

2.3. Dynamic Shear Rheometer (DSR)

The rheological properties (complex modulus and phase angle) of the SBS-modified asphalt binders before and after UV aging with different UV aging times and intensities were investigated by the DSR test. The DSR test instrument was Physica MCR 301 by Anton Paar (Ostfildern, Germany). Temperature sweep tests were conducted under the strain-controlled mode with a constant frequency of 10 rad/s, and the temperature ranged from 40 to 90 °C, with an increment of 2 °C/min. The test specimens were prepared with a set of parallel plates of 25 mm diameter, with a 1 mm gap between each plate according to the AASHTO T315-05 [22].

2.4. Bending Beam Rheometer (BBR)

BBR tests were carried on RTFOT- and UV-aged specimens by a Cannon thermoelectric rheometer at two different temperatures of −12 and −18 °C, following the procedures of American Association of State Highway and Transportation Officials (AASHTO) T313-12 [23]. The load and deformation of all specimens at 8, 15, 30, 60, 120, and 240 s were collected, and the stiffness modulus S and creep rate m were calculated. Three duplicated samples were prepared for each test. The values of S and m at 60 s were used to characterize the low-temperature properties of SBS-modified asphalt binders with different UV aging degrees.

2.5. Fourier Transform Infrared Spectroscopy (FTIR)

FTIR is a quantitative and qualitative technique for analyzing the contents of organic components based on the changes of functional groups, which has been demonstrated to be an effective tool to describe the aging characterization of asphalt binders [17,24,25]. The principle of FTIR is that the molecules vibrate when organic compounds absorb infrared light at the corresponding wavelength, and the molecular vibration can be divided into stretching vibration and bending vibration (see Figure 2). Since those two vibrational types have their corresponding absorption peaks in infrared spectra, the existence of functional groups and the contents of components in the molecules can be identified according to the positions and intensities of the absorption peaks.

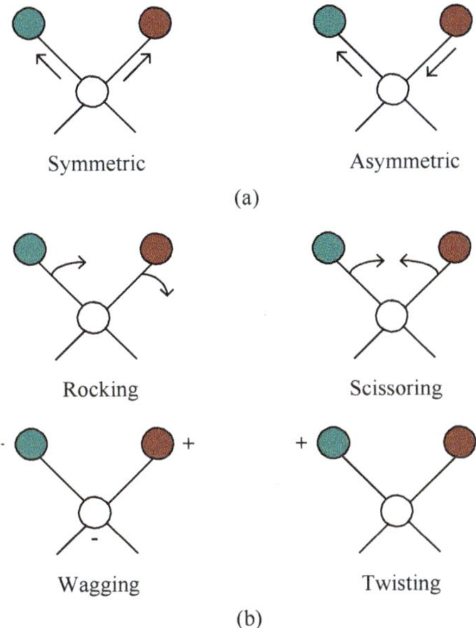

Figure 2. Molecular vibration form: (**a**) Stretching vibration and (**b**) bending vibration.

In this study, the changes of specific functional groups of asphalt binders with different aging degrees were evaluated by FTIR (Tensor 27 of Bruker manufactured by Germany). The absorbance spectra of asphalt binders were recorded, ranging from 4000 to 400 cm^{-1}, and the spectral resolution was 1–0.4 cm^{-1}. The main functional group band assignments and positions are summarized in Table 3. Furthermore, the main aging indexes, including carbonyl group (C=O) and sulfoxide group (S=O), were introduced for a more intuitive result [16,26], as shown in Equations (1)–(4):

$$I_{C=O} = \frac{A_{1700}}{A_{1460} + A_{1375}} \tag{1}$$

$$I_{S=O} = \frac{A_{1030}}{A_{1460} + A_{1375}} \tag{2}$$

$$I_{PB} = \frac{A_{966}}{A_{1460} + A_{1375}} \tag{3}$$

$$I_{PS} = \frac{A_{699}}{A_{1460} + A_{1375}} \tag{4}$$

where $I_{C=O}$ is the aging index of the carbonyl group (C=O), $I_{S=O}$ is the aging index of the sulfoxide group (S=O), I_{PB} is the aging index of butadiene (PB), and I_{PS} is the aging index of styrene (PS). A is the area of the corresponding absorption peak. For example, A_{1700} refers to the absorption peak at a wavenumber of 1700 cm^{-1}. Figure 3 shows a typical FTIR spectra and the functional groups of petroleum asphalt and biobinders [27].

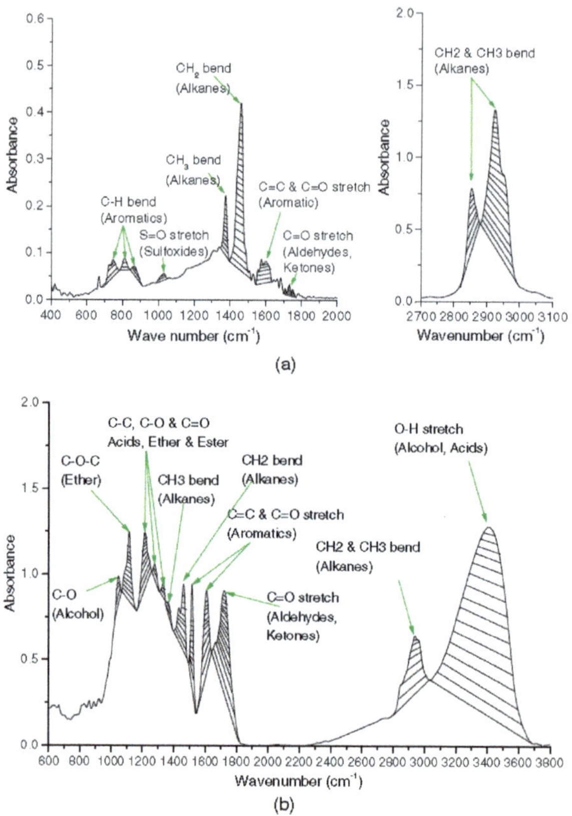

Figure 3. Spectra and functional groups: (**a**) Spectra and functional groups for the asphalt binder; (**b**) spectra and functional groups for the biobinder [27].

Table 3. Functional groups and corresponding wavenumber in infrared spectra [28].

Wavenumber (cm^{-1})	Assignment
2951	ν as CH$_3$-aryl
2921	ν as CH$_3$, CH$_2$
2852	ν s CH$_3$, CH$_2$
1710	ν C=O
1600	ν C=C
1456	δ as CH$_3$, CH$_2$
1376	δ s CH$_3$
1311	ν SO$_2$, or ester groups
1168	ν C–O–C (anhydrides), –S–C
1021	ν O-C=O, C–S=O, ether or ester groups
865	δ CH aromatic (out-of-plane bending)
813	δ CH aromatic (out-of-plane bending)
744	δ CH aromatic (out-of-plane bending)
723	r CH$_2$ (rocking CH$_2$ groups chains (CH$_2$)$_n$)
966	ν butadiene(PB)
699	ν styrene (PS)

ν = stretching; δ = bending; s = symmetric; as = asymmetric.

2.6. Scanning Electron Microscope (SEM)

The scanning electron microscope (SEM) is an important tool which can be used to observe the micro-morphology and microstructure of materials. The principle is that by emitting high-energy electron beams on a sample, when the electron beams come into contact with the surface of the sample, they will produce information such as secondary electron, scattered back electron, and so on, and the information will be changed from optical signals to electric signals through systematic processing. After the video amplifier is amplified, the image of the sample surface can be formed on the screen of the picture tube. In this paper, the aging asphalt binder samples were scanned by SEM under different conditions to observe the surface cracking characteristics of the asphalt binder samples after UV aging.

3. Results

3.1. High-Temperature Performance

The temperature sweep data from the DSR testing was used to obtain the indexes of the complex modulus and phase angle, as shown in Figures 4 and 5. It was concluded from Figure 4 that, together with the UV aging time or UV intensity increment, the complex modulus of SBS-modified asphalt binders also increased. With the prolongation of aging time and the increase of UV intensity, asphalt binder aging became worse, and the light component of the asphalt binder was transferred to the recombination fraction. In addition, the more serious the aging, the more volatile the light components. After aging, the viscosity of the asphalt binder decreased, the elastic component increased, and the asphaltene content increased, which show that the asphalt binder hardens, so the composite modulus gradually becomes larger. It can be observed that SBS-modified asphalt aging kept approximately the same complex modulus under the same aging time when the UV intensity was 15 mW/cm^2 or less. However, when the UV intensity was greater than 15 mW/cm^2, the complex modulus was greatly increased. In addition, the curves have different slopes, indicating different temperature susceptibilities between virgin, RTFOT-aged, and UV-aged asphalt.

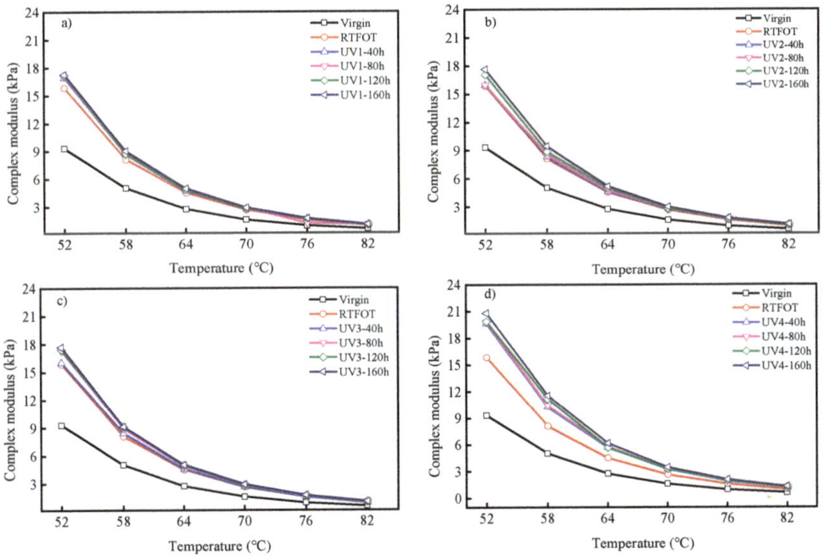

Figure 4. Complex modulus of SBS-modified asphalt binders under different UV intensities: (**a**) UV = 5 mW/cm^2; (**b**) UV = 10 mW/cm^2; (**c**) UV = 15 mW/cm^2; (**d**) UV = 20 mW/cm^2.

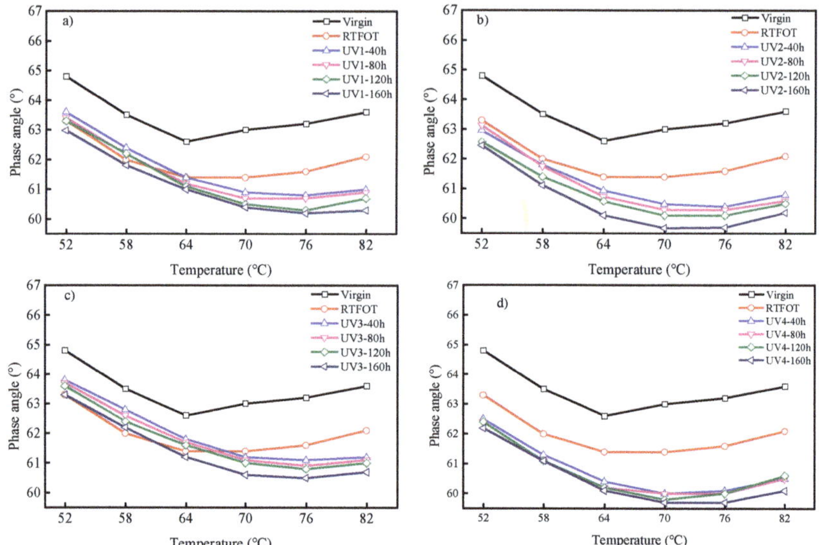

Figure 5. Phase angle of SBS-modified asphalt binders under different UV intensities: (**a**) UV = 5 mW/cm^2; (**b**) UV = 10 mW/cm^2; (**c**) UV = 15 mW/cm^2; (**d**) UV = 20 mW/cm^2.

The phase angle represents the ratio of viscosity to elastic composition of the asphalt binder. The smaller the phase angle, the greater the proportion of the elastic component. It can be seen from Figure 5 that the phase angle of SBS-modified asphalt binders gradually decreases with increasing test temperature at the beginning, and then increases regardless of aging type and aging time, showing different characteristics to virgin asphalt binders. Generally speaking, with the increase of temperature, asphalt binders will gradually change from viscoelastic to viscosity properties, and the phase angle increases gradually. The reduction of phase angle at the beginning is mainly attributed to the existence of SBS. At a certain temperature, SBS has high elasticity and fatigue resistance. When the temperature continues to rise, the polystyrene phase gradually softens and flows to make the SBS plastic. Under the same conditions, the longer the aging time or the greater the UV intensity is, the smaller the phase angle and the greater the level of asphalt binder aging is. Compared with the original asphalt binder and the short-term aging asphalt binder, the turning point temperature of the UV-aging asphalt binder decreases first and then increases, which is due to the serious destruction of the network structure formed by the crosslinking effect in the SB-modified asphalt binder by UV radiation, and the aggravated degradation of SBS. Regardless of UV intensity and aging time, the turning point of the phase angle of SBS-modified asphalt binders was approximately 70 °C, which was about 6 °C higher than that of the virgin asphalt binder and the one after short-term aging.

3.2. Low-Temperature Performance

Strategic Highway Research Program (SHRP) demonstrated that asphalt binder properties greatly impact the low temperatures performance of asphalt mixture [29]. According to the mechanism of low-temperature cracking of asphalt pavement, low-temperature rheology is the most important factor affecting low-temperature cracking of asphalt pavement [30]. Therefore, the aging behavior of asphalt binders can be theoretically evaluated from the perspective of the rheological properties of asphalt binders at low temperature. The test results show that at −18 °C, the SBS-modified asphalt binders with different aging conditions have $S > 300$ MPa and $m < 0.3$. This indicates that the low-temperature performance of SBS-modified asphalt binders has failed at −18 °C, according to the specifications. Therefore, only the BBR test results of asphalt binder samples at −12 °C are listed in this paper, as shown in Figures 6 and 7. The stiffness modulus S basically increases with the increase of aging time or UV radiation intensity. However, when the UV radiation intensity reaches a certain level, the stiffness modulus decreases as the strength increases. The creep rate m of UV-aged samples shows a slight degree of a decreasing trend, however, the trend is not significant. The reason is mainly due to the fact that UV only affects the surfaces of asphalt binders, and the aging part of the asphalt binder distribution in the beam is different.

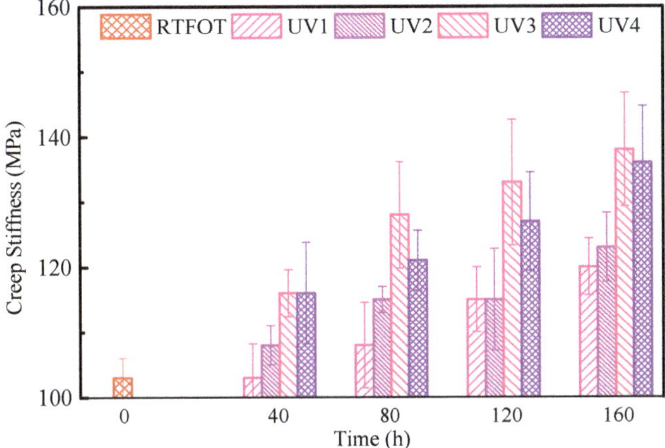

Figure 6. Creep stiffness (S value) of SBS-modified asphalt binders under UV and Rolling Thin Film Oven Test (RTFOT) aging.

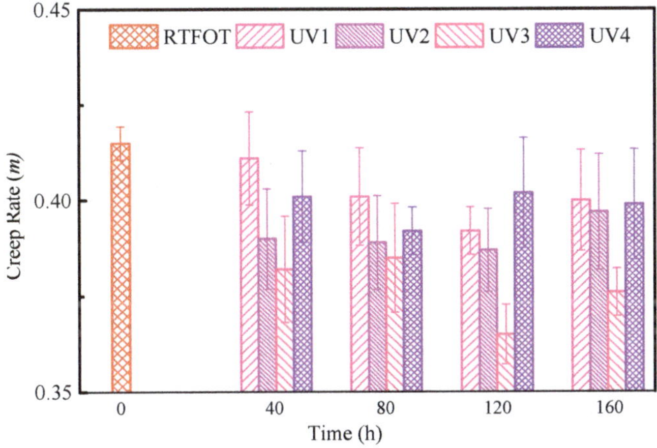

Figure 7. Creep rate (m value) of SBS-modified asphalt binders under UV and RTFOT aging.

3.3. Functional Group Change

Due to the influences of coating thickness, environmental noise, instrument, and background interference on the infrared spectrum test, the spectrum will drift and tilt to a certain extent, so it is necessary to optimize the processing of the original spectral data [31]. In this paper, the normalized preprocessing method is used to eliminate the adverse effects of interference information on the test results and to enhance the differences between spectra. The corresponding functional group aging index is shown in Figure 8.

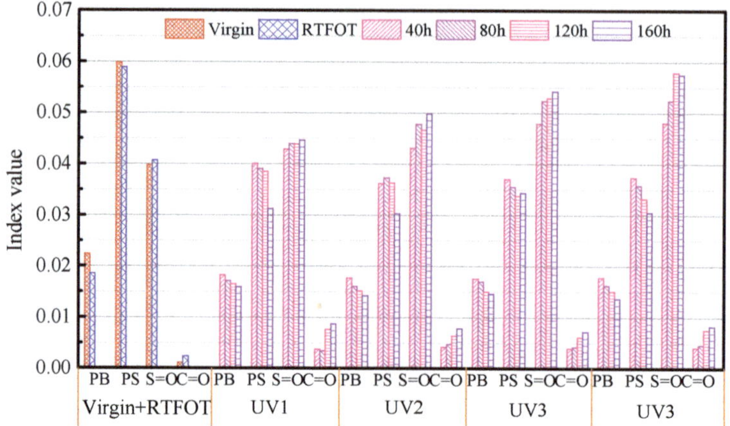

Figure 8. Aging indices of different functional groups in SBS-modified asphalt binders.

It can be seen that, after UV aging, the intensities of the stretching vibration absorption peaks of C=O at 1700 cm^{-1} and S=O at 1030 cm^{-1} fluctuate slightly, but the amplitudes are not large. This is due to the C–O oxygenation to C=O and the C–O and C–S breakage to produce S=O during asphalt binder aging. Studies have shown that the degree of aging can be characterized by the $I_{C=O}$ and $I_{S=O}$ indices. However, the amplitudes of the C=O and S=O bands in the spectrogram caused by the aging effect are too weak to calculate the peak area accurately. In addition, as the content of S in asphalt binders is fixed, the $I_{C=O}$ and $I_{S=O}$ indices are inappropriate to represent the degree of asphalt binder aging.

The symmetric and asymmetric bending vibration absorption peaks of methyl and methylene at 1460 and 1375 cm^{-1}, respectively, are the most stable in the spectrum, and therefore, they are often used as the reference bands for quantitative analysis of infrared spectra. The absorption peaks of virgin and SBS-modified asphalt binders at 1460 and 1375 cm^{-1} are basically the same as those of short-term aging asphalt binders. However, after UV aging, the absorption peaks of these two sites increase slightly, and the amplitudes of absorption spectra increase with the increase of UV aging time. This is mainly due to the destruction of C=C bonds of unsaturated olefins in asphalt binders irradiated by UV radiation, and the formation of C–H bonds. The absorption peaks at 966 and 699 cm^{-1} are the differences between SBS-modified asphalt binders and base asphalt binders. The peak at 966 cm^{-1} represents the bending vibration absorption peak of PB C=C in SBS, and the 699 cm^{-1} peak represents the bending vibration peak of PS C–H in SBS. Compared with the virgin asphalt binder and the short-term aging asphalt binder, the amplitudes of these two bands decrease slightly after UV aging, and the decrease in amplitude became greater with the prolongation of UV aging time. It is believed that this is due to the breakdown of SBS segment structure and the decrease of molecular weight with the increase of aging time and UV radiation intensity, which is the main reason for SBS degradation in the aging process of SBS-modified asphalt binders. The results show that the greater the degree of degradation, the stronger the intermolecular conjugation effect.

3.4. Apparent Topography

In general, asphalt pavement cracking is mostly in the form of fatigue and low-temperature cracking caused under driving loads and temperature stress, or in the other form of reflective cracking due to the existence of cracks in the base layer. The results of SEM show that UV radiation can also cause top-down cracking of the asphalt binder surfaces at a room temperature of 25 °C. Based on the UV aging level, the asphalt binder surface cracking process can be roughly divided into four stages (as shown in Figure 9).

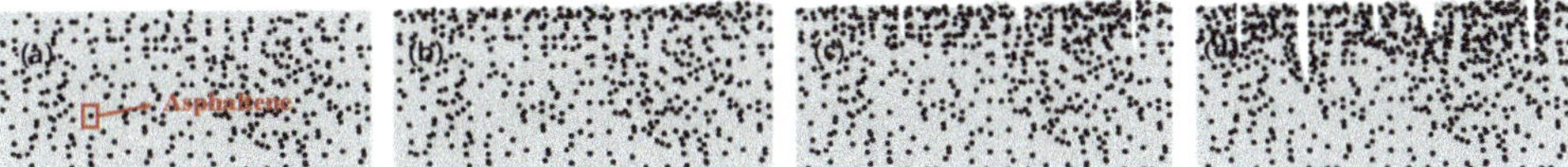

Figure 9. The four stages of asphalt binder base topography change under UV radiation: (**a**) The asphalt binder surface is smooth and flat, (**b**) the asphalt binder surface is wrinkled, (**c**) micro-cracks occur in the asphalt binder surface, and (**d**) the cracks are further extended and expanded to be wider and deeper.

Figure 10 shows the apparent topography of SBS-modified asphalt binders under different UV radiation (365 nm, 10 mW/cm^2) times (magnification 300 times). After 40 h of UV radiation, the surfaces of SBS-modified asphalt binders appeared wrinkled, and there were micro-cracks beginning to be produced. When the UV radiation time reached 80 h, the small cracks extended to become dense cracks on the surfaces of SBS-modified asphalt binders. When the UV radiation time increased to 120 h, the surface cracks further developed and extended, and the surfaces began to fragment. When the UV radiation time reached 160 h, the micro-cracks in the surfaces were further connected and expanded, and the asphalt binder surfaces were completely fragmented and the corners became warped.

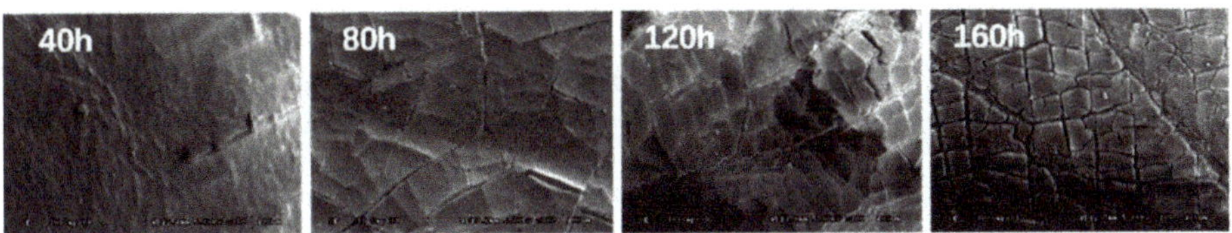

Figure 10. SEM scanning of SBS-modified asphalt binders under different UV radiation times.

According to the cracking criterion of pavements, when the stress of an asphalt pavement reaches the tensile strength of asphalt concrete under load, it will crack. When the stress intensity factor of the crack tip exceeds the fracture toughness of the material, the crack will expand. The cracking criterion of pavement considers that when the stress of an asphalt pavement reaches the tensile strength of asphalt concrete under the action of load, the crack will expand when the stress intensity factor at the crack tip exceeds the fracture toughness of the material [32]. Therefore, the author considers that under UV radiation, the asphalt binder surface will produce a kind of shrinkage stress similar to the temperature stress. In addition, the UV aging will cause the stiffness modulus of the asphalt binders to increase, causing the stress relaxation performance and the ultimate strength of the asphalt binders to decrease. When the shrinkage stress is greater than the limit stress, cracking will occur, and with the increase of the aging level, the crack will expand further.

4. Conclusions

In this paper, rheology tests and spectroscopy tests were conducted to investigate the UV aging of asphalt binders. Based on the test results of SBS-modified asphalt binders at different aging states, the following conclusions can be drawn:

(1) UV seriously destroyed the network structure formed by the crosslinking effect in SBS-modified asphalt binders, and the degradation of SBS was aggravated, which resulted in the SBS-modified asphalt binders becoming more homogeneous after aging. The nature of the UV aging includes the change of component content and degradation of SBS, which leads to the transformation of colloidal structure, manifested by the increase of asphaltene, the dynamic stability of resin, and a reduction in the molecular weight of SBS.

(2) The complex modulus of SBS-modified asphalt binders increases continuously, and the phase angle decreases with the increase of aging. Although this means that the rutting resistance is

improved, the aging asphalt binders are more rigid and brittle, which can easily lead to the pavement cracking at low temperature. With the increase of temperature, the phase angle decreases at the beginning and then increases, and the turning point temperature of the phase angle increases after UV aging.

(3) Continuous UV radiation can induce a kind of shrinkage stress similar to the temperature stress in the surface asphalt binders at mild constant temperature. When the shrinkage stress exceeds the limit stress of the material, it will lead to the formation of top-down cracks in the surface layers of the asphalt binders.

Author Contributions: Conceptualization, H.Y. and G.Q.; data curation, X.B., H.W., X.G., and Z.L.; formal analysis, X.B., H.W., X.G., and Z.L.; funding acquisition, G.Q.; investigation, X.B., H.W., X.G., and J.J.; project administration, H.Y. and G.Q.; resources, H.Y.; software, H.W., X.G., and Z.L.; supervision, G.Q.; validation, J.J.; visualization, J.J.; writing—original draft, X.B.; writing—review & editing, H.Y.

Funding: The authors gratefully acknowledge funding support by the National Key R&D Program of China (2018YFB1600100); the National Natural Science Foundation of China (51778071 and 51808058); and open fund (kfj180103) of the National Engineering Laboratory of Highway Maintenance Technology (Changsha University of Science and Technology).

Conflicts of Interest: The authors declare no conflict of interest.

References

1. Jin, J.; Tan, Y.; Liu, R.; Zheng, J.; Zhang, J. Synergy Effect of Attapulgite, Rubber, and Diatomite on Organic Montmorillonite-Modified Asphalt. *J. Mater. Civ. Eng.* **2019**, *31*, 04018388. [CrossRef]

2. Jin, J.; Xiao, T.; Zheng, J.; Liu, R.; Qian, G.; Xie, J.; Wei, H.; Zhang, J.; Liu, H. Preparation and thermal properties of encapsulated ceramsite-supported phase change materials used in asphalt pavements. *Constr. Build. Mater.* **2018**, *190*, 235–245. [CrossRef]

3. Zhang, Y.; Leng, Z. Quantification of bituminous mortar ageing and its application in ravelling evaluation of porous asphalt wearing courses. *Mater. Des.* **2017**, *119*, 1–11. [CrossRef]

4. Behnood, A.; Gharehveran, M.M. Morphology, rheology, and physical properties of polymer-modified asphalt binders. *Eur. Polym. J.* **2019**, *112*, 766–791. [CrossRef]

5. Behnood, A.; Olek, J. Rheological properties of asphalt binders modified with styrene-butadiene-styrene (SBS), ground tire rubber (GTR), or polyphosphoric acid (PPA). *Constr. Build. Mater.* **2017**, *151*, 464–478. [CrossRef]

6. Zaumanis, M.; Mallick, R.B.; Frank, R. Evaluation of different recycling agents for restoring aged asphalt binder and performance of 100% recycled asphalt. *Mater. Struct.* **2015**, *48*, 2475–2488. [CrossRef]

7. Airey, G.D. State of the Art Report on Ageing Test Methods for Bituminous Pavement Materials. *Int. J. Pavement Eng.* **2003**, *4*, 165–176. [CrossRef]

8. Durrieu, F.; Farcas, F.; Mouillet, V. The influence of UV aging of a Styrene/Butadiene/Styrene modified bitumen: Comparison between laboratory and on site aging. *Fuel* **2007**, *86*, 1446–1451. [CrossRef]

9. Zheng, N.; Xiaoping, J.I.; Hou, Y. Nonlinear Prediction of Attenuation of Asphalt Performance after Ultraviolet Aging. *J. Highw. Transp. Res. Dev.* **2009**, *26*, 33–36. [CrossRef]

10. Wang, J.; Xue, Z.; Tan, Y. Influence of Ultraviolet Aging on Mechanical Behavior and Aggregated State of Asphalt. *China J. Highw. Transp.* **2011**, *24*, 14–19. [CrossRef]

11. Xiao, F.; Newton, D.; Putman, B.; Punith, V.S.; Amirkhanian, S.N. A long-term ultraviolet aging procedure on foamed WMA mixtures. *Mater. Struct.* **2013**, *46*, 1987–2001. [CrossRef]

12. Kemp, G.R.; Predoehl, N.H. A comparison of field and laboratory environments on asphalt durability. In *Association of Asphalt Paving Technologists Proceedings*; Asphalt Institute: Washington, DC, USA, 1981.

13. Ran, L. Research on Aging Mechanism and High Performance Regenerant of SBS Modified Asphalt under Coupling Condition of Light, Heat, Water. Ph.D. Thesis, Chongqing Jiaotong University, Chongqing, China, 2016.

14. Wei, H.; Bai, X.; Qian, G.; Wang, F.; Li, Z.; Jin, J.; Zhang, Y. Aging Mechanism and Properties of SBS Modified Bitumen under Complex Environmental Conditions. *Materials* **2019**, *12*, 1189. [CrossRef] [PubMed]

15. Zeng, W.; Wu, S.; Pang, L.; Chen, H.; Hu, J.; Sun, Y.; Chen, Z. Research on Ultra Violet (UV) aging depth of asphalts. *Constr. Build. Mater.* **2018**, *160*, 620–627. [CrossRef]

16. Mouillet, V.; Farcas, F.; Besson, S. Ageing by UV radiation of an elastomer modified bitumen. *Fuel* **2008**, *87*, 2408–2419. [CrossRef]

17. Hou, X.; Lv, S.; Chen, Z.; Xiao, F. Applications of Fourier transform infrared spectroscopy technologies on asphalt materials. *Measurement* **2018**, *121*, 304–316. [CrossRef]

18. Hu, J.; Wu, S.; Liu, Q.; Hernández, M.I.G.; Wang, Z.; Nie, S.; Zhang, G. Effect of ultraviolet radiation in different wavebands on bitumen. *Constr. Build. Mater.* **2018**, *159*, 479–485. [CrossRef]

19. Zeng, W.; Wu, S.; Wen, J.; Chen, Z. The temperature effects in aging index of asphalt during UV aging process. *Constr. Build. Mater.* **2015**, *93*, 1125–1131. [CrossRef]

20. China, MoTotPsRo. *Standard Test Methods of Bitumen and Bituminous Mixtures for Highway Engineering, JTG E20-2011*; China Communications Press Co., Ltd.: Beijing, China, 2011.

21. Qian, G.; Yu, H.; Gong, X.; Zhao, L. Impact of Nano-TiO2 on the NO2 degradation and rheological performance of asphalt pavement. *Constr. Build. Mater.* **2019**, *218*, 53–63. [CrossRef]

22. *AASHTO T315-05, Standard Method of Test for Determining the Rheological Properties of Asphalt Binder Using a Dynamic Shear Rheometer*; American Association of State Highway and Transportation Officials: Washington, DC, USA, 2005.

23. *AASHTO T313-12, Determining the Flexural Creep Stiffness of Asphalt Binder Using the Bending Beam Rheometer (BBR)*; American Association of State Highway and Transportation Officials: Washington, DC, USA, 2012.

24. Qian, G.; Wang, K.; Bai, X.; Xiao, T.; Jin, D.; Huang, Q. Effects of surface modified phosphate slag powder on performance of asphalt and asphalt mixture. *Constr. Build. Mater.* **2018**, *158*, 1081–1089. [CrossRef]

25. Hou, X.; Xiao, F.; Wang, J.; Amirkhanian, S. Identification of asphalt aging characterization by spectrophotometry technique. *Fuel* **2018**, *226*, 230–239. [CrossRef]

26. Mouillet, V.; Farcas, F.; Chailleux, E.; Sauger, L. Evolution of bituminous mix behaviour submitted to UV rays in laboratory compared to field exposure. *Mater. Struct.* **2014**, *47*, 1287–1299. [CrossRef]

27. Yang, X.; You, Z.; Mills-Beale, J. Asphalt Binders Blended with a High Percentage of Biobinders: Aging Mechanism Using FTIR and Rheology. *J. Mater. Civ. Eng.* **2015**, *27*, 04014157. [CrossRef]

28. Wei, J. Study on Surface Free Energy of Asphalt, Aggregate and Moisture Diffusion in Asphalt. Ph.D. Thesis, China University of Petroleum, Qingdao, China, 2008.

29. Asphalt Institute. *Performance Graded Asphalt Binder Specification and Testing*; Asphalt Institute: Lexington, KY, USA, 2003.

30. Tan, Y.; Shao, X.; Zhang, X. Research on evaluations approach to low-temperature performance of asphalt based on rheology characteristic of low temperature. *China J. Highw. Transp.* **2002**, *15*, 1–5. [CrossRef]

31. Wang, Y.; Luo, A.; Lu, W.; Yuan, H. Determination of Wax Content in Bitumen Using NIR Spectrometric Method. *Acta Pet. Sin. Pet. Process. Sect.* **2001**, *17*, 68–72.

32. Che, F.; Chen, S.; Li, Z.; Liu, W.; Wang, A. Analysis of Cracks Propagation on Asphalt Pavement Surface under Load. *J. Highw. Transp. Res. Dev.* **2010**, *27*, 26–29.

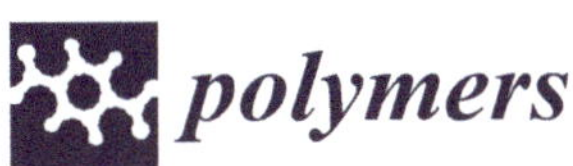 

Article

Examination of Poly (Styrene-Butadiene-Styrene)-Modified Asphalt Performance in Bonding Modified Aggregates Using Parallel Plates Method

Xiangbing Gong [1,2,*]**, Zejiao Dong** [2,*]**, Zhiyang Liu** [2]**, Huanan Yu** [1] **and Kaikai Hu** [1]

[1] State Engineering Laboratory of Highway Maintenance Technology, Changsha University of Science and Technology, Changsha 410114, China; huanan.yu@csust.edu.cn (H.Y.); kaikaihu@stu.csust.edu.cn (K.H.)

[2] School of Transportation Science and Engineering, Harbin Institute of Technology, Harbin 150090, China; liuzhiyanghit@163.com

[*] Correspondence: xbgong@csust.edu.cn (X.G.); hitdzj@hit.edu.cn (Z.D.); Tel.: +86-17603658650 (X.G.); +86-15846533166 (Z.D.)

Received: 5 June 2019; Accepted: 10 December 2019; Published: 14 December 2019

Abstract: Although asphalt-aggregate bonding provides contacting strength for hot mix asphalt (HMA), it is still ignorant in dynamic shear test, due to the only use of metal parallel plate. Modified parallel plates cored from different types of aggregate were provided to simulate aggregate-asphalt-aggregate (AAA) sandwich in HMA, aiming at the comprehensive interpretation on bonding's influence. This study began with an experimental design, aggregate plates, and joint clamps were processed to be installed into the rheometer. Aggregate type and loading conditions were set as essential variables. Subsequently, microscopic tests were utilized to obtain chemical components of aggregate, micro morphology of interface, and roughness of plates. The shearing tests for poly (styrene-butadiene-styrene)-modified asphalt were conducted in bonding with aggregate plates. Meanwhile, contrasting groups adopting metal plates followed the same experimental procedures. The results indicate that the influence of aggregate type on binder's rheological characteristics is dependent on the experimental variables, and microscopic characteristics and component differences should be taken into consideration when selecting aggregates in designing asphalt mixtures.

Keywords: rheology; microscopic characteristic; poly (styrene-butadiene-styrene)-modified asphalt; modified clamps; adhesion

---

1. Introduction

Adhesive bonding [1] has been a major concern for many decades. The corresponding theory is being employed as a cutting-edge technique in the study for composite materials [2], and current research related to adhesion characteristics of asphalt-aggregate bonding has been taken account into hot mix asphalt (HMA). The corresponding methodologies can be divided into three major categorizations that are dependent on physical scales: micro-scale [3], meso-scale [4], and macro-scale [5]. Numerical simulations and analytical approaches are commonly applied at these scales, such as molecular dynamics (MD) [6], finite element method (FEM) [7], and discrete element method (DEM) [8]. A fact was known, as that asphalt-aggregate bonding illustrates an important impact on the performance of HMA. However, several assumptions contribute to be a trend analysis, rather than an accurate value for numerical methods.

Except for numerical methods, microscopic characteristics and mechanical tests were used to evaluate the impact of factors associated with the bonding interaction at the asphalt-aggregate bonding interface. Force-displacement curves that were provided by atomic force microscopy (AFM) were developed to obtain adhesive energy for thin asphalt films [9]. Instead of the silicon AFM tip, a calcium carbonate tip was modified to evaluate the bonding strength by simulating the microscopic interaction between asphalt and limestone [10]. Furthermore, a chemical model was provided to express a fundamental explanation of asphalt-aggregate bonding characteristics [11]. A former study found that the modulus of asphalt film fluctuated with the distance between asphalt and aggregate while using AFM [12]. Additionally, the size limitation of scanning at the micro-scale is insufficient for comprehensively understanding the impact of several factors at a larger scale, such as mineralogy [13], texture [13], and film thickness [14]. Currently, mechanical tests are generally conducted to exhibit adhering responses that were subjected to a given loading condition [15]. However, asphalt pavement is directly exposed to the environment, and temperature, moisture, and stress/strain are dominant factors. Thus, complicated filed conditions that are associated with multiple scales are challenges to completely understand the bonding interaction at the asphalt-aggregate interface.

The rheological characterization of asphalt is generally measured by a dynamic shear rheometer (DSR). Complex modulus ($|G^*|$), phase angle (δ), and viscosity (η) are the typical outputs related to performance evaluation for binder [16], which are widely used to predict the stability [17] and the durability of polymer [18]. The variables for dynamic shear test consist of sample geometry, temperature, stress or strain level [19], and loading duration [20]. Thus, an appropriate experimental design is essential prior to the material preparation in this study. Moreover, instead of metal parallel plates, rock parallel plates were cored according to the protocols that were described in the Superpave binder specification. Several similar experimental designs have been provided to evaluate the asphalt-aggregate bonding behavior under various loading conditions, such as aging [21], changing film thicknesses [22], and moisture damage [23]. The rock plates were applied to test the shearing properties of asphalt [24] and asphalt mastic effectively [25], and aggregate-mastic interaction was finally illustrated based on a modulus model [26]. However, the obvious issue for those studies is that only one plate size was adopted, which limits the temperature range and loading form for the dynamic shear test.

The primary motivation of this study is to design rock cylinders that can cover a wider temperature range based on a rectangle clamp of DHR-II (Discovery Hybrid Rheometer, TA Instruments, Lindon, UT, USA). One significant advantage for DHR-II is that the air bath has less interference rather than the water bath. An Φ 25 mm rock cylinder and an Φ 8 mm rock cylinder were both included in the experimental design. Therefore, the rheological properties of asphalt could be characterized at high, intermediate, and low temperatures that were subjected to different types of modified parallel plates. Another motivation is to conduct several tests by changing the variables to simulate the binder layer in real asphalt pavement (Figure 1), such as temperature, loading, aggregate type, and frequency. It aims at explaining how those fundamental variables affect the rheological characteristics of aggregate-asphalt-aggregate (AAA) system. This setup is designed to simulate the real conditions in asphalt pavement engineering partially. In addition, this study tries to provide a preliminary protocol to evaluate the influence of the aggregate plates on binder performances, and it will be improved based on the future works.

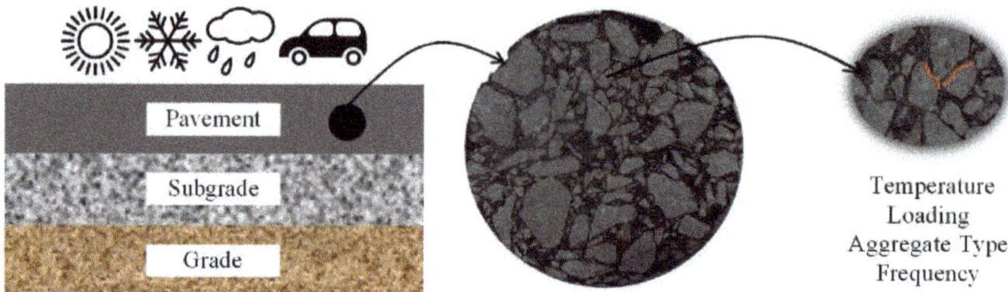

Figure 1. Schematic image to illustrate the purpose of aggregate-asphalt-aggregate dynamic shear rheometer (DSR) tests.

2. Materials and Experimental Design

2.1. Materials

Strategic Highway Research Program (SHRP) developed Superpave PG (performance gradate) asphalt binder specification. It defines a series of standard experiments that characterize the viscoelastic behavior of binder at multiple temperatures [27]. In this specification, the Φ 8 mm parallel plate is selected at intermedium temperatures, even low temperatures applying to a sample with 2000 μm thickness, Φ 25 mm plate is the corresponding clamps for oscillation experiments at high temperatures applying to a sample with 1000 μm thickness. Two types of rock were cored into Φ 8 mm and Φ 25 mm cylinders with a thickness of 5 mm. Two types of aggregate plate (limestone and basalt) were chosen, because they are used as the universal aggregate type in China. While the asphalt type was not designed to be a variable in this study, it is significant to select poly (Styrene-Butadiene-Styrene, SBS)-modified asphalt, because it is widely used. Based on several physical properties tests, the testing results are shown, as follows: penetration value is 58.3 (0.1 mm, 100 g, and 25 °C), softening point is 56.0 °C (ball and ring) while using the ring and ball method, and ductility is 40.1 mm (25 °C). Its fundamental properties meet the specification requirements of China.

2.2. Metal Base and Rock Cylinder Design

The first step was to process a base that is used as a joint part to connect a rock cylinder and a rectangle clamp of DHR. The base is divided into two parts: one part is an upper cylinder and other part is a lower polygon. The rock cylinder was fixed onto the upper part through the super glue [24]; the polygon bottom was fastened into the rectangle clamp of DHR through bolts and shims (Figure 2). A key procession was to ensure that the cylinder top should be concentric with the loading top of DHR, and thus several precise procedures were presented to generate the metal base. Ultimately, modified clamps were designed to have the same axis of symmetry with the loading system of DHR (Figure 2), which avoids eccentric force.

The next step was to core the rock cylinder into different diameters (8 mm and 25 mm), and two types of aggregate (limestone and basalt) were selected. Two types of upper cylinder were designed to be with 8 mm or 25 mm diameter, correspondingly. The rock cylinders were cored from aggregate slabs with a 5 mm height, and those cubic slabs were initially polished to create the parallel surfaces with a height tolerance of 0.02 mm [24]. However, there is no further polishing to reach a smooth surface, thus the microscopic roughness is still kept as the variable. Furthermore, the microscopic roughness of the rock cylinders was inspected by AFM and the chemical components were identified by X-ray fluorescence (XRF) and X-ray diffraction (XRD). Finally, gel epoxy was utilized to fix the rock cylinders onto the upper cylinder of the base. Super glue must be effective at temperatures ranging from −30 °C to 250 °C.

2.3. Experimental Description

Instead of metal plates, a pair of modified clamps can simulate an asphalt film that is sandwiched between two pieces of aggregate in HMA. It is essential to conduct the glue operation in the zero-gapped pattern [24]. Consequently, the upper and lower aggregate plate can be easily assembled to be concentric. When the epoxy is soft, the rock cylinders must be adjusted to be concentric with the upper cylinder. After 12 hours' curing time, the modified parallel plates can then be installed into the rectangle clamps of DHR (Figure 2).

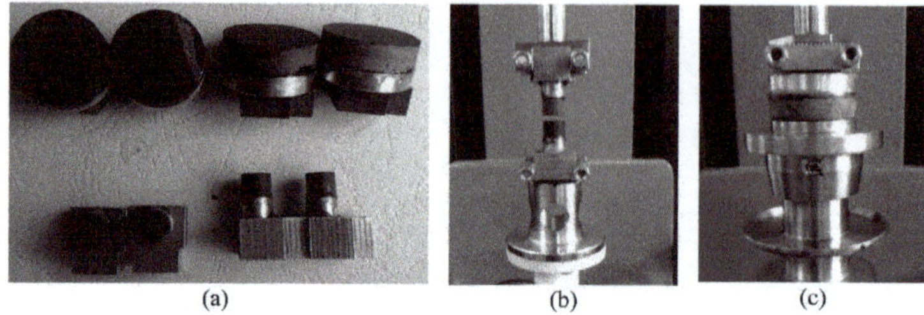

Figure 2. Experimental procedures: (**a**) rock cylinders (Φ 8 mm and Φ 25 mm) and bases after the glue operation, (**b**) installation in the Discovery Hybrid Rheometer (DHR) for the Φ 8 mm cylinder, and (**c**) installation in the DHR for the Φ 25 mm cylinder.

DHR is widely applied in the rheological research for poly-modified asphalt. Precise displacement/force control is the biggest advantage for DHR. The zero-gapping and trimming operations follow the same protocol that was described in the Superpave specifications, except for an additional curing process to generate the adhesive strength for the asphalt-aggregate interface (60 °C, 10 min). The aggregate plates must be heated to approximate 160 °C in an oven prior to the curing process and binder film installation. The function of this step is to dry out moisture and simulate the preheating process while blending for HMA.

In this study, the first test is the oscillatory sweep at a wide temperature range by adopting Φ 8 mm and Φ 25 mm plates, the second test is the relaxation test, and the last one is the multiple stress creep recovery test (MSCR). Each test has several groups: (a) three types of parallel plates, called basalt (BS), limestone (LS), and metal, respectively; (b) three pairs of parallel plates, named aggregate-aggregate (AA), aggregate-metal (AM), and metal-metal (MM), respectively; and, (c) several kinds of loading form, plate type, and frequency.

3. Results and Discussion

All of the tests were conducted within the upper limitation of LVE (Linear Viscoelasticity) range, and DHR was applied to measure the rheological outputs of asphalt film under different conditions. Complex modulus, relaxation modulus, and creep-recovery responses were the main outputs in this study. Afterwards, major results and discussion are described, as follows.

3.1. Microscopic Characteristics

3.1.1. Morphology Analysis

The microscopic characteristics of aggregate surfaces cannot be ignored with respect to the asphalt-aggregate interface; SEM (Scanning Electron Microscope) and AFM are the effective observation methods [28]. SEM was used to investigate the morphology of asphalt-aggregate interface, and AFM determined the microscopic roughness of the cutting surface of aggregate.

Images that were taken by SEM and AFM in Figure 3 show conditions of the asphalt-aggregate bonding boundary at micro-scale. The samples were prepared by following the same procedures adopted in the experimental design in order to control the variable. Additionally, the average roughness (R_a) of plates was calculated via *NanoScope Analysis Software*, which is known as a specific tool for AFM. In this paper, it mainly aims at micro roughness, except for macro shape of aggregate. Consequently, the aggregate plate surface was polished under the same polishing conditions. It is concluded that DHR metal plate (R_a, 7.5 nm) has the smallest average roughness, followed by limestone (R_a, 206.4 nm), and the largest one is basalt (R_a, 264.8 nm). Micro roughness differences should be directly related to mineral components of various plates. The AFM phase images differentiate the surface morphology for BS and LS at micro-scale. Figure 3c exhibits that the valley is narrow and the peak is granular for the rough surface of BS. On the contrary, the valley is broad and the peak is flat for LS, as shown in Figure 3d.

Moreover, those differences can also be depicted from the SEM images. The microscopic defects in basalt is bigger than limestone, which could be clearly seen in Figure 3a,b. Therefore, hot asphalt can flow into the voids and easily fill the defects for basalt. Consequently, the boundary along the asphalt-basalt interface in Figure 3a was coated well. However, the boundary of asphalt-limestone interface in Figure 3b seems like an accumulation of binder without flowing into defects. The SEM images exhibit that the particles in limestone appear to be large cubic, which should block the flowing of hot asphalt. On the contrary, hot asphalt might coat those flat particles in basalt well. Therefore, it is concluded that basalt shows a better coating at the micro-scale due to the better distribution of microscopic morphology.

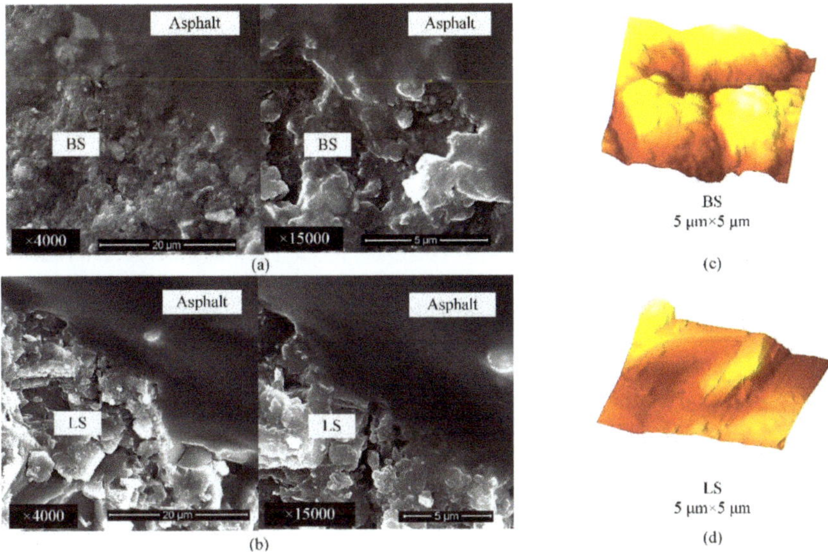

Figure 3. Microscopic characteristics: (**a**) Scanning Electron Microscope (SEM) images of basalt (BS), (**b**) SEM images of limestone (LS), (**c**) atomic force microscopy (AFM) image of BS, and (**d**) AFM images of LS.

3.1.2. Component Analysis

Not only microscopic morphology, but also the chemical component, are decisive factors that affect the microscopic interaction, because different chemical minerals must have a significant impact on absorbed asphalt proportion. In this part, the main objective is to differentiate mineral components. Besides, chemical component analysis can be used to check that the type of aggregate is correct. Consequently, XRF (X-Ray Fluorescence) and XRD (X-Ray Diffraction) were utilized to investigate chemical components of aggregate. XRF first determined the chemical elements of aggregate, and XRD was then applied into the elementary analysis of chemical components. For basalt, silicon takes the largest proportion, followed by iron and calcium. For limestone, silicon and calcium are the primary elements, followed by aluminum.

On the basis of the XRD results, the main compositions in aggregates were achieved. Calcite and quartz are the primary minerals for the limestone, which are mainly composed of $CaCO_3$ and SiO_2. Diopside, nepheline, and forsterite are the primary minerals for the basalt, being known as typical silicates. Based on the contact angle test [29], limestone has a smaller angle than basalt with respect to asphalt. A former study described that limestone exhibited a greater adsorption proportion of asphalt than quartzite and granite [30]. Additionally, numerical studies conclude that the limestone-asphalt samples have a better adhesive capability after boiling in the water bath. Therefore, it is estimated that limestone has the stronger adhesive capability that is associated with asphalt than basalt while accounting for chemical component differences.

3.2. Shearing Testing Results

Aiming at comprehensively evaluating the factors mentioned above, experimental plans focus on a wide temperature range that covers high, intermediate, and low temperatures. The outputs that were obtained through strain sweep, frequency sweep, relaxation test, and MSCR test are discussed, as follows.

3.2.1. Strain Sweep

The low temperature cracking is a predominant issue that results in several early distresses of asphalt pavement. As the temperature drops, the pavement shrinks, then the aggregate-asphalt bonding zone tends to easily fracture because of the increasing tensile stress. In this part, the strain sweep experiment attempts to investigate the influence of strain on $|G^*|$ at a low temperature undergoing different bonding conditions. In addition, the test temperature was 0 °C with a constant frequency of 10 rad/s; Figure 4 shows the results.

The initial $|G^*|$ shows significant differences especially for different plate types. The AA plates have the greatest modulus, and followed by the AM plate, the MM plate has the smallest value. As strain increases, $|G^*|$, as obtained from the AM plate significantly decreases, and followed by the AA plates and the MM plate. The largest decreasing percentage is clearly found in the AM plate, the MM plate shows the smallest reduction rate. Additionally, the LS plate generates the better resistance to the increasing strain than the BS plate at 0 °C, no matter what pair type is. These findings indicate that limestone-asphalt-limestone system is the strongest interaction among those plates at 0 °C. However, the $|G^*|$ that was measured from aggregate plates are easier to decrease with the increase of strain than metal plates. This finding implies that smooth aggregate surface could be concluded as an advantage when asphalt pavement is located in the cold region.

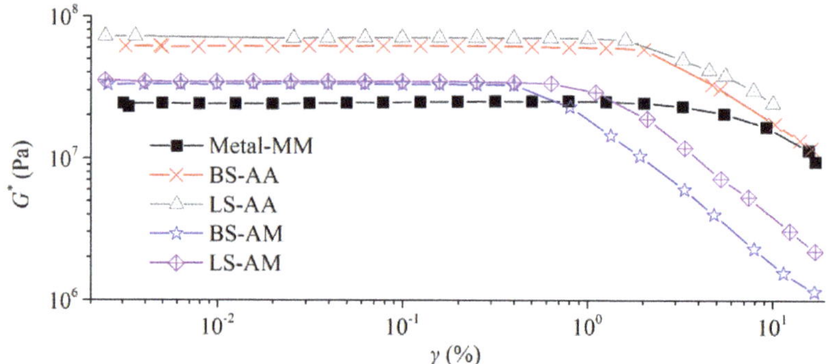

Figure 4. Strain sweep curves at 0 °C with different parallel plates (2000 μm, gap).

3.2.2. Frequency Sweep Using Φ 8 mm Plates

Modified 8 mm parallel plates were used to obtain master curves at intermediate temperatures and low temperatures. The shearing modulus was measured by DHR with various frequencies ranging from 0.01 Hz to 30 Hz, and the tests were conducted at four temperatures (−15 °C, 0 °C, 15 °C, and 30 °C). Figure 5 shows the master curves of modulus at 15 °C, and the results and discussion are presented, as follows.

It is clear that the AA plates contribute to the largest modulus with respect to low-frequency shearing. While, the master curve of the MM plate moves upward in Figure 5 as the frequency increases via contrasting with the AA plate and the AM plate. However, the modulus curves for the AA and AM plates tend to be smaller than the MM plate as the frequency increases. The modulus differences between the AA plates and the MM plate become larger with the increase of gap. It is summarized that $|G^*|$ for the MM plate is nearly the largest when the frequency is around 10^4 Hz. It could be concluded that smooth surface might be a good property for the AAA system associated with high-frequency loading, as well as low-frequency loading at intermediate temperature.

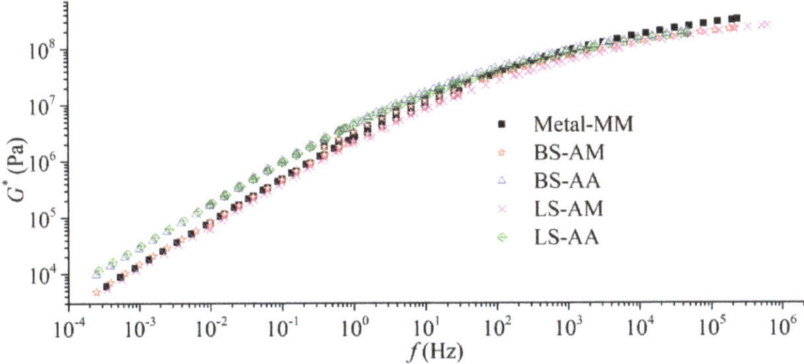

Figure 5. Master curves at 15 °C with different types of aggregates and plates (2000 μm, gap).

Table 1 and Figure 5 present a comparison between the AM plate and the MM plate. The modulus between the AM plate and MM plate are close when the frequency is low. As the frequency gets higher, $|G^*|$ for the AM plate becomes smaller than the MM plates. In Table 1, SBS binder modulus are significantly affected by the type of aggregate verified by experimental results from the BS and LS groups, and the basalt generate a stiffer AAA system than limestone at an intermediate temperature. This finding demonstrates that aggregate chemical components should be treated as factors when designing intermediate temperature properties of mixtures, such as fatigue.

Table 1. Dynamic modulus at 15 °C with different frequencies and plates (2000 μm, gap).

Frequency (Hz)	Dynamic Modulus (Pa)				
	Metal-MM	**BS-AM**	**BS-AA**	**LS-AM**	**LS-AA**
0.01	8.04×10^4	7.21×10^4	1.72×10^5	6.27×10^4	1.74×10^5
1	2.40×10^6	2.24×10^6	4.61×10^6	2.06×10^6	4.21×10^6
10	9.86×10^6	9.00×10^6	1.65×10^7	8.27×10^6	1.51×10^7
30	1.74×10^7	1.59×10^7	2.73×10^7	1.44×10^7	2.52×10^7

3.2.3. Frequency Sweep Using Φ 25 mm Plates

Modified Φ 25 mm plates were employed into frequency sweep at high temperatures. In this section, the frequency sweep tests were generally conducted at four temperatures (45 °C, 60 °C, 75 °C, and 90 °C). The dynamic frequency changed from 0.1 Hz to 30 Hz. Figure 6 illustrates the aster curves at 60 °C; the results can be concluded, as follows.

All of the solid plates are less sensitive to the frequency when comparing Figure 6 with Figure 5. As seen from Figure 6 and Table 2, $|G^*|$ for the AA plates is greater than the MM plates, and the smallest $|G^*|$ is found in AM plates. Therefore, the homogeneous aggregate property is also an index in selecting aggregates that are subjected to high temperatures. The largest $|G^*|$ for the AA plates and AM plates are both found in limestone. In Figure 3, the microscopic roughness of limestone is not the greatest one, but the well-coating boundary might result in the largest $|G^*|$ at high temperature. Consequently, the strong bonding could predominate the shearing flow behavior of AAA system, being associated with the high-temperature stability of HMA.

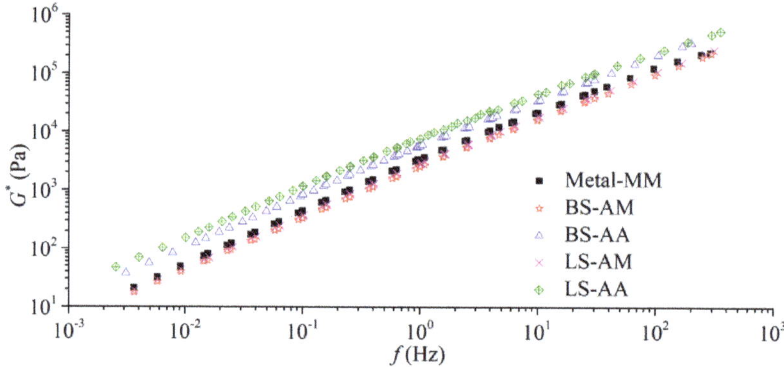

Figure 6. Master curves at 60 °C with different types of aggregates and plates (1000 μm, gap).

Table 2. Dynamic modulus at 60 °C with different frequencies and plates (1000 μm, gap).

Frequency (Hz)	Dynamic Modulus (Pa)				
	Metal-MM	BS-AM	BS-AA	LS-AM	LS-AA
0.01	4.35×10^2	3.31×10^2	8.42×10^2	3.64×10^2	1.17×10^3
1	3.34×10^3	2.55×10^3	5.68×10^3	2.82×10^3	7.66×10^3
10	2.14×10^4	1.63×10^4	3.41×10^4	1.80×10^4	4.51×10^4
30	5.07×10^4	3.87×10^4	7.95×10^4	4.29×10^4	1.02×10^5

3.2.4. Relaxation Test

The relaxation phenomenon is defined as modulus decreases undergoing a constant strain. It is a typical factor that is related to the resistance to the cracking of asphalt pavement at low temperatures. A rapid decrease of relaxation modulus ($G(t)$) refers to the strong resistance to the thermal cracking. The relaxation test of AAA system was conducted under strain mode, based on the modified parallel plates. The experimental temperature was 0 °C, and the shear strain was kept as a constant (1%). $G(t)$ versus time data was measured by DHR for the AA and the MM plates, Figure 7 presents the findings depicted, as below.

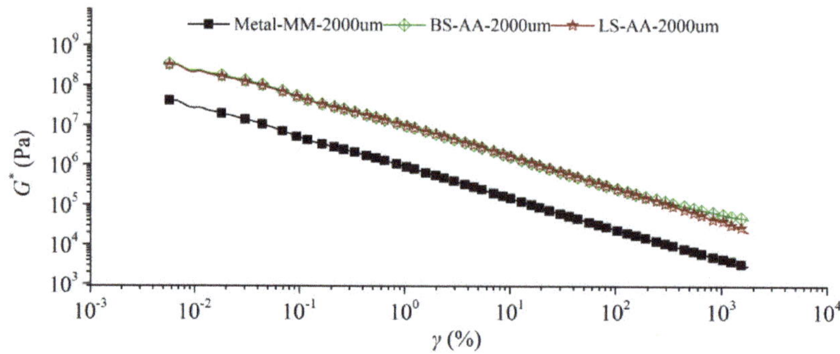

Figure 7. Relaxation modulus curves at 0 °C with different parallel plates: metal-metal (MM) plates, basalt-aggregate-aggregate (BS-AA) plates, and limestone-AA (LS-AA) plates (2000 μm, gap).

In Figure 7, the initial SBS binder stiffness is related to the plate type. The smooth metal plate leads to the smallest initial $G(t)$, which means that the thermal stress would be reduced in the metal-binder-metal system. Although the $G(t)$ curves of the BS-AA plate and the LS-AA plate seem to be similar, the curve of the BS-AA plate tends to be flat when comparing with the LS-AA plate and MM plate. Thus, it could be summarized that the smooth surface results in easy thermal stress relaxation. This observation denotes that rough aggregate surface at the micro-scale could degrade the resistance to thermal cracking of asphalt mixture, especially suffering a long-term cold climate or a sharp temperature dropping climate.

3.2.5. MSCR Test

Instead of the rutting factor, the multiple stress creep recovery test (MSCR) is widely used to amend the evaluation index at high temperatures for poly-modified asphalt [31]. This test generally covers two stresses (100 Pa, 3200 Pa), and each test repeats ten cycles. Each cycle endures one second creep loading and then recovers for nine seconds without loading. The creep strain at the end of one second' loading is StrainA$_i$, the recoverable strain at the end of nine seconds' recovery is StrainB$_i$. Namely, the non-recoverable strain (StrainC$_i$) is the difference between StrainA$_i$ and StrainB$_i$. The recoverable rate (R, %) and non-recoverable creep compliance (J_{nr}, kPa^{-1}) are effective parameters for analyze results [32].

According to the strain illustration (Figure 8), all of the parameters could be calculated via formulas exhibited as below, and i means the cycle number. The MSCR test for SBS-modified asphalt utilizing different plates was conducted at four temperatures (45 °C, 50 °C, 55 °C, and 60 °C) with a gap of 1000 μm. The strain versus time was recorded by DHR, and strain can be drawn and found in Figure 9. In Figure 10, the bars represent R and the plots represent J_{nr}, the results and discussion are shown, as follows.

$$R100 = (1/10)\left\{\sum_{i=1}^{10} StrainB_i / StrainA_i\right\} * 100 \tag{1}$$

$$R3200 = (1/10)\left\{\sum_{i=1}^{10} StrainB_i / StrainA_i\right\} * 100 \tag{2}$$

$$J_{nr}100 = (1/10)\left\{\sum_{i=1}^{10} StrainC_i / 0.1\right\} \tag{3}$$

$$J_{nr}3200 = (1/10)\left\{\sum_{i=1}^{10} StrainC_i / 3.2\right\} \tag{4}$$

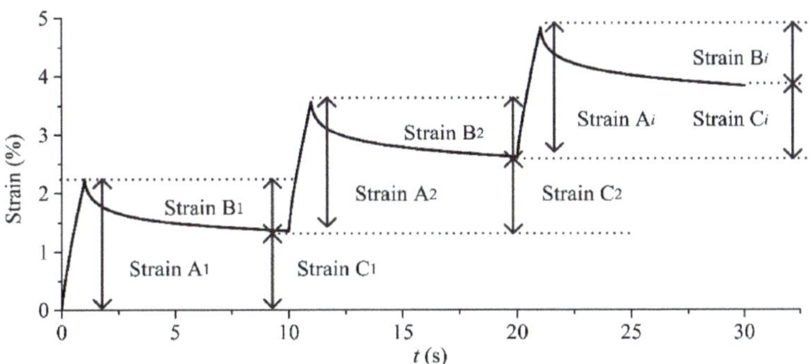

Figure 8. Strain illustration for the multiple stress creep recovery test (MSCR) test and the initial strain is assumed to be zero for the first cycle.

It can be found that the J_{nr} increases as the temperature goes up no matter what kind of plate and stress were applied. In Figure 10a, the curves of J_{nr} for different stress are similar. However, the differences between $J_{nr}100$ and $J_{nr}3200$ present an undeniable increase when the temperature changes from 45 °C to 60 °C, and the greatest gap is approximately 60 kPa^{-1}. The AA plate has a smaller J_{nr} than others, and the AA plate with limestone delivers a better recoverable performance than the basalt. The AM plate with basalt exhibits the highest J_{nr} while excluding the $J_{nr}100$ at 55 °C, which indicates that the microscopic roughness could be a critical factor affecting the recoverable resistance of the AAA system in HMA at high temperatures.

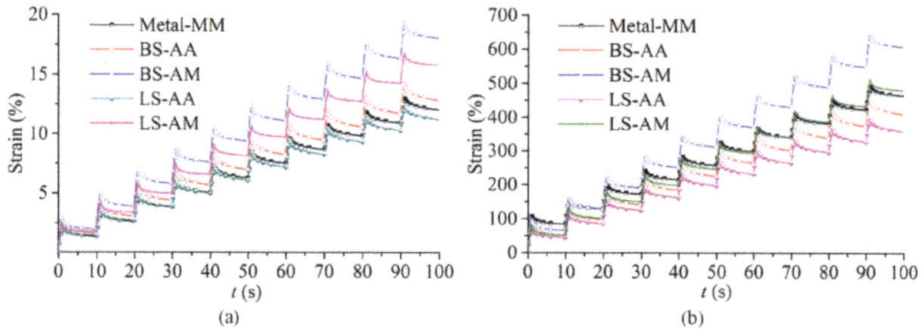

Figure 9. Strain versus time plots of different types of aggregates and plates (1000 μm, gap): (**a**) 100 Pa, and (**b**) 3200 Pa.

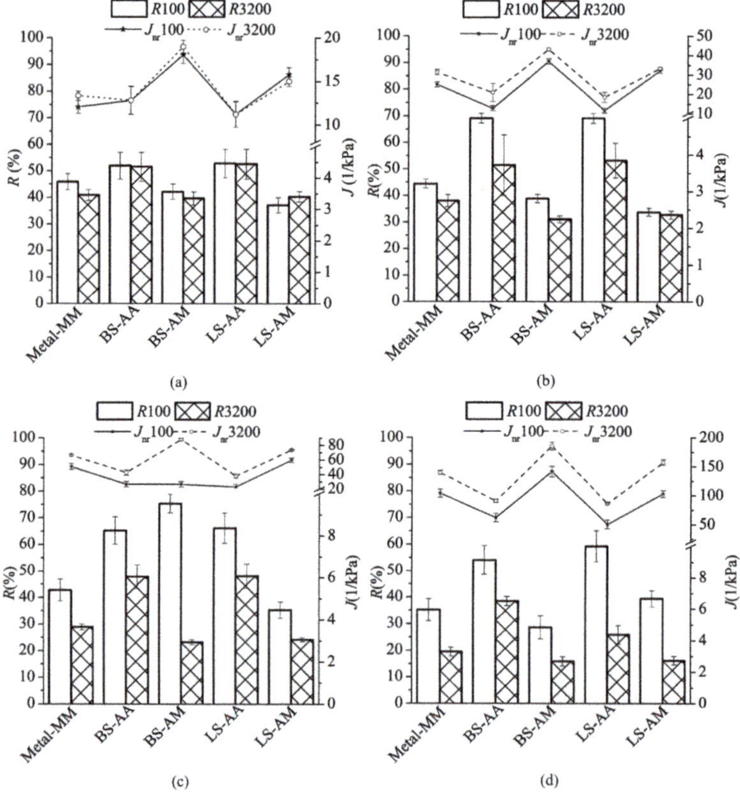

Figure 10. Two parameters calculated from the MSCR test: (**a**) 45 °C, (**b**) 50 °C, (**c**) 55 °C, and (**d**) 60 °C, the error bar is equal to the standard deviation, 100 means 100 Pa, and 3200 represents 3200 Pa.

Accounting for R bars, one finding is that value gap between $R100$ and $R3200$ becomes larger with the increase of temperature. It is summarized like that a high temperature combined with a heavy traffic would significantly degrade the displacement recoverable capability of the asphalt mixture. Another finding can be concluded as that the AA plate shows the best resistance to the creep deformation. However, the $R100$ at 55 °C for the AM plate with basalt rises from 37% to 75%, the gap ranks as the highest increasing rate. This observation might be related to a better binder coating at the micro-scale for the basalt-asphalt interface zone that is shown in Figure 3.

The $R3200$ decreases as the temperature increases, but $R100$ presents more complicated fluctuation when it comes to the AA plate. First, the $R100$ of two types of plate increases to around 70% at 50 °C. Subsequently, it decreases by approximate 5% at 55 °C. Finally, the $R100$ of the AA plate with basalt and limestone reduces to 54% and 60%. The microscopic roughness could be a major index that results in the fluctuation of R, just when the shearing loading is not too heavy. At the highest temperature of 60 °C, the maximum $R100$ occurs in the AA plate with limestone, and the maximum $R3200$ occurs in the AA plate with basalt, which might be owed to its microscopic roughness. When considering the rutting issue at high temperatures, limestone would be a good choice while undergoing regular traffic, but the basalt should be the preferred aggregate type subjected to a heavy traffic.

4. Conclusions

This study aims at understanding how different aggregate-modified plates affect the rheological characteristics of poly (SBS)-modified asphalt, and a development of this study is to conduct modified shearing tests at a wide temperature range as compared with former studies. Several conclusions can be summarized based on the analysis mentioned above.

Aggregate selection in the asphalt mixture design is a dominant procedure for generating a strong asphalt mixture. Additionally, the aggregate-plate-modified DSR testes developed in this study could provide instructions to choose an optimum aggregate type. The testing methods would be further used to predict HMA performances under various conditions, such as different temperatures and traffic volumes.

One obvious fact is that the MM plate is less sensitive to DSR testes than other plates, which lacks the capability to illustrate AAA properties in HMA. Aggregate-modified DSR tests are effective in investigating the influence of AAA bonding and simulate a real AAA system in asphalt mixtures, which involves binder coating at the micro-scale and aggregate chemical components.

The effects of microscopic roughness and chemical components are dependent on temperature, frequency, and loading form. Rough aggregates combined with strong adhesive capability could improve the asphalt mixture performances. In this study, similarity among aggregate particles is also known as a critical index, because it might contribute to homogeneous interaction within the AAA system. Therefore, the source of stone must be seriously controlled to avoid large property variability.

5. Future Research

More aggregate types and asphalt types will be included in future works. Additionally, the macro shape of an aggregate needs to be involved to interpret the shearing behavior of the AAA system. Next, the experimental design will focus on the thickness effect as well as the compliance verification of DSR testing subjected to difference film thinness.

Author Contributions: G.X. and Z.D. organized the research; Z.L. and K.H. performed DSR tests; X.G. and H.Y. wrote the manuscript; Z.D. checked the manuscript.

Funding: The study was supported by National Key R&D Program of China (grant No.: 2018YFB1600100) and National Natural Science Foundation of China (grant No.: 51808058, grant No.: 51908072).The project was also supported by Open Fund of State Engineering Laboratory of Highway Maintenance Technology (Changsha University of Science and Technology) (grant No.: kfj170102) and Scientific Research Project of Education Department of Hunan Province (grant No.: 18C0192).

Acknowledgments: This study was completed at School of Transportation Science and Engineering in Harbin Institute of Technology, as well as School of Traffic and Transportation Engineering in Changsha University of Science and Technology.

Conflicts of Interest: The authors declare no conflict of interest.

References

1. Lacombe, R. *Adhesion Measurement Methods: Theory and Practice*; CRC Press Inc.: Boca Raton, FL, USA, 2010.
2. Necasova, B.; Liska, P.; Kelar, J.; Slanhof, J. Comparison of adhesive properties of polyurethane adhesive system and wood-plastic composites with different polymers after mechanical, chemical and physical surface treatment. *Polymers* **2019**, *11*, 397. [CrossRef] [PubMed]
3. Lu, Y.; Wang, L. Nanoscale modelling of mechanical properties of asphalt-aggregate interface under tensile loading. *Int. J. Pavement Eng.* **2010**, *11*, 393–401. [CrossRef]
4. Wang, P.; Zhai, F.; Dong, Z.; Wang, L.; Li, G. Micromorphology of asphalt by polymer and carbon nanotubes modified through molecular dynamics simulation and experiments: Role of the strengthened interfacial interactions. *Energy Fuel* **2018**, *32*, 1179–1187. [CrossRef]
5. Mo, L.; Huurman, M.; Wu, S.; Molenaar, A.A.A. Ravelling investigation of porous asphalt concrete based on fatigue characteristics of bitumen-stone adhesion and mortar. *Mater. Des.* **2009**, *77*, 170–179. [CrossRef]
6. He, L.; Li, W.; Chen, D.; Zhou, D.; Lu, G.; Yuan, J. Effects of amino silicone oil modification on properties of ramie fiber and ramie fiber/polypropylene composites. *Mater. Des.* **2015**, *77*, 142–148. [CrossRef]
7. Dong, Z.; Gong, X.; Zhao, L.; Zhang, L. Mesostructural damage simulation of asphalt mixture using microscopic interface contact models. *Constr. Build. Mater.* **2014**, *53*, 665–673. [CrossRef]
8. Hossain, M.I.; Faisal, H.M.; Tarefder, R.A. Characterisation and modelling of vapour-conditioned asphalt binders using nanoindentation. *Int. J. Pavement Eng.* **2015**, *16*, 382–396. [CrossRef]
9. Pauli, T.; Grimes, W.; Wang, M.; Lu, P.; Huang, S.C. Development of an adherence energy test via force-displacement atomic force microscopy (FD-AFM). In *Multi-Scale Modeling and Characterization of Infrastructure Materials*; Springer: Dordrecht, The Netherlands, 2013; Volume 8, pp. 273–284.
10. Al-Rawashdeh, A.S.; Sargand, S. Performance assessment of a warm asphalt binder in the presence of water by using surface free energy concepts and nanoscale techniques. *J. Mater. Civ. Eng.* **2014**, *26*, 803–811. [CrossRef]
11. Cho, D.W.; Kim, K.; Lee, M.-J. Chemical model to explain asphalt binder and asphalt–aggregate interface behaviors. *Can. J. Civ. Eng.* **2010**, *37*, 45–53. [CrossRef]
12. Zhu, X.; Yuan, Y.; Li, L.; Du, Y.; Li, F. Identification of interfacial transition zone in asphalt concrete based on nano-scale metrology techniques. *Mater. Des.* **2017**, *129*, 91–102. [CrossRef]
13. Raisanen, M.; Kupiainen, K.; Tervahattu, H. The effect of mineralogy, texture and mechanical properties of anti-skid and asphalt aggregates on urban dust. *Bull. Eng. Geol. Environ.* **2005**, *64*, 247–256. [CrossRef]
14. Wang, D. Binder film thickness effect on aggregate contact behavior. Ph.D. Thesis, Virginia Polytechnic Institute and State University, Blacksburg, VA, USA, 2007.
15. Ban, H.; Kim, Y.; Pinto, I. Integrated experimental-numerical approach for estimating material-specific moisture damage characteristics of binder-aggregate interface. *Transp. Res. Rec.* **2011**, *2209*, 9–17. [CrossRef]
16. Wen, G.; Zhang, Y.; Zhang, Y.; Sun, K.; Fan, Y. Rheological characterization of storage-stable SBS-modified asphalts. *Polym. Test.* **2002**, *21*, 295–302. [CrossRef]
17. Yang, Z.; Zhang, X.; Zhang, Z.; Zou, B.; Zhu, Z.; Lu, G.; Xu, W.; Yu, J.; Yu, H. Effect of aging on chemical and rheological properties of bitumen. *Polymers* **2018**, *10*, 1345. [CrossRef] [PubMed]
18. Miao, Y.; Wang, T.; Wang, L. Influences of interface properties on the performance of fiber-reinforced asphalt binder. *Polymers* **2019**, *11*, 542. [CrossRef]
19. Motamed, A.; Bahia, H.U. Influence of test geometry, temperature, stress level, and loading duration on binder properties measured using DSR. *J. Mater. Civ. Eng.* **2011**, *23*, 1422–1432. [CrossRef]
20. Newman, J.K. Dynamic shear rheological properties of polymer-modified asphalt binders. *J. Elastomers Plast.* **1998**, *30*, 245–263. [CrossRef]
21. Scholz, T.V.; Brown, S.F. Rheological characteristics of bitumen in contact with mineral aggregate. In Proceedings of the 1996 Conference of the Association of Asphalt Paving Technologies: Asphalt Paving Technology, Baltimore, MD, USA, 18–20 March 1996; pp. 357–384.

22. Huang, S.C.; Branthaver, J.F.; Robertson, R.E.; Kim, S.S. Effect of film thickness on the rheological properties of asphalts in contact with aggregate surface. *Transp. Res. Rec.* **1998**, *1638*, 31–39. [CrossRef]

23. Rottermond, M.P.; Williams, R.C.; Bausano, J.P.; Scholz, T.V. A new test protocol to evaluate the moisture susceptibility of asphalt binders. *J. Appl. Asph. Bind. Technol.* **2005**, 31–53.

24. Cho, D.W.; Bahia, H.U. Effects of aggregate surface and water on rheology of asphalt films. *Transp. Res. Rec.* **2007**, *1998*, 10–17. [CrossRef]

25. Dong, Z.; Liu, Z.; Wang, Y.; Gong, X. Effects of aggregate mineralogy on rheology of asphalt mastics. TRB 96th Annual Meeting Compendium of Papers. In Proceedings of the Transportation Research Board 96th Annual Meeting, Washington, DC, USA, 8–12 January 2017.

26. Dong, Z.; Liu, Z.; Wang, P. Modelling asphalt mastic modulus considering substrate-mastic interaction and adhesion. *Constr. Build. Mater.* **2018**, *166*, 324–333. [CrossRef]

27. American Association of State Highway and Transportation Officials (AASHTO). *Determining the Rheological Properties of Asphalt Binder Using a Dynamic Shear Rheometer (DSR)*; T 315-10; American Association of State Highway and Transportation Officials: Washington, DC, USA, 2010.

28. Lu, Y.; Wang, L. Nano-mechanics modelling of deformation and failure behaviours at asphalt-aggregate interfaces. *Int. J. Pavement. Eng.* **2011**, *12*, 311–323. [CrossRef]

29. Xiao, Q.Y.; Wei, L.Y. A precise evaluation method for adhesion of asphalt aggregate. *Int. J. Pavement Res. Technol.* **2010**, *2*, 270–274.

30. Petersen, J.; Plancher, H. Model studies and interpretive review of the competitive adsorption and water displacement of petroleum asphalt chemical functionalities on mineral aggregate surfaces. *Pet. Sci. Technol.* **1998**, *6*, 89–131. [CrossRef]

31. Zoorob, S.E.; Castro-Gomes, J.P.; Pereira Oliveira, L.A.; O'Connell, J. Investigating the multiple stress creep recovery bitumen characterisation test. *Constr. Build. Mater.* **2012**, *30*, 734–745. [CrossRef]

32. Soenen, H.; Blomberg, T.; Pellinen, T.; Laukkanen, O.V. The multiple stress creep-recovery test: A detailed analysis of repeatability and reproducibility. *Road Mater. Pavement Des.* **2013**, *14*, 2–11. [CrossRef]

Article

Hydrophilic Molecularly Imprinted Chitosan Based on Deep Eutectic Solvents for the Enrichment of Gallic Acid in Red Ginseng Tea

Guizhen Li and Kyung Ho Rwo *

Department of Chemistry and Chemical Engineering, Inha University, Incheon 402751, Korea
* Correspondence: rowkho@inha.ac.kr

Received: 3 July 2019; Accepted: 30 August 2019; Published: 1 September 2019

Abstract: Hydrophilic molecularly imprinted chitosan (HMICS) were synthesized based on hydrophilic deep eutectic solvents (DESs) and the DESs was used as both a template and functional monomer for the enrichment of gallic acid (GA) from red ginseng tea using a solid phase microextraction (SPME) method. Using the response surface methodology (RSM) strategy, the optimal extraction amount (8.57 mg·g^{-1}) was found to be an extraction time of 30 min, a solid to liquid ratio of 20 mg·mL^{-1}, and five adsorption/desorption cycles. Compared to traditional methods, the produced HMICS-SPME exhibited the advantages of simplicity of operation, higher recovery and selectivity, improved analytical characteristics and reduced sample and reagent consumption, and it is expected to promote the rapid development and wide applications of molecular imprinting.

Keywords: hydrophilic molecularly imprinted chitosan; deep eutectic solvents; solid phase microextraction; gallic acid; response surface methodology

---

1. Introduction

Gallic acid (GA) is one of the major polyphenolic compounds in plants, such as green tea, vegetables and fruits [1,2]. GA is used in the pharmaceutical and biomedical industries owing to their anti-oxidant, anti-inflammatory, anti-cancer, and anti-viral activities [3,4].

Chitosan (CS) is the deacetylated form of chitin and has been used widely in material-formulations because of its distinct advantages, such as biocompatible, biodegradable, less toxic, and bioactive chemistry properties [5–7]. Moreover, CS is an eco-friendly and cost-effective biopolymer that can be modified easily by various chemical reactions to improve its physicochemical properties [8]. CS can be modified mainly using its amine group, by crosslinking reactions to make it insoluble in acidic pH or by grafting new functional groups to the amine and hydroxyl groups to add new chemical properties and improve the selectivity for the targets [9]. With the addition of new functional groups on CS, it is possible to increase the number of adsorption sites [10].

Currently, the molecular imprinting technique has attracted increasing attention for the production of artificial materials with selective recognition for the target molecules [11]. Molecular imprinting is an approach of artificially generating recognition sites in polymer structures to specifically rebind the target molecules. These materials are obtained by polymerizing functional and cross-linking monomers around a template molecule, leading to a highly cross-linked three-dimensional network polymer. With different kinds of imprinting methods, many products of the molecular imprinting technique have been shown with excellent selectivity and unique structural predictability, such as molecularly imprinted polymers (MIPs) [12], molecular imprinted resins (MIRs) [13,14], and molecular imprinted nanoparticles (MINs) [15]. Practical molecular imprinting materials (MIMs) have become a rapidly evolving research area, providing key factors for understanding the separation, recognition, and regenerative properties toward biological molecules [16].

For further applications, however, MIMs should have good water compatibility because many targeted molecules are present in the aqueous biological matrix [17]. Therefore, imprinting materials need to be modified for use in the aqueous phase system. In molecular imprinting techniques, deep eutectic solvents (DESs) has been used as the solvent for template elution and additive in the preparation of imprinting materials for higher adsorption capacity, acting as monomers [18], solvents [19], or others [20]. DESs are a eutectic mixture of a quaternary ammonium salt as a hydrogen bond acceptor (HBA) with either the organic amine, alcohol, or organic acid as the hydrogen bond donor (HBD) [21], which are characterized by a melting point lower than those of each individual component. Recently, some efforts have been made to introduce DESs in the fields of molecular imprinting [22] and for the separation of bioactive compounds because of their unique properties, including low vapor pressure, easy preparation, and benign biodegradability [23].

In this study, a novel hydrophilic molecularly imprinted chitosan (HMICS) was synthesized using DESs for the selective enrichment of GA in red ginseng tea leaves via a solid phase microextraction (SPME) method. The DESs were used as both the template and functional monomer. The materials obtained were characterized by scanning electron microscopy (SEM), Fourier transform infrared spectroscopy (FT-IR), thermogravimetric analysis (TGA), and nuclear magnetic resonance spectroscopy (NMR). The adsorption kinetics and isotherms for the adsorption models were also investigated. The optimal extraction conditions for GA were optimized using a response surface methodology (RSM).

2. Experimental

2.1. Chemical and Reagents

Red ginseng tea was purchased from a local market (Incheon, Korea). Methanol, GA, acetic acid, sodium hydroxide, protocatechuic acid, 3-hydroxybenzoic acid, 4-hydroxybenzoic acid, protocatechuic acid, and 3,5-dihydroxybenzoic acid were supplied by Sigma-Aldrich. Co, Ltd. (St Louis, MO, United Stated). CS powder (molecular weight (M_w), 75 kDa; degree of deacetylation (DD), 80 mol %) was obtained from Seafresh Chitosan (Lab, Incheon, Korea) Co. Acetonitrile, acetone, hexane, ethyl acetate, liquid paraffin, span-80, glutaraldehyde, isopropanol, choline chloride, and azobisisobutyronitrile (AIBN) were acquired from Daejung Chemicals & Metals Co., Ltd. (Gyonggido, Yongin, Korea). The other chemicals used were of high performance liquid chromatography (HPLC) grade.

2.2. Instrument

The surface morphology was examined by field emission SEM (Hitachi S-4200, Hitachi, Toronto, ON, Canada). FT-IR (Vertex 80 V, Bruker, Billerica, MA, USA) spectroscopy was conducted to examine the functional groups between 4000 and 400 cm^{-1} using KBr pellet samples. The formation of particles was confirmed by TGA (TG 209 F3, Netzsch, Selb, Germany) under a nitrogen atmosphere from room temperature to 800 °C at a heating rate of 10 °C·min^{-1}. The chromatographic measurements were performed on a high performance liquid chromatography (HPLC) system using YL9110 equipment from Younglin Co. Ltd. (Daegu, Korea) consisting of a Rheodyne injector (20 μL sample loop), Waters 600 s Multi solvent Delivery System, Waters 1515 liquid chromatography (Waters, Bedford, MA, USA), and variable wavelength 2489 UV dual channel detector. EmpowerTM 3 software (Waters, Bedford, MA, USA) was used as the data acquisition system. The analysis was performed on an OptimaPak C$_{18}$ column (5 μm, 250.0 mm × 4.6 mm, i.d., RStech Corporation, Daejeon, Korea). All the 1H NMR experiments were carried out with a Bruker DMX 300 spectrometer (Bruker, Karlsruhe, Germany), equipped with a diffusion probe capable of producing magnetic field gradient pulses up to 11.76 T/m in the z-direction. The measurements were carried out in a temperature range between 293.8 and 300 K. A Bruker Variable Temperature unit, BVT 3000 (Bruker, Karlsruhe, Germany), was used to set the required temperature for each experiment. The sample was placed in a 5 mm NMR glass tube with a height of approximately 20 mm and left for 20 min at the desired temperature in order to reach thermal equilibrium.

2.3. Preparation of the Materials

2.3.1. Synthesis of Hydrophilic DESs

Three types of hydrophilic DESs were prepared by mixing choline chloride with GA (1:1, *n/n*, DES-1), (1:2, *n/n*, DES-2), (1:3, *n/n*, DES-3) heated under 80 °C with constant stirring. After 8 h, the resulting DESs were obtained until a homogeneous colorless liquid had formed. Table S1 lists the basic physicochemical properties of the obtained hydrophilic DES. Only DES-2 (1:2, *n/n*) had a stable form prepared and was used in the following procedure.

2.3.2. Preparation of HMICS and NICS

CS microspheres were obtained using an emulsion polymerization technique considering the requirements of high adsorptive surface areas and controllable preparation route [24]. A 10.0 g sample of CS was dissolved in 50.0 mL of a 5.0% (*w/v*) acetic acid solution and 25.0 mL isopropanol with constant stirring. A 2.0 mL sample of the above CS liquid mixture was then dispersed as the aqueous phase into 2.0 mL of liquid paraffin containing 10% (*v/v*) Span 80. Subsequently, 2.0 mL of the obtained hydrophilic DES-2 (both as a template and monomer) was added followed by the dropwise addition of (5%, 1.0 mL) glutaraldehyde solution (as a crosslinker). Following this, 0.5 g of AIBN was added as an initiator, and the homogeneous mixture was stirred at 60 °C for 3 h. The reacting mixture was left to stand at room temperature overnight for complete polymerization. The obtained materials were washed sequentially with petroleum ether and deionized water, and dried in a vacuum oven at 60 °C until a constant weight was reached. Figure 1 shows a schematic diagram of the preparation of HMICS with DES. Non-imprinted chitosan (NICS) was obtained in an identical manner, except that the template was not added during the preparing process of HIMCS, and Figure S1 shows the NMR spectroscopy of HMICS (a) and NICS (b).

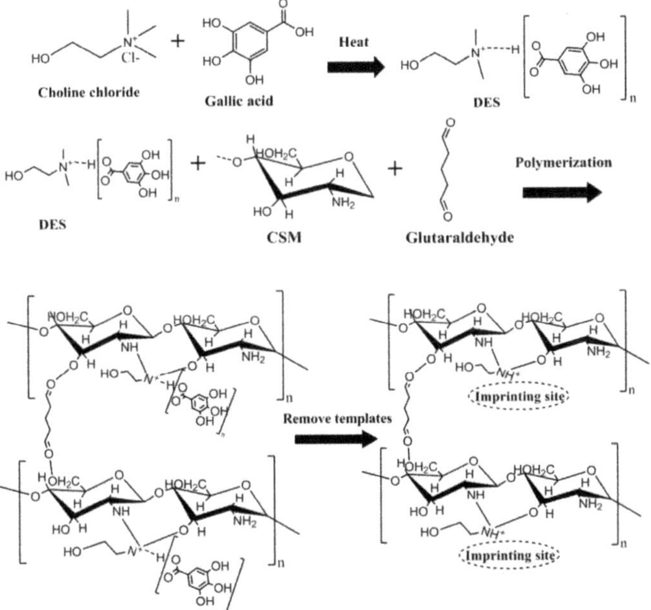

Figure 1. Schematic diagram of the preparation of hydrophilic molecularly imprinted chitosan (HMICS) with deep eutectic solvents (DES).

2.3.3. Adsorption Properties of HMICS and NICS

Static and dynamic adsorption experiments were conducted to evaluate the adsorption properties of the obtained HMICS and NICS. The adsorption capability for HMICS and NICS was determined by the following procedure and model. Briefly, 20 mg of HMICS and NICS were suspended in 10 mL of a GA solution at concentrations ranging from 5 to 200 $\mu g \cdot mL^{-1}$. The series of mixtures were shaken for 240 min under 25 °C to ensure the equilibrium.

The adsorption capacity (Q) was calculated according to the following equation:

$$Q = \frac{(C_0 - C_{free}) \times V}{W} \tag{1}$$

where C_{free} ($\mu g \cdot mL^{-1}$) is the free concentration of GA; C_0 ($\mu g \cdot mL^{-1}$) is the initial concentration; V (mL) is the volume of the GA solution; and W (mg) is the mass of the materials.

For the adsorption kinetics study, 20 mg of HMICS and NICS was suspended in 10 mL of a 100 $\mu g \cdot mL^{-1}$ GA solution and shaken at 25 °C. The concentrations of GA from 30 to 360 min at a certain interval (30 min) were centrifuged and calculated using the following equation:

$$\ln(Q_e - Q_t) = \ln(Q_e) - k_1 t \tag{2}$$

where Q_t ($mg \cdot g^{-1}$) and Q_e ($mg \cdot g^{-1}$) are the amount adsorbed at the given time and equilibrium, respectively; k_1 (min^{-1}) is the rate constant of the adsorption.

2.3.4. Selectivity and Reusability Experiments

To estimate the selectivity of the obtained imprinted materials, selectivity experiments were conducted on GA along with protocatechuic acid, 3,5-dihydroxybenzoic acid, 3-hydroxybenzoic acid, and 4-hydroxybenzoic acid as competitive compounds. A 2 mg sample of HMICS or NICS was added to 1 mL of the mixture solution containing 100 $\mu g/mL$ of the above five compounds. After shaking for 4 h, the solutions were collected by centrifugation and analyzed by HPLC.

The imprinting factor (α) and selectivity factor (β) were used to evaluate the properties of selectivity of HMICS and NICS toward the template molecule (GA) and analogs (protocatechuic acid, 3,5-dihydroxybenzoic acid, 3-hydroxybenzoic acid, and 4-hydroxybenzoic acid). The α and β were calculated from the following equations:

$$\alpha = \frac{Q_{HMICS}}{Q_{NICS}} \tag{3}$$

$$\beta = \frac{\alpha_{template}}{\alpha_{analog}} \tag{4}$$

where Q_{HMICS} and Q_{NICS}, and $\alpha_{template}$ and α_{analog} are the sorption capacity and imprinting factor toward GA (template) or the analog on the HMICS and NICS, respectively.

To test the regeneration capability, a 2 mg sample of HMICS or NICS mixed in the 1 mL of GA standard solution were evaluated by ten sequential cycles of adsorption-regeneration.

2.3.5. Optimization of the SPME Conditions

Red ginseng tea was cleaned, dried in an oven at 60 °C, and ground to a powder. A 10 g sample dried powder was ultrasonicated in 200 mL MeOH/water (80:20, *v/v*) at room temperature for 6 h. The suspension was then filtered as the extraction samples. The miniature SPME procedure was performed using SPE unit. A 20.0 mg sample of the obtained adsorbents was packed into SPE cartridges and connected to a conventional syringe to ensure a suitable and constant flow rate, and capped with decreased cotton in the middle of the SPE cartridges and the syringe. Figure 2 presents a schematic diagram of the miniature SPME procedure.

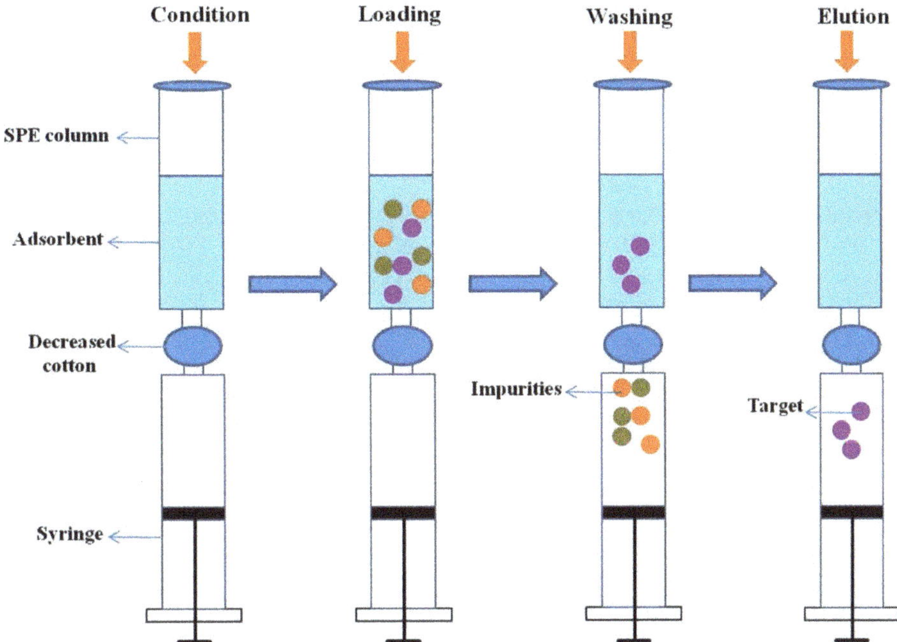

Figure 2. Schematic illustration of the miniature solid phase microextraction (SPME) procedure.

To remove interferents from matrix, different washing solvents, including methanol, acetonitrile, acetone, hexane, and ethyl acetate solutions, with different volumes (0.2–2.0 mL) were tested. The washing solvent was forced to pass through the system by regulating the vacuum of approximately 20 kPa to obtain a flow rate of 0.5 mL·min^{-1}.

To select the most appropriate eluent to desorb GA from the prepared HMICS, several eluent solvents (methanol, methanol-acetic acid (85:15, *v/v*), methanol-ammonia water (85:15, *v/v*), acetonitrile-acetic acid (85:15, *v/v*), and acetone-acetic acid (85:15, *v/v*)) were tested. The washing solvent was forced to pass through the system by regulating the vacuum of approximately 20 kPa to obtain a flow rate of 0.5 mL·min^{-1}.

2.3.6. SPME Procedure with Real Samples Using the RSM

The RSM was applied to determine the optimal levels of the three variables having a significant effect on the extraction efficiency. After determining the preliminary range of the analysis variables through a single-factor test, the experimental variables were designed to optimize the adsorption efficiency of GA. The effects of the three independent variables, namely extraction time (X_1, min), solid to liquid ratio (X_2, mg·mL^{-1}), and number of adsorption/desorption cycles (X_3) on the extraction yields of analytes were investigated using a Box-Behnken design (BBD) of 17 experimental points.

Each variable coded at its three levels (−1, 0, 1) represents the lower, middle and higher value (Table S2). The generalized second-order polynomial Equation (5) used in response surface analysis is as follows:

$$Y = A_0 + \sum_{i=1}^{3} A_i X_i + \sum_{i=1}^{3} A_{ii} X_i^2 + \sum_{i=1}^{2} \sum_{j=i+1}^{3} A_{ij} X_i X_j \tag{5}$$

where Y is the measured response; A_0 is a constant; A_i, A_{ii}, and A_{ij} are linear, quadratic, and interaction coefficients, respectively; and X_i and X_j are the levels being studied. Data analysis was performed using Design-Expert software (v.7.1.6, Stat-Ease, Inc., Minneapolis, MN, USA) and evaluated by an analysis of variance (ANOVA). The fitness of the polynomial equation to the responses was estimated using the coefficient of determination (R^2), and the differences with a p-value less than 0.05 were considered significant.

3. Results and Discussion

3.1. Adsorption Properties

The binding capacity of HMICS and NICS increased with increasing GA concentration, and the adsorption capacity of HMICS with DES-2 was much higher than NICS without DES (Figure 3a). The additional GA bound to HMICS compared to NICS could be attributed to the binding of GA to the imprinting sites with higher specificity.

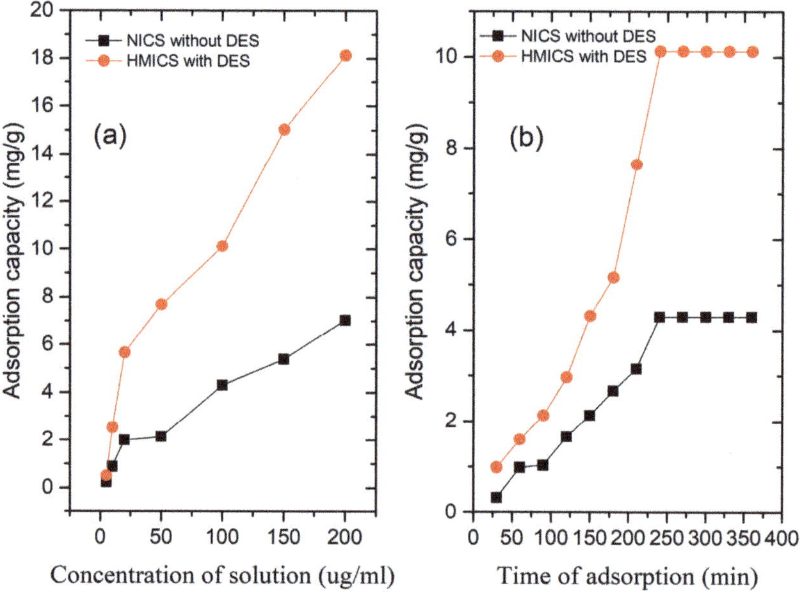

Figure 3. Equilibrium adsorption isotherms (**a**) and kinetic adsorption curves (**b**) of HMICS with DES, and non-imprinted chitosan (INCS) without DES.

The adsorption kinetics of GA onto HMICS and NICS was investigated by varying the adsorption time from 30 to 360 min (Figure 3b). The adsorption capacity increased rapidly in the first 0–100 min, and then the increment from 125–225 min until the process approximately reached equilibrium after 225 min. At the beginning of the adsorption process, the GA could enter many empty specific binding sites easily and rapidly and mass-transfer resistance was significantly small. With time prolonging, it became difficult to find an imprinted site for target. Therefore, the adsorption rate decelerated up to reaching equilibrium. A sharp increase in the adsorption amounts towards GA by HMICS and NICS occurred within 240 min. The adsorption capacity of GA on HMICS with DES was 10.13 mg·g^{-1}, which is approximately 2.36 times as high as that (4.30 mg·g^{-1}) of NICS without DES. The fast and greater dynamics of HMICS adsorption were attributable to the imprinted sites.

3.2. Characterization

Figure 4a presents the FT-IR spectra of HMICS and NICS. The absorbance at 3400–3500 cm^{-1} was assigned to the overlapping stretching vibration of the O–H bonds and N–H bonds on the NICS surface. The absorption at approximately 1620 cm^{-1} in HMICS was attributed to the bending vibration of NH$_2$, which provides evidence of the presence of DES in the HMICS skeleton. In addition, a characteristic absorption band at approximately 1154 cm^{-1} is related to a C−N stretching vibration. The absorbances at 2900 and 1650 cm^{-1} were assigned to the overlapped stretching vibration of CH groups and CO groups, respectively. These results demonstrate the successful formation of HMICS.

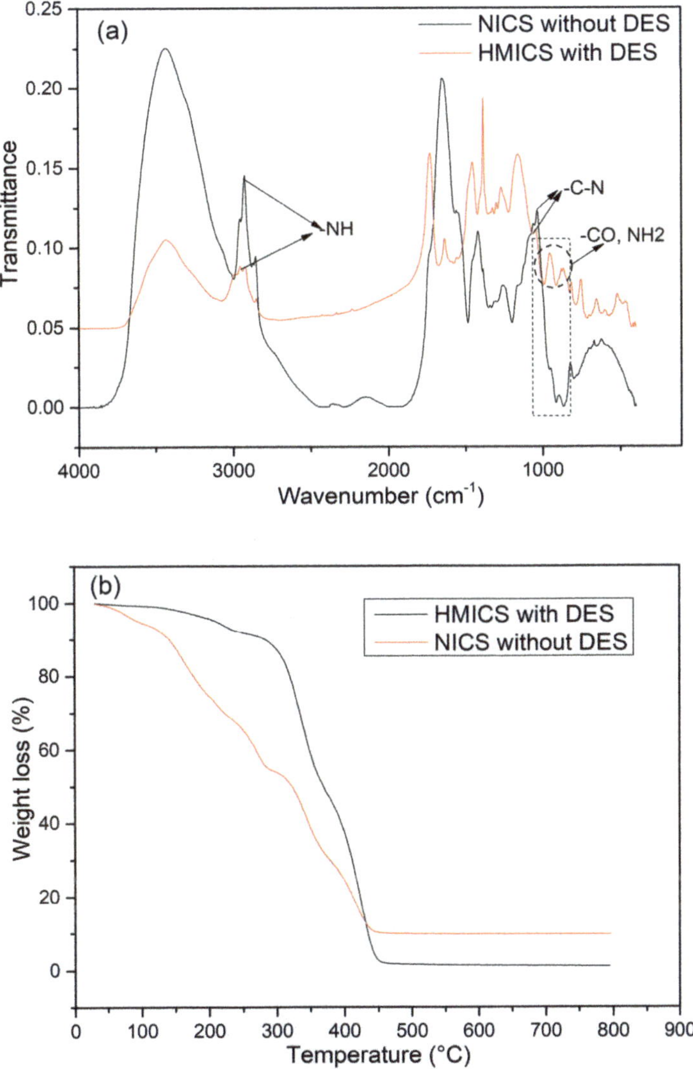

Figure 4. FT-IR spectra (**a**) and thermogravimetric analysis (**b**) of HMICS with DES, and NICS without DES.

TGA was conducted to estimate the thermal stability of the prepared HMICS and NICS, as shown in Figure 4b. HMICS showed 9.2% weight loss between 25 and 300 °C, which was attributed to the elimination of free water and structural water molecules, whereas NICS showed a small mass loss due to the evaporation of residual water from 0 °C and 50 °C. When the temperature was increased to more than 200 °C, the organic shell gradually lost weight and decomposed completely at 400 °C. TGA of HMICS showed significant mass loss from 300 °C to 450 °C, whereas NICS without DES showed significant mass loss from 50 to 420 °C. These weight losses were attributed to the decomposition and vaporization of a grafted macromolecular microsphere. The weight loss of both of HMICS and NICS remained relatively constant from 450 °C to 800 °C. Therefore, HMICS had good thermal stability below 300 °C, whereas NICS was stable below 50 °C.

SEM images were obtained to observe the shape and morphology of HMICS and NICS; the images revealed a uniform structure with a spherical morphology. As shown in Figure 5, HMICS had a more uniform structure with a more regular spherical morphology than NICS. Moreover, the HMICS possessed a slightly spherical structure with a relatively greater size distribution, which facilitated mass transfer and rapid sorption kinetics.

Figure 5. SEM images of HMICS with DES, (**a**), and NICS without DES (**b**).

3.3. Optimization of the Extraction Conditions

Owing to the complex of the tea sample matrices, it is essential for further validation and optimization of the washing and elution conditions. A washing step was required to remove the other constituents of the tea samples bound nonspecifically to the imprinted sorbent. Regarding the chemical structure of the GA, solubility of the target analyte and its compatibility with chromatographic system, the washing solvents (acetonitrile/acetone/hexane/ethylacetate), showed efficient, which may be due to the similarity between the polarity of the target and the washing solvents could break hydrogen bonds between trapped analyte and sorbent easily. The cleanest extract with the highest recovery was obtained using hexane as the washing solvent. Different hexane volumes (0.2–2.0 mL) were tested and the optimal value (91.4%) was 1.0 mL (Figure 6a). Because of the hydrophilicity of HMICS and high potential of hexane to dissolve non-polar compounds, most of the matrix interfaces were washed with hexane without interfering with the interactions between the GA and sorbent.

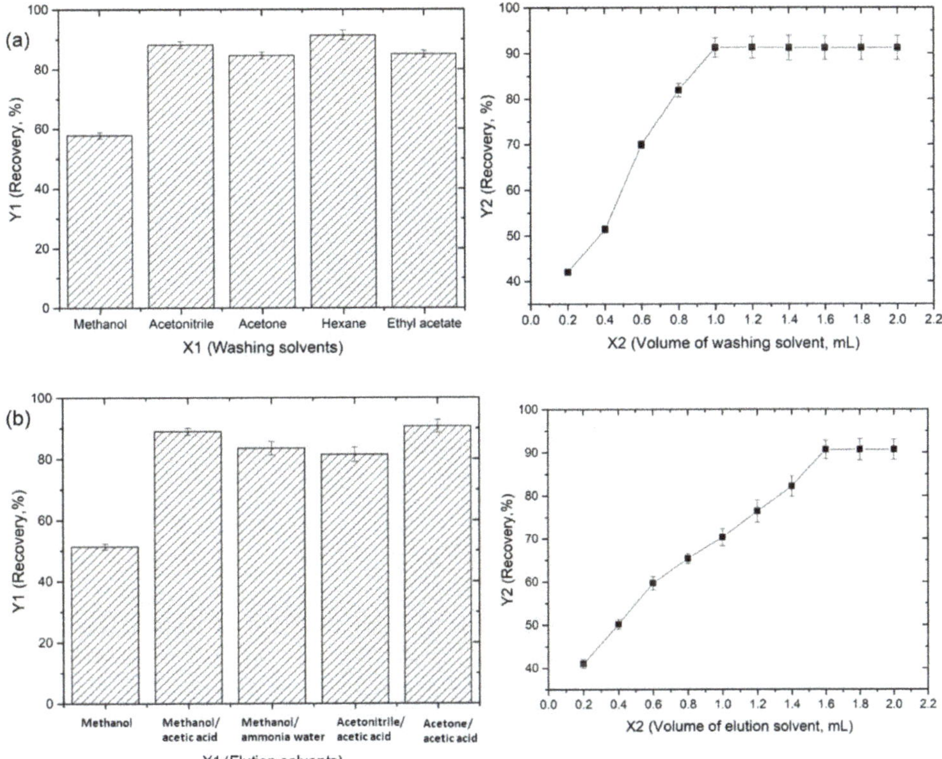

Figure 6. Optimization of the HMICS in the SPME procedure for GA ((**a**): Washing solvents, (**b**): Elution solvents).

The selection of a suitable eluent solute has a great influence on the extraction recovery of a target molecule. According to the experimental results, the best recovery was achieved using acetone-acetic acid (85:15, *v/v*) as the eluent (Figure 6b). Different eluent volumes (0.2–2.0 mL) were tested and a volume of 1.6 mL of acetone-acetic acid (85:15, *v/v*) was found to be sufficient to desorb the analyte. The addition of acetic acid could impede hydrogen bonding between GA and the stationary phase, leading to the easier removal of GA.

3.4. Selectivity and Reusability of HMICS and NICS

Imprinting factor and selectivity factor are the chief and prominent advantages of imprinted materials. The imprinting and selectivity capability of HMICS were evaluated by comparing the imprinting factor (α) and selectivity (β) of GA in the presence of competitive compounds. As shown in Figure 7a, α values of HMICS for GA, protocatechuic acid, 3,5-dihydroxybenzoic acid, 3-hydroxybenzoic acid, and 4-hydroxybenzoic acid were 3.67, 1.62, 1.53, 1.09, and 1.07, respectively, and the β values of HMICS for GA, protocatechuic acid, 3,5-dihydroxybenzoic acid, 3-hydroxybenzoic acid, and 4-hydroxybenzoic acid were 0.70, 0.22, 0.20, 0.14, and 0.13, respectively. It was shown that the HMICS had the highest selectivity value toward GA, and it was further verified that GA was adsorbed onto the HMICS by means of specified imprinted sites. Moreover, these results show that HMICS had highly selective recognition capability toward GA.

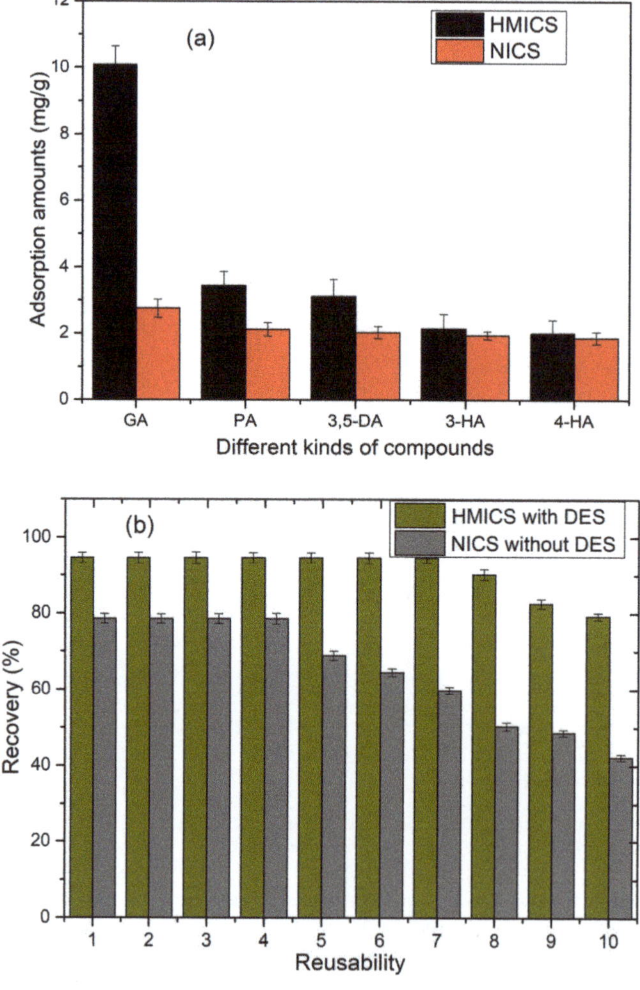

Figure 7. Selectivity (**a**) and reusability (**b**) of HMICS with DES, and INCS without DES. (GA: Gallic acid; PA: Protocatechuic acid; 3,5-DA: 3,5-Dihydroxybenzoic acid. 3-HA: 3-Hydroxybenzoic acid; 4-HA: 4-Hydroxybenzoic acid)

Reproducibility and reusability are very important for designing an advanced and effectual sorbent. As shown in Figure 7b, the extraction efficiency of the HMICS remained at a relatively high level, even after seven cycles, whereas NICS showed an obvious decrease at the fourth cycle, indicating that the HMICS sorbent can be employed frequently as an effective sorbent for GA recovery.

3.5. RSM Model Fitting and Statistical Analysis

The extraction variables optimized for the extraction efficiency for GA were the extraction time, solid to liquid ratio, and number of adsorption/desorption cycles. Table S3 lists the central composite experimental design with the independent variables. The quadratic response surface regression model was used to predict the extraction efficiency in terms of the extraction parameters (coded factors), as expressed in Equation (6):

$$Y = 92.92 + 0.29X_1 - 1.34X_2 + 1.72X_3 + 0.025X_1X_2 + 0.75X_1X_3 + 0.25X_2X_3$$
$$- 0.70X_1{}^2 + 1.00X_2{}^2 - 4.42X_3{}^2 \tag{6}$$

Table S4 lists the ANOVA results for the quadratic response surface regression model and the significance of the regression coefficients to maximize the extraction recovery. The F-value of the model was 13.47 and the p-value was less than 0.0001, indicating that the model was significant in predicting the extraction yield. The Model *F*-value of 13.47 suggested that the model was not significant relative to the noise, and there was only a 0.12% probability that a "Model F-Value" could occur due to noise. The "Lack of Fit F-value" of 5.46 suggested that the Lack of Fit was not significant relative to the pure error, and there was a 6.74 % chance that a "Lack of Fit F-value" could occur due to noise. Moreover, its corresponding p-value was 0.0674, indicating that the model fitted the experimental data well.

The coefficient of determination (R^2 = 0.9454) indicated that 94.54% of the variability in the extraction recovery could be explained by this model (Table S5). Moreover, the predicted extraction recovery values were close to those from the BBD, which confirmed the reliability of the model. The difference between the predicted $R^2{}_{Pred}$ (0.2814) and adjusted $R^2{}_{Adj}$ (0.8752) indicates there is reasonable agreement in the regression polynomial model. The signal to noise ratio was measured using "Adeq. Precision"; a ratio of greater than four was normally desirable. The "Adeq. Precision" of 11.623 suggested that this model could be used to navigate the design space.

The contour plot and three-dimensional (3D) response surfaces were plotted to investigate the interactions among the variables (Figure 8). The extraction yield decreased with increasing extraction time ranged from 30 min to 50 min, increased with increasing solid to liquid ratio from 20 mg·mL^{-1} to 30 mg·mL^{-1}, and then decreased from 30 mg·mL^{-1} to 40 mg·mL^{-1}. Regarding the number of adsorption/desorption cycles, in the designed ranges of one to nine, the extraction recovery increased and then decreased. The theoretical maximum extraction recovery (94.6%) for GA was obtained at an extraction time of 30 min, solid to liquid ratio of 20 mg·mL^{-1}, and five adsorption/desorption cycles. Under the optimal extraction conditions, the actual extraction recovery was 93.9%, which is close to 94.6%, highlighting the suitability and accuracy of the suggested models.

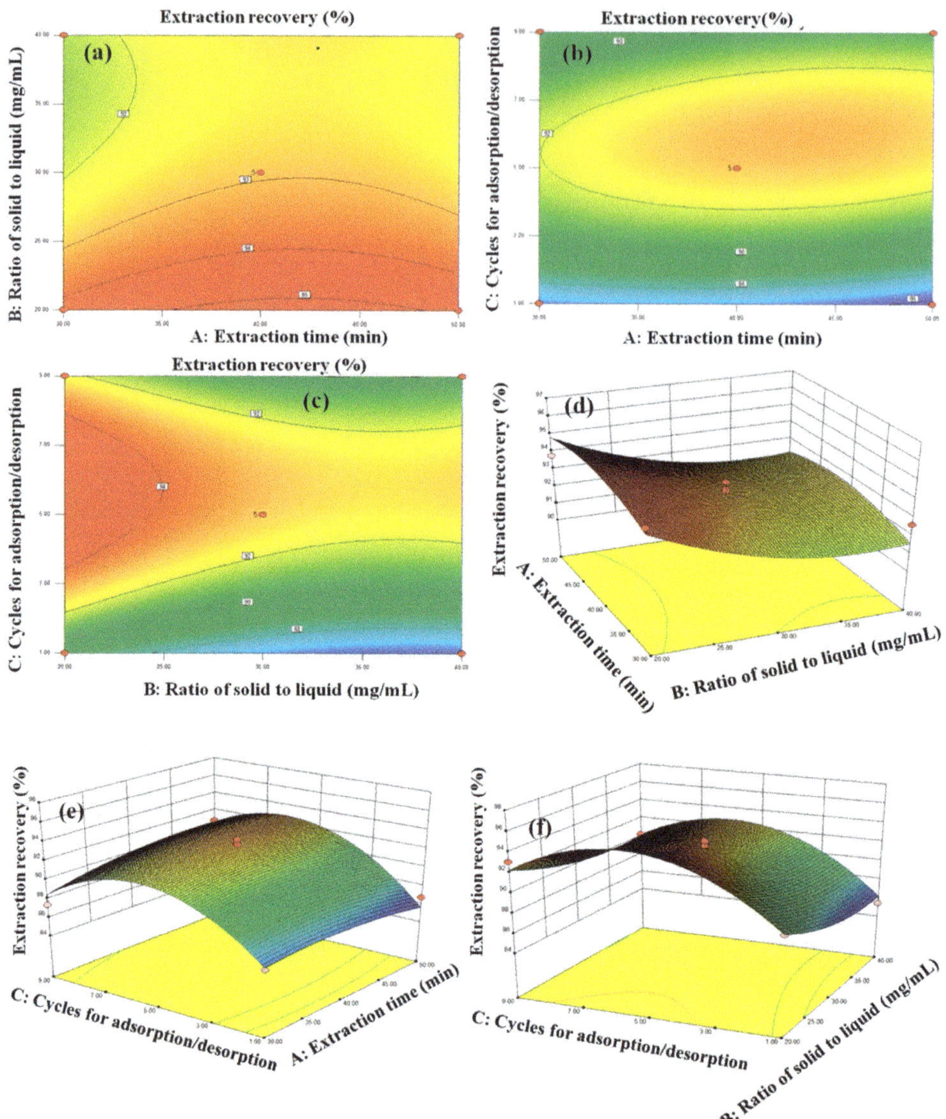

Figure 8. The contour plot (**a–c**), and reciprocal 3D response interaction (**d–f**) on the HMICS extraction recovery yield of three variables (effects of extraction time, ratio of solid to liquid, cycles for adsorption/desorption).

3.6. Method Validation and Real Sample Analysis

The standard curve for GA was linear over the range, 5.00–100.00 μg·mL^{-1}, by assaying five data points in triplicate (Y = 3.78 × 10^4 + 2.54 × 10^{4}X, R^2 = 0.9997). To validate the developed method for the determination of GA, further experiments with regard to the calibration linearity range, limit of detection (LOD), limit of quantification (LOQ), recovery, and relative standard deviation (RSD) were conducted under the optimized experimental conditions (Table 1). The LOD (0.32–0.41 μg·mL^{-1}) and LOQ (0.22–0.36 μg·mL^{-1}) were calculated as three and ten times the standard deviation of the noise signal, respectively. The recoveries of the targets extracted did not differ significantly and the RSD was no higher than 4.16%, demonstrating the high selectivity of the proposed method.

Table 1. Intra-day and Inter-day precisions and accuracy of GA.

Targets	Concentration ($\mu g \cdot mL^{-1}$)	Limit of Detection ($\mu g \cdot mL^{-1}$)	Limit of Quantification ($\mu g \cdot mL^{-1}$)	Inter-day Recovery (%)	Inter-day Relative Standard Deviation (%, n = 4)	Intra-day Recovery (%)	Intra-day Relative Standard Deviation (%, n=4)
	50	0.21	0.24	90.13	4.16	90.04	3.68
Gallic Acid	100	0.15	0.18	94.82	3.84	93.68	3.15
	200	0.08	0.13	102.68	2.65	100.84	3.06

As shown in Figure 9, the red ginseng tea extract is a complex matrix and there were some peaks other than that for GA. After pretreatment with HMICS, the interfering peaks weakened. A significant peak for GA was observed, confirming that GA in the tea extract sample could be extracted selectively by HMICS, and the GA content in red ginseng tea was 8.57 mg·g^{-1}.

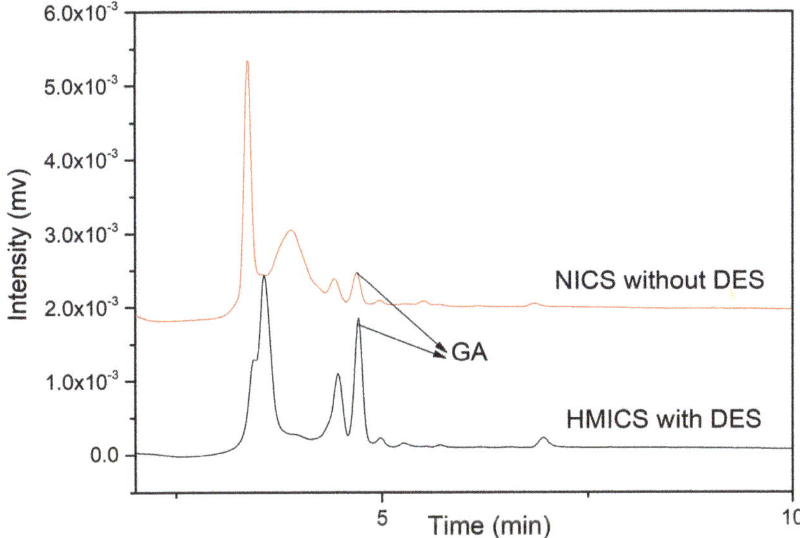

Figure 9. Extraction chromatograms of red ginseng tea extracts with HMICS with DES, and INCS without DES. (Column: C$_{18}$ column, mobile phase: (Acetonitrile-0.05% phosphoric acid solution = 5:95, *v/v*), flow rate: 1.0 mL·min^{-1}, UV: 270 nm, injection: 20 μL).

3.7. Comparison with other Methods

To highlight the distinct merits of the HMICS and present method, a comparison with other reported studies was made, as shown in Table 2. The obtained HMICS had lower LODs than previous imprinted materials for the extraction of GA.

The developed method provided a wide linear range and a much lower LOD than the other methods, whereas the recovery and precision of this method were comparable to or better than the other methods. Therefore, the developed HMICS-SPME method can be used as an effective method for the simple, rapid, cost-effective, sensitive, and selective determination of GA in red ginseng tea samples.

Table 2. Comparison of different materials for extraction of GA.

Materials Types	Function Monomer	Source	Recovery (%)	Reference
Molecularly imprinted microparticles	Methacrylic acid	Olive mill wastewaters	85.0–97.0	[25]
Molecularly imprinted microspheres and nanoparticles	Acrylic acid	Emblica officinalis	75.0–83.4	[26]
Hydrophilic molecularly imprinted chitosan microsphere	DES	Tea sample	90.0–102.7	This research

4. Conclusions

A novel HMICS was synthesized based on a hydrophilic DES used as both the template and functional monomer for the enrichment of GA from red ginseng tea using the SPME method. The optimal extraction amount (8.57 mg·g^{-1}) was found at an extraction time of 30 min, solid to liquid ratio of 20 mg·mL^{-1}, and five adsorption/desorption cycles using the RSM strategy. Compared to the traditional CS microspheres, the HMICS produced using the hydrophilic DESs exhibited higher extraction capacity. Such improvements will allow an extension of the field of application of the new imprinted CS, which could become a new tool used routinely in analytical laboratories in the future.

Supplementary Materials: The following are available online at http://www.mdpi.com/2073-4360/11/9/1434/s1, Table S1 Density (ρ), viscosity (μ), and conductivity (σ) of the three kinds of DESs at 293.15 K and atmospheric pressure (1.01 bar); Table S2 Independent variables their levels used for BBD; Table S3 Central composite experimental design with the independent variables; Table S4 Analysis of variance of the experimental results of the BBD; Table S5 Analysis of variance for the fitted quadratic polynomial model of extraction of GA; Figure S1. 1H NMR spectra of the HMICS (a) and NICS (b).

Author Contributions: Conceptualization, G.L. and K.H.R.; methodology, G.L.; software, G.L.; validation, G.L.; formal analysis, G.L.; investigation, G.L.; resources, K.H.R.; data curation, G.L.; writing-original draft preparation, G.L.; writing-review and editing, K.H.R.; visualization, G.L.; supervision, K.H.R.; project administration, K.H.R.; funding acquisition, K.H.R.

Funding: This study was supported by the National Research Foundation of Korea (NRF) grant funded by Korea Government (MSIT) (No. NRF-2019R1A2C1010032).

Conflicts of Interest: The authors declare no conflict of interest.

References

1. Zhang, J.J.; Li, B.Q.; Yue, H.J.; Zheng, Y.S. Highly selective and efficient imprinted polymers based on carboxyl-functionalized magnetic nanoparticles for the extraction of gallic acid from pomegranate rind. *J. Sep. Sci.* **2018**, *41*, 540–547. [CrossRef] [PubMed]

2. Karimi-Khouzani, O.; Sharifi, A.; Jafari, A. A Systematic Review of the Potential Gallic Acid Effective in Liver Oxidative Stress in Rats. *J. Med. Res.* **2018**, *4*, 5–12.

3. Kahkeshani, N.; Farzaei, F.; Fotouhi, M.; Alavi, S.S.; Bahramsoltani, R.; Naseri, R.; Momtaz, S.; Abbasabadi, Z.; Rahimi, R.; Farzaei, M.H.; et al. Pharmacological effects of gallic acid in health and disease: A mechanistic review. *Iran J. Basic Med. Sci.* **2019**, *22*, 225–237. [PubMed]

4. Dludla, P.; Nkambule, B.; Jack, B.; Mkandla, Z.; Mutize, T.; Silvestri, S.; Orlando, P.; Tiano, L.; Louw, J.; Mazibuko-Mbeje, S.E. Inflammation and oxidative stress in an obese state and the protective effects of gallic acid. *Nutrients* **2019**, *11*, 23. [CrossRef] [PubMed]

5. Ali, A.; Ahmed, S. A review on chitosan and its nanocomposites in drug delivery. *Int. J. Biol. Macromol.* **2018**, *109*, 273–286. [CrossRef] [PubMed]

6. Elgadir, M.A.; Uddin, M.S.; Ferdosh, S.; Adam, A.; Chowdhury, A.J.K.; Sarker, M.Z.I.S. Impact of chitosan composites and chitosan nanoparticle composites on various drug delivery systems: A review. *J. Food Drug Anal.* **2015**, *23*, 619–629. [CrossRef] [PubMed]

7. Khan, A.; Aqil, M.; Imam, S.S.; Ahad, A.; Sultana, Y.; Ali, A.; Khan, K. Temozolomide loaded nano lipid based chitosan hydrogel for nose to brain delivery: Characterization, nasal absorption, histopathology and cell line study. *Int. J. Biol. Macromol.* **2018**, *116*, 1260–1267. [CrossRef] [PubMed]

8. Moreira, A.L.S.L.; de Souza Pereira, A.; Speziali, M.G. Bifunctionalized chitosan: A versatile adsorbent for removal of Cu (II) and Cr (VI) from aqueous solution. *Carbohydrate Polym.* **2018**, *201*, 218–227. [CrossRef]

9. Wang, B.J.; Bai, Z.S.; Jiang, H.R.; Prinsen, P.; Luque, R.; Zhao, S.L.; Xuan, J. Selective heavy metal removal and water purification by microfluidically-generated chitosan microspheres: Characteristics, modeling and application. *J. Hazard. Mater.* **2019**, *364*, 192–205. [CrossRef]

10. Chen, X.; Zhao, Y.; Zhang, Y.S.; Lu, A.G.; Li, X.R.; Liu, L.Z.; Qin, G.M.; Fang, Z.L.; Zhang, J.L.; Liu, Y.X. A novel design and synthesis of multifunctional magnetic chitosan microsphere based on phase change materials. *Mater. Lett.* **2019**, *237*, 185–187. [CrossRef]

11. Zhang, N.; Zhang, N.; Xu, Y.; Li, Z.L.; Yan, C.R.; Mei, K.; Ding, M.L.; Ding, S.C.; Guan, P.; Qian, L.W.; et al. Molecularly Imprinted Materials for Selective Biological Recognition. *Macromol. Rapid Comm.* **2019**, *1900096*, 1–21. [CrossRef] [PubMed]

12. Li, G.; Row, K.H. Ternary deep eutectic solvent magnetic molecularly imprinted polymers for the dispersive magnetic solid-phase microextraction of green tea. *J. Sep. Sci.* **2018**, *41*, 3424–3431. [CrossRef] [PubMed]

13. Li, H.; Long, R.Q.; Tong, C.Y.; Li, T.; Liu, Y.G.; Shi, S.Y. Shell thickness controlled hydrophilic magnetic molecularly imprinted resins for high-efficient extraction of benzoic acids in aqueous samples. *Talanta* **2019**, *194*, 969–976. [CrossRef] [PubMed]

14. Zhou, T.Y.; Ding, J.; He, Z.Y.; Li, J.Y.; Liang, Z.H.; Li, C.Y.; Li, Y.; Chen, Y.H.; Ding, L. Preparation of magnetic superhydrophilic molecularly imprinted composite resin based on multi-walled carbon nanotubes to detect triazines in environmental water. *Chem. Eng. J.* **2018**, *334*, 2293–2302. [CrossRef]

15. Abbasi Ghaeni, F.; Karimi, G.; Mohsenzadeh, M.S.; Nazarzadeh, M.; Motamedshariaty, V.S.; Mohajeri, S.A. Preparation of dual-template molecularly imprinted nanoparticles for organophosphate pesticides and their application as selective sorbents for water treatment. *Sep. Sci. Technol.* **2018**, *53*, 2517–2526. [CrossRef]

16. Li, G.; Row, K.H. Recent applications of molecularly imprinted polymers (MIPs) on micro-extraction techniques. *Sep. Purif. Rev.* **2018**, *47*, 1–18. [CrossRef]

17. Yuan, Y.N.; Yang, C.L.; Lv, T.W.; Qiao, F.X.; Zhou, Y.; Yan, H.Y. Green synthesis of hydrophilic protein-imprinted resin with specific recognition of bovine serum albumin in aqueous matrix. *Anal. Chim. Acta* **2018**, *1033*, 213–220. [CrossRef]

18. Zhang, Y.D.; Cao, H.W.; Huang, Q.W.; Liu, X.Y.; Zhang, H.X. Isolation of transferrin by imprinted nanoparticles with magnetic deep eutectic solvents as monomer. *Anal. Bioanal. Chem.* **2018**, *410*, 6237–6245. [CrossRef]

19. Li, G.; Row, K.H. Selective extraction of 3,4-dihydroxybenzoic acid in Ilex chinensis Sims by meticulous mini-solid-phase microextraction using ternary deep eutectic solvent-based molecularly imprinted polymers. *Anal. Bioanal. Chem.* **2018**, *410*, 7849–7858. [CrossRef]

20. Wang, R.; Li, W.; Chen, Z. Solid phase microextraction with poly (deep eutectic solvent) monolithic column online coupled to HPLC for determination of non-steroidal anti-inflammatory drugs. *Anal. Chim. Acta* **2018**, *1018*, 111–118. [CrossRef]

21. Liu, Y.; Friesen, J.B.; McAlpine, J.B.; Lankin, D.C.; Chen, S.N.; Pauli, G.P. Natural deep eutectic solvents: Properties, applications, and perspectives. *J. Nat. Prod.* **2018**, *81*, 679–690. [CrossRef] [PubMed]

22. Xu, K.J.; Wang, Y.Z.; Wei, X.X.; Chen, J.; Xu, P.L.; Zhou, Y.G. Preparation of magnetic molecularly imprinted polymers based on a deep eutectic solvent as the functional monomer for specific recognition of lysozyme. *Microchim. Acta* **2018**, *185*, 146–153. [CrossRef] [PubMed]

23. Fu, N.J.; Liu, X.L.; Li, L.T.; Tang, B.K.; Row, K.H. Ternary choline chloride/caffeic acid/ethylene glycol deep eutectic solvent as both a monomer and template in a molecularly imprinted polymer. *J. Sep. Sci.* **2017**, *40*, 2286–2291. [CrossRef] [PubMed]

24. Zhang, W.J.; Li, Q.; Mao, Q.; He, G.H. Cross-linked chitosan microspheres: An efficient and eco-friendly adsorbent for iodide removal from waste water. *Carbohyd. Polym.* **2019**, *209*, 215–222. [CrossRef] [PubMed]

25. Puoci, F.; Scoma, A.; Cirillo, G.; Bertin, L.; Fava, F.; Picci, N. Selective extraction and purification of gallic acid from actual site olive mill wastewaters by means of molecularly imprinted microparticles. *Chem. Eng. J.* **2012**, *198*, 529–535. [CrossRef]

26. Pardeshi, S.; Dhodapkar, R.; Kumar, A. Molecularly imprinted microspheres and nanoparticles prepared using precipitation polymerisation method for selective extraction of gallic acid from Emblica officinalis. *Food Chem.* **2014**, *146*, 385–393. [CrossRef] [PubMed]

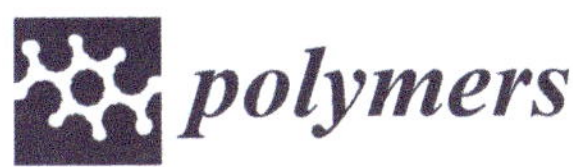 

Article

Ion-Imprinted Polypropylene Fibers Fabricated by the Plasma-Mediated Grafting Strategy for Efficient and Selective Adsorption of Cr(VI)

Zhengwei Luo [1], Jiahuan Xu [2], Dongmei Zhu [2], Dan Wang [2], Jianjian Xu [2], Hui Jiang [1], Wenhua Geng [1], Wuji Wei [2,*] and Zhouyang Lian [2,*]

[1] College of Biotechnology and Pharmaceutical Engineering, Nanjing Tech University, 30# Puzhu South Road, Nanjing 211816, China; luozw1989@163.com (Z.L.); jianghuinjtech2018@163.com (H.J.); gengwenhua2018@163.com (W.G.)

[2] School of Environmental Science and Engineering, Nanjing Tech University, 30# Puzhu South Road, Nanjing 211816, China; xu_jiahuan2018@163.com (J.X.); dongmei_zhunjtech@163.com (D.Z.); wangdannjtech@163.com (D.W.); xujianjian@163.com (J.X.)

* Correspondence: wjwei@njtech.edu.cn (W.W.); lianzy1985@163.com (Z.L.); Tel.: +86-25-83588831 (W.W.); Fax: +86-25-58139382 (W.W.)

Received: 14 August 2019; Accepted: 12 September 2019; Published: 16 September 2019

Abstract: To improve the adsorption selectivity towards hexavalent chromium anion (Cr(VI)), surface Cr(VI)-imprinted polypropylene (PP) fibers were fabricated by the plasma-mediated grafting strategy. Hence, a non-thermal Rradio frequency discharge plasma irradiation followed by a gaseous phase grafting was used to load acrylic acid (AA) onto PP fibers, which was afterwards amidated with triethylenetetramine and subjected to imprinting with a Cr(VI) template. The plasma irradiation conditions, i.e., gas species, output power, pressure, and time, were optimized and then the influence of grafting time, pressure, and temperature on the grafting degree of AA was investigated. Scanning electron microscopy and Fourier transform infrared spectroscopy were used for the characterization of pristine and modified fibers and to confirm the synthesis success. The hydrophilicity of modified fibers was greatly improved compared with pristine PP fibers. The adsorption thermodynamics and kinetics of Cr(VI) were investigated, as well as the elution efficiency and reusability. The prepared imprinted fibers showed superior adsorption selectivity to Cr(VI) compared with non-imprinted fibers. Finally, the stability of the imprinted fibers against the oxidation ability of Cr(VI) is discussed.

Keywords: polypropylene; nonwoven fibers; plasma; imprinted polymer; chromium

1. Introduction

Electroplating, metallurgy, and leather tanning industries produce a large amount of waste water containing hexavalent chromium anions (Cr(VI)), mainly in the form of oxyanions (e.g., $HCrO_4^-$). Cr(VI) has a strong oxidation character and mobility, as well as strong carcinogenic, teratogenic, and mutagenesis effects on humans [1,2]. Therefore, industrial wastewater containing a high concentration of Cr(VI) must be treated before being discharged. The usual methods include chemical reduction [3], membrane separation [4], the photocatalytic method [5], microbial removal [6], and adsorption. Among the chemical methods, the most commonly used are reduction and precipitation, but the cost is high and the sludge treatment is difficult, frequently causing secondary pollution. The recycling and utilization of Cr(VI) is considered to be the most environmentally friendly and economical approach [7]. Due to its advantages, such as high removal efficiency, low treatment cost, and reusability, the adsorption is one of the most efficient methods for the removal of Cr(VI) [8,9]. Although many adsorbents have been used for the adsorption of Cr(VI), there are still shortcomings to overcome, such as the adsorption capacity, adsorption rate, stability, and selectivity [10]. For instance, in the presence of competing ions, e.g., sulfate and phosphate, it is difficult to achieve a high adsorption capacity for Cr(VI) with the currently used adsorbents. Hence, there is a real demand for development of novel adsorbents.

Polypropylene (PP) melt-blown non-woven fabric has been widely used as an adsorbent due to its excellent chemical stability and good flexibility [11]. Grafting modification allows the polymerization of other monomers with polarity and functionality to the molecular chain of PP, which is the most effective way to improve the surface chemistry of PP. The plasma method requires mild process conditions, it is environmentally friendly, and it does not damage the internal structure of polymer materials. The principle of plasma grafting consists of using ionized high-energy particles to bombard the surface of the material to produce active sites (e.g., free radicals), triggering the grafting reaction of monomers on the surface. Grafting modification of PP fibers can greatly improve the hydrophilicity, functionality, and other physical and chemical properties [12]. Basarir et al. found that the electrochemical properties of PP films were improved as a result of activation with plasma followed by polymerization of the acrylic monomers on the surface [13]. Chen et al. grafted glycidyl methacrylate (GMA) on the surface of PP fibers, and the chelated fibers effectively adsorbed Cd(II) from water [14]. Tseng et al. used plasma to treat PP fibers on which GMA was grafted [15]. Then, the chelating fibers were modified with diethanolamine for Ag^+ adsorption. Generally, gaseous plasma treatment and grafting of a monomer from the liquid phase are applied into two steps [16–18] or the plasma discharge treatment and grafting process occur simultaneously in monomer solution [19,20]. If the monomer gas is directly used for plasma pretreatment and grafting, the preparation steps can be simplified and the grafting efficiency can be improved [21].

Surface grafting modification of PP fibers for the adsorption of heavy metal ions can take advantage of the excellent properties of the polymer fibers, and high efficiency and selectivity of ion adsorption can be further achieved by using the surface ion-imprinted polymers. The essence of ion-imprinted polymers is to create specific adsorption sites on the surface, so as to improve the adsorption selectivity and capacity. The preparation steps generally include the preparation of polymeric adsorbent, adsorption of template metal ions, crosslinking, and elution of heavy metal templates [22]. Li et al. grafted the polyacrylic acid onto PP fiber surfaces followed by the reaction with polyethylene imine to prepare a Cu^{2+}-imprinted polymer [23]. The resulted adsorbent exhibited good selectivity and regenerability. Zhang et al. prepared surface imprinted PP fibers induced by ^{60}Co radiation to selectively adsorb uranyl carbonate from the seawater [24,25].

Plasma modification has been used to improve the adsorption performance of PP fibers for the removal of heavy metal cations. However, few studies have been conducted on the preparation of PP fiber-imprinted material by plasma grafting and its application for the adsorption of high valent oxyanions. Hence, this paper reports on the use of the gaseous plasma grafting method to load an acrylic acid (AA) group on the surface of a melt-blown non-woven PP fiber, which was further amidated to prepare a Cr(VI)-imprinted fiber. The parameters of the plasma grafting process were optimized, and the morphology, hydrophilicity, and surface structure of the fiber were analyzed. The adsorption performance and reuse efficiency of the ion-imprinted fiber toward the Cr(VI) anion are also discussed.

2. Experimental

2.1. Reagents and Instruments

A PP melt-blown non-woven fabric with a fiber diameter of 2–10 μm and specific surface area of 1.50 m^2/g was homemade using a previously reported method [26]. Argon and air (purity of 99.99%) were supplied by Nanjing Shangyuan Industrial Gas Company (Nanjing, China). Other reagents, i.e., acrylic acid (AA), triethylenetetramine (TETA), HCl, acetone, ethanol, N, N-dimethylformamide (DMF), 1-hydroxybenzotriazole (HOBT), N, N′-dicycloethyl carbamide (DCC), formaldehyde, glutaral, epichlorohydrin, and sodium hydroxide were of analytic grade and used as received.

A pulsed radio frequency (RF) power source (500 W, MSY–I) and radio frequency matcher (2000 W, SP-II) were supplied by the Institute of Microelectronics, Chinese Academy of Sciences (Beijing, China). A rotary-vane vacuum pump (2X-8B, 1.1 kW) was supplied by Nanjing Vacuum Pump Co., Ltd (Nanjing, China).

2.2. Preparation of Ion-Imprinted Fibers

The reactions involved in the preparation of ion-imprinted fibers are shown in Scheme 1. First, AA was grafted onto the surface of a PP fiber [14,17]. As shown in Figure 1, three gases, namely AA, air, and argon, were used for the radio frequency discharge plasma irradiation to generate free radicals on PP fibers, respectively. Afterwards, AA gas was induced into the reactor at a certain pressure and temperature and grafted onto PP fibers through reaction with the free radicals. Then, the resultant AA-grafted PP fiber (PA) was amidated using 0.1 mol/L TETA in the medium of 50 mL DMF solution with DCC and HOBT concentrations of 0.1 and 0.15 mol/L, respectively. After grafting, the sample was washed three times with acetone and deionized water to remove the remaining reactants on the fiber surface. The amidated fiber (PAT) was obtained after drying in vacuum at 50 °C to a constant mass.

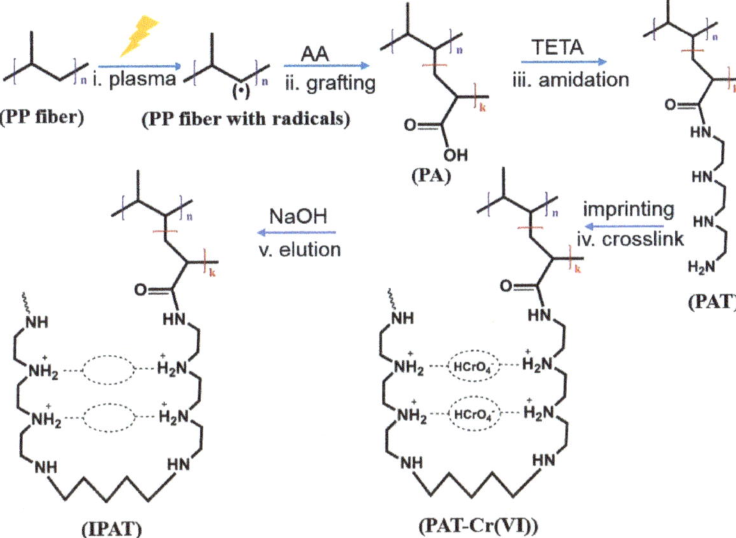

Scheme 1. Schematic representation of the preparation process of ion-imprinted PP fibers (IPAT) (**i**: plasma treatment of PP fibers to form radicals; **ii**: grafting of acrylic acid (AA) on PP fibers to form AA-grafted PP (PA); **iii**: amidation of AA with triethylene tetramine (TETA) to form PAT; **iv**: imprinting of Cr(VI) template and crosslinking; **v**: elution of Cr(VI) template using a NaOH solution).

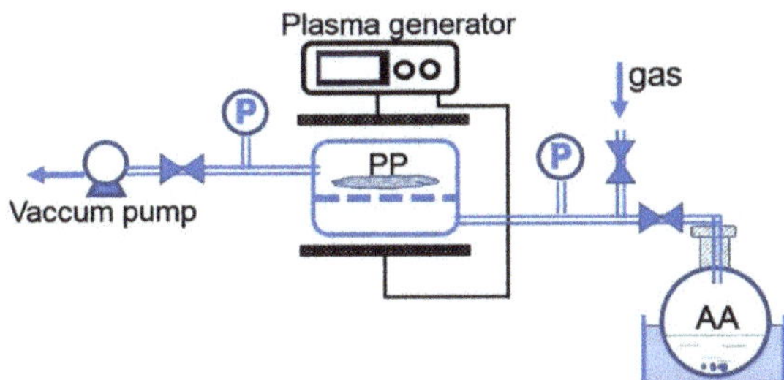

Figure 1. Illustration of the experiment setup of the plasma-induced polymerization of AA.

The PAT was weighed and soaked in 100 mL potassium dichromate solution (200 mg/L, pH = 3), followed by shaking and adsorption at 300 rpm for 2 h, and then immersed in a 50 mL aqueous solution of 30% TETA (pH = 3) for 30 min. The resulting fibers were crosslinked using 30% aqueous epoxy chloropropane, formaldehyde, and glutaraldehyde-ethanol solution at constant temperature, respectively. To remove template ions, the fibers were washed with 0.1 mol/L NaOH until Cr(VI) was not detected in the washing solution. After washing with deionized water and acetone, and drying at 50 °C to a constant mass, the Cr(VI)-imprinted fibers, denoted as IPAT, were obtained. At the same time, non-imprinted fibers (NIPAT) were prepared following the same preparation steps as for IPAT, except for the templating step, and used as a reference.

2.3. Analytic Methods

2.3.1. Grafting Degree

The grafting degree of the PA fiber was determined by the titration method. The PA fiber was weighed and dissolved in 100 mL of xylene, then 20 mL of 0.05 mol/L NaOH-ethanol solution was added and refluxed for 30 min. After cooling to room temperature, titration with 0.05 mol/L HCl-isopropanol solution to the end point was performed using bromothymol blue as a pH indicator. The grafting degree, Gr (mmol/g), was calculated as:

$$G_r = (M_1 \times V_1 - M_2 \times V_2)/m \times 100\% \tag{1}$$

where V_1 (mL) and V_2 (mL) are the consumed volumes of NaOH and HCl solutions, respectively, M_1 (mol/L) and M_2 (mol/L) are the molar concentrations of NaOH and HCl solutions, respectively, and m (g) is the mass of PA. Each experiment was performed in triplicate.

2.3.2. Amino Content

The PAT was weighed and placed in a 100 mL conical flask with a stopper. An HCl-ethanol solution (50 mL) was added and shaken well. The flask was tightly sealed and immersed in a water bath at 40 °C for 2 h. After cooling to room temperature, 10 mL of solution was taken and placed in a new conical flask. After adding 10 mL of ethanol and 2 drops of phenolphthalein indicator solution, titration with the NaOH-ethanol solution to the end point was carried out. The amine group content, E (mmol/g), in the sample was calculated with Equation (2).

$$E = (50 \times C_1 - 5 \times C_2 \times V_1)/m \tag{2}$$

where C_1 (mol/L) is the concentration of the HCl-ethanol solution, C_2 (mol/L) is the concentration of the NaOH–ethanol solution, V_1 (mL) is the volume of the NaOH-ethanol solution consumed for the titration, and m (g) is the mass of PAT.

2.3.3. Crosslink Degree of IPAT

The IPAT fiber sample was weighed, immersed in 50 mL acetic acid, and then soaked for 24 h. Then, it was washed three times with deionized water and dried to a constant mass at 50 °C. The crosslinking degree, C (%), was calculated as:

$$C = m_1/m_0 \times 100\% \tag{3}$$

where m_1 (g) is the mass of IPAT after soaking and drying, and m_0 (g) is the mass of IPAT before soaking.

2.4. Characterizations

The fiber morphology was observed by scanning electron microscopy (SEM, S-3400N II, Hitachi, Japan). The surface functional groups of fibers were analyzed by Fourier transform infrared spectrometry (FTIR, NEXUS870, Nicolet, Japan). Elemental analyzer (Vario MACRO cube, Elementar, Germany) was used to evaluate the elemental composition (C, H, N, O) of the fibers. Non-woven fabric samples were cut into pieces of about 1 cm × 1 cm, pasted on the glass slide, and placed in the water contact angle tester (JC2000C, Shanghai Zhongchen Digital Technic Apparatus Co., Ltd, Shanghai, China). The water contact angle was measured by dripping about 1 L ultrapure water on the fiber surface.

2.5. Adsorption and Regeneration

The fibers were immersed into 100 mL of Cr(VI) solution (400 mg/L) in a shaker (300 rpm) and allowed to adsorb for 60 min. The Cr(VI) concentration before and after adsorption was determined on a inductively coupled plasma emission spectrometer (iCAP 6300, Thermo Fisher Scientific, Cambridge, UK). The adsorption capacity, Q (mg/g), was calculated as:

$$Q = 0.05 \times (C_0 - C_1)/m \tag{4}$$

where C_0 (mg/L) and C_1 (mg/L) are the concentrations of Cr(VI) before and after adsorption, respectively, and m (g) is the mass of fibers.

The adsorption kinetics were fitted with pseudo-first-order (Equation (5)) and pseudo-second-order (Equation (6)) models.

$$\ln(q_e - q_t) = \ln q_e - k_1 t \tag{5}$$

$$\frac{t}{q_t} = \frac{1}{k_2 q_e^2} + \frac{t}{q_e} \tag{6}$$

where q_e and q_t are the amount of adsorbed Cr(VI) at equilibrium and at an arbitrary time t, respectively, and k_1 (min^{-1}) and k_2 (mg g^{-1} min^{-1}) are the pseudo-first-order and pseudo-second-order rate constants of adsorption, respectively.

The adsorption isotherm was analyzed with the Langmuir and Freundlich models:

$$\frac{C_e}{Q_e} = \frac{C_e}{Q_m} + \frac{1}{K_L Q_m} \tag{7}$$

$$\ln Q_e = \ln K_F + \frac{1}{n} \ln C_e \tag{8}$$

where C_e (mg/L) is the concentration of Cr(VI) in solution at equilibrium, Q_e (mg/g) is the equilibrium adsorption capacity, Q_m (mg/g) is the maximum adsorption capacity, K_L (L/mg) is the Langmuir constant, K_F is the Freundlich constant indicative of the adsorption capacity, and n is a parameter related to the sorption intensity.

After the adsorption of Cr(VI), the fibers were washed with 50 mL of NaOH solution (0.2%) for 30 min (150 rpm) to remove the adsorbed Cr(VI). The cleaning process was repeated until no Cr(VI) was detected in the washing solution. Elution efficiency, D_n (%), was calculated as:

$$D_n = M_n/Q_m \times 100\% \tag{9}$$

where n is the number of washes, M_n (mg) is the total amount of eluted Cr(VI) after n times elution, and Q_m (mg) is the adsorption capacity of fibers.

The regeneration efficiency, R_n, of fibers was calculated as:

$$R_n = Q_n/Q \times 100\% \tag{10}$$

where Q_n (mg/g) is the adsorption capacity of fibers after n cycles, and Q (mg/g) is the adsorption capacity of fibers in the first cycle.

To assess the adsorption selectivity of IPAT for Cr(VI), the fibers were placed into a mixed solution containing Cr(VI) (200 mg/L, pH = 3) and competing ions (phosphate, sulfate, and carbonate), and allowed to adsorb for 30 min. The adsorption selectivity, S (%), was calculated as:

$$S = Q_n/Q_0 \times 100\% \tag{11}$$

where Q_n (mg/g) and Q_0 (mg/g) are the adsorption capacity of fibers towards Cr(VI) in the presence and absence of competing ions, respectively.

3. Results and Discussion

3.1. Optimization of Preparation Conditions

3.1.1. Plasma Irradiation

The etching effect of plasma irradiation on PP fibers and the effects of the applied power, plasma irradiation pressure, and irradiation time on the grafting degree of AA monomers are depicted in Figure 2. Figure 2a shows that the PP fibers lose weight under the three atmospheres (i.e., AA, Ar, and air) due to the plasma etching. After 13 min of irradiation, the high-energy particles in the plasma are consumed and the energy of particles in the atmosphere is reduced, so the etching effect on the PP fiber surface is weakened and no change in mass of fibers is noticed. The weight loss of PP fibers observed in the three gases at the same applied power (120 W) and pressure (10 Pa) follows the order of air > Ar > AA. This may be explained by the fact that the high-energy particles generated by air ionization have the maximum energy, while the particles generated by the AA monomer have the minimum energy. Particles with higher energy tend to bombard the material surface more violently and lead to higher mass loss. In general, the weight loss within 19 min was less than 2.0%, which could be ignored in the calculation of the grafting degree [17].

Figure 2b shows the effect of different powers on the AA grafting degree under the three gases. As can be seen, the grafting degree follows the order of AA > Ar > air. Unlike the latter two, when AA gas is used, active free radicals are generated by the ionization and fragmentation of the AA monomers, which are more likely to bind to the surface of the PP fibers and create active free radical sites. On the contrary, air and Ar atmospheres are more prone to etch the fiber surface, which is not favorable for the formation of free radicals. In addition, the oxygen in the air may remove the active free radical sites due to oxidation, which makes the grafting degree in the air atmosphere slightly lower than that in the inert Ar atmosphere [27]. As for grafting in an AA atmosphere, the applied power has an obvious effect. Hence, the grafting is favored until a power of 120 W, then the efficiency of grafting strongly decreases as the power increases from 120 to 240 W. The rationale for this behavior is related to the increase in dissociation degree of AA for a power between 70 and 120 W when the active particles move fast, increasing the number of free radicals on the surface of the PP fiber and favoring grafting of AA. When the discharge power exceeds 120 W, the etching effect generated by bombarding the PP surface is too large, so the grafting degree gradually decreases [28].

Figure 2c shows the effect of plasma irradiation pressure on the grafting degree efficiency. Compared with the applied power, the irradiation pressure has a slight influence on the grafting degree. Under an AA atmosphere, a grafting degree of almost 50% is seen between 5 and 12 Pa, which then decreases to 40% with the increase in irradiation pressure to 20 Pa. Figure 2d shows the effect of the irradiation time on the grafting degree of the PA fiber at 120 W and 10 Pa. Irrespective of the atmosphere used, an increase is noticed within the first 5–6 min of irradiation, but it then decreases. This behavior is explained by the saturation of the fiber surface with free radicals, which corresponds to the highest degree of grafting. When the discharge time is too long, the adjacent radicals are easily coupled and annihilated, a phenomenon associated with a decrease in grafting degree.

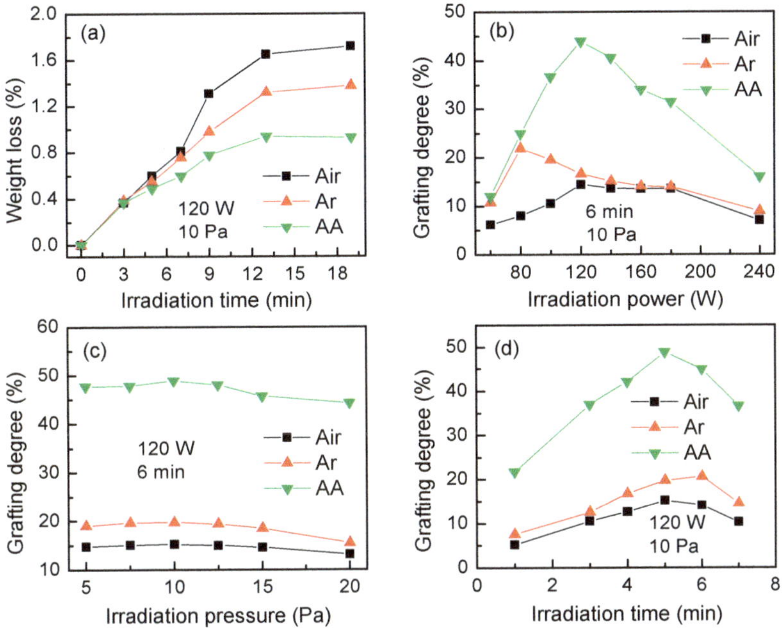

Figure 2. (**a**) The effect of the irradiation time on the weight loss of fibers and the effects of (**b**) the irradiation power, (**c**) pressure, and (**d**) time on the grafting degree of AA (grafting conditions: 150 Pa, 90 min, and 45 °C).

3.1.2. Grafting Conditions

The effects of the gas pressure, duration and temperature on the grafting degree were studied, as shown in Figure 3a–c, respectively. As displayed in Figure 3a, the grafting degree of PA decreases with the increase in gas pressure. This is explained by the high concentration of AA in the reaction system when the pressure is low and thus, a higher number of monomers can be grafted on the surface active sites of PP fibers. The lowest pressure in this reaction system is controlled at 150 Pa using an AA atmosphere. The grafting degree of modified fibers increases with the extension of grafting time from 0 to 90 min, and then it is stable, as illustrated in Figure 3b. As the reaction progresses, the active sites on the surface of the fibers are consumed and the grafting reaction is gradually completed. As can be seen from Figure 3c, the reaction temperature increase results in an increase in grafting degree of PA in the temperature range of 25–45 °C, afterwards, it remains stable. In the first temperature range, when the temperature increases while the pressure is constant, the saturated vapor pressure of the AA monomer increases in the reaction system, as well as the concentration, enhancing the reaction rate and thus, the grafting degree. However, the grafting degree is stopped at temperatures higher than 45 °C because the number of active sites on the surface of the PP fiber is limited and the high temperature favors the self-polymerization of a fraction of the AA monomer.

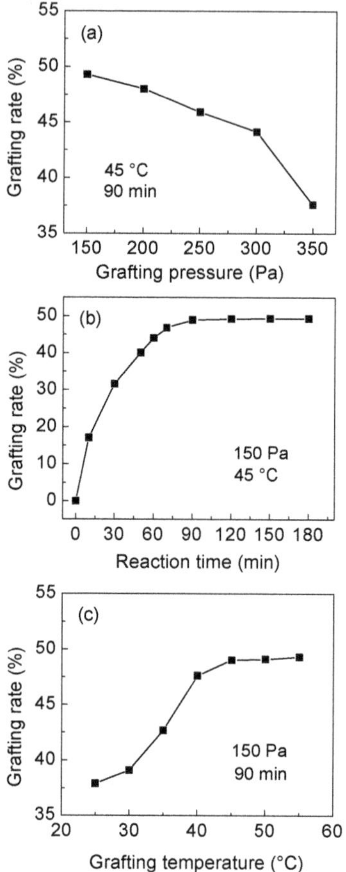

Figure 3. The effects of (**a**) the pressure, (**b**) reaction time, and (**c**) temperature on the grafting degree of PA (plasma irradiation conditions: 120 W, 10 Pa, and 5 min).

3.1.3. Amidation Efficiency of PAT and Grafting Degree of PA

Amidation of PA was carried out with different AA grafting degrees, and the relationship between the AA grafting degree of PA and the amidation efficiency of PAT is shown in Figure 4. From the first-order equation obtained by fitting, the concentration of the amino groups was 2.78 times higher than that of AA. According to the theoretical calculation, the highest number of amine groups on the polymer should be three times higher than that of AA. Hence, the amidation efficiency of the reaction is 92.7%.

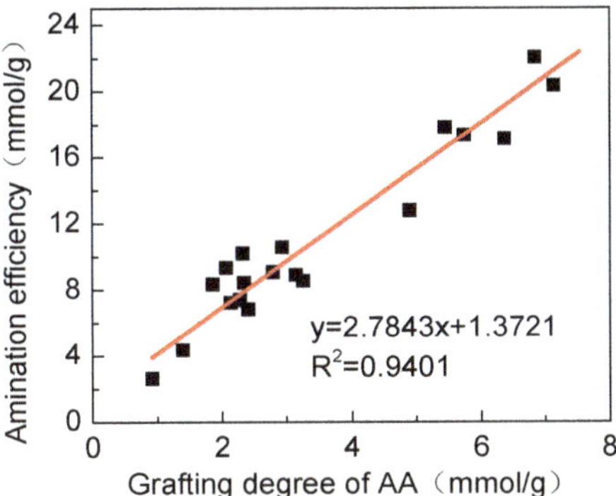

Figure 4. Relationship between the amidation efficiency of PAT and the grafting degree of PA.

3.1.4. Crosslinking Reaction

First, the effect of the crosslinking agent on the crosslinking degree was studied. Figure 5a shows the weight gain of fibers when formaldehyde, epichlorohydrin, and glutaraldehyde were used as crosslinking agents. When formaldehyde was used, the change in fiber weight was very small; only about 10% after 180 min of reaction. When epichlorohydrin was used as the crosslinking agent, the crosslinking effect was poor, and the fiber mass decreased by 7.8% after 180 min of crosslinking reaction. This effect is attributed to the ability of epichlorohydrin to desorb amines or AA chains on the surface of fibers. Among the three crosslinking agents, glutaraldehyde was the most appropriate, since an obvious gain of 214% was noticed after 180 min of reaction. According to the reaction scheme shown in Figure 1, two TETA molecules can be connected by glutaraldehyde during the cross-linking condensation reaction to bridge the polymeric spatial network.

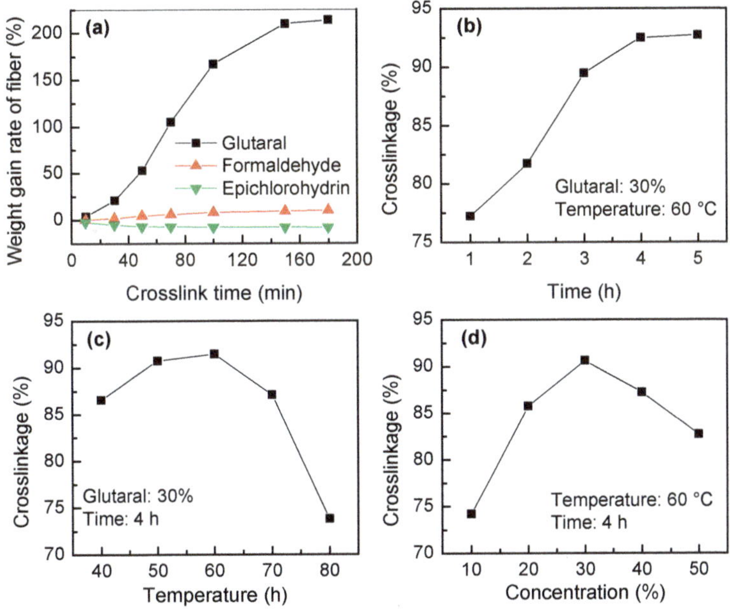

Figure 5. The effects of (**a**) the crosslinking agents, (**b**) time, (**c**) temperature, and (**d**) crosslinking agent concentration on the crosslinking degree of IPAT.

Once the cross-linking agent was identified, i.e., glutaraldehyde, the factors affecting the crosslinking degree of IPAT were further investigated. According to Figure 5b, the increase in crosslinking time results in an increase in the crosslinking degree of IPAT, so that an optimum time of 4 h was established. As Figure 5c shows, increasing the temperature from 40 to 60 °C leads to an increase in the degree of crosslinking. At higher temperatures (60–80 °C), the cross-linking degree decreases significantly due to the partial loss of fiber structural organization. According to Figure 5d, the optimum amount of glutaraldehyde concentration is 30%. Therefore, the optimal crosslinking conditions for IPAT are 30% glutaraldehyde at 60 °C for 4 h of reaction.

3.2. Characterization

3.2.1. SEM

Figure 6 shows SEM images of pristine and modified PP fibers. It can be seen from Figure 6 that the pristine PP fiber, whose diameter is in the micron range, displays a smooth surface. After gas phase grafting, the PA fiber surface is slightly rough, and corrugated layers can be seen. After amidation, the surface of PAT became rougher. After cross-linking with glutaraldehyde, the entire surface of the fiber is covered with a large number of granular polymers, thus denoting the formation of large local agglomerates. These results show that the fiber surface was successfully modified by grafting, amidation, and crosslinking.

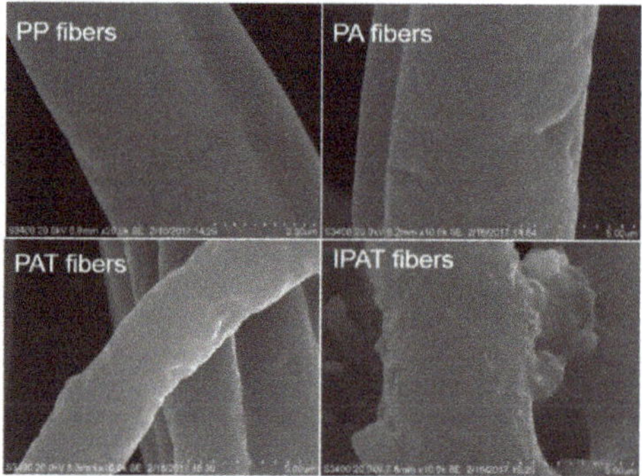

Figure 6. SEM images of PP, PA, PAT, and IPAT fibers.

3.2.2. FTIR

Figure 7 shows FTIR spectra of the fibers. The characteristic vibration bands of C–C, –CH$_3$, and –CH$_2$ could be observed only in the infrared spectrum of the pristine PP fiber. The vibration bands at 3171.1, 1708.7, and 1250.9 cm^{-1} displayed by the spectrum of PA are attributed to the stretching vibrations of –OH, C=O, and C–O, respectively, in the carboxyl group. After amidation, the band that appeared at 3346 cm^{-1} is the telescopic vibration of the –NH in the amide group. The bending vibration of –NH$_2$ and the stretching vibration of C–N were identified at 1648.5 and 1403.8 cm^{-1}, respectively. However, the strong vibration band at 1560.7 cm^{-1} is attributed to C=O in the amide group, which shifted to the right relative to the vibration band of C=O in the carboxyl group. This is because the π electrons in the amide group and the P electron conjugation on the nitrogen atom lead to the increase of C–N stretching vibration frequency. Compared with PAT, the intensity of the C–N telescopic vibration at 1051.6 cm^{-1} (mostly in the fingerprint area) increased in the spectrum of IPAT. On the other hand, the elemental analysis showed a large amount of the N element in the PAT fibers, which indicates successful amidation of the fiber surface. The content of the N element in IPAT increased significantly, pointing out that a larger amount of TETA was fixed on the surface of imprinted fibers by the crosslinking reaction. In addition, the content of the O element significantly increased as a result of the hydroxyl groups generated on the fiber by the acetal reaction between glutaraldehyde and TETA.

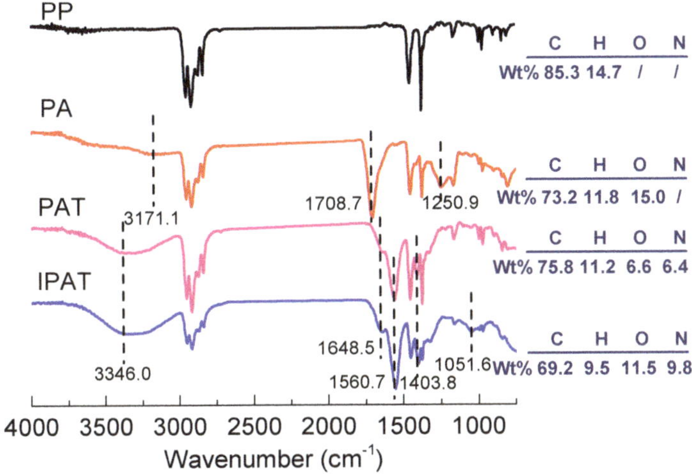

Figure 7. FTIR spectra and elemental analyses of PP, PA, PAT, and IPAT.

3.2.3. Water Contact Angle

The change in water contact angle of fibers is depicted in Figure 8. Compared with PP fibers, the water contact angle of PA fibers decreased significantly from 123.1° to 10.7°, indicating that grafting with the AA monomer significantly improved the hydrophilicity of the PP fibers. It can be seen from the illustration inset that PP fibers float on the water surface due to hydrophobicity, while PA fibers can be immersed in water after grafting with the AA monomer. The hydrophilicity of PAT and IPAT was further increased, and the water contact angle was reduced from 10.7° to nearly 0°, i.e., water droplets were immediately adsorbed on the fabric surface. The amines grafted on the surface of the fiber through the amidation reaction enhanced the surface polarity of the fibers, resulting in an enhanced hydrophilicity, which is conducive to the adsorption process in an aqueous solution.

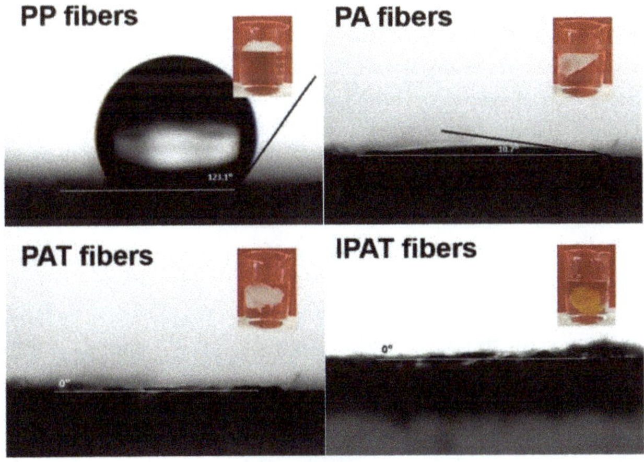

Figure 8. Water contact angle and hydrophilicity analysis of PP, PA, PAT, and IPAT.

3.3. Adsorption of Cr(VI)

3.3.1. Effect of pH on the Adsorption Capacity of IPAT

The effect of the pH on the adsorption of Cr(VI) on IPAT is shown in Figure 9. In the pH range of 1–3, the adsorption capacity of IPAT to Cr(VI) is high with a slight increase tendency, reaching the maximum adsorption capacity of about 167 mg/g at pH = 3. As the pH continues to increase, the adsorption capacity of the fiber decreases rapidly, and when the pH value is 8, the adsorption capacity drops to about 10 mg/g. When the solution is acidic, the protons in solution easily combine with the lone pair electrons of the N in the amino groups on the surface of IPAT, making the amine groups positively charged. As Cr(VI) is an anion in solution, the adsorption on the protonated amine groups is facilitated by electrostatic forces [29]. Meanwhile, IPAT was prepared at pH = 3, so the microstructure of imprinted areas is favorable for the adsorption of Cr(VI). When the pH is less than 3, part of $HCrO_4^-$ is converted into non-ionic H_2CrO_4, which is difficult to bind to the protonated amine groups, so the adsorption capacity of the fiber is slightly reduced. When the pH is higher than 3 and increases gradually, the amine groups on the fiber surface are deprotonated and negatively charged, which negatively impacts the adsorption of Cr(VI).

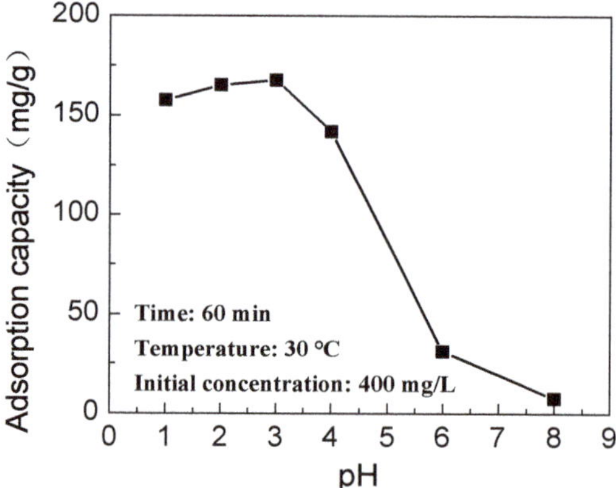

Figure 9. The effect of the pH on the adsorption capacity of IPAT towards Cr(VI).

3.3.2. Adsorption Kinetics

The kinetics for the adsorption of Cr(VI) onto IPAT was studied and the results are illustrated in Figure 10a. The adsorption of Cr(VI) was very fast in the first 10 min of adsorption. As the functional groups of fiber materials are evenly distributed on the surface and all available for the adsorption, the saturation is achieved in a short time. After 35 min, the surface amine groups are all occupied by Cr(VI), and the sorption process reaches equilibrium. The linear fitting results and parameters are shown in Figure 10b,c. Generally, the adsorption process that conforms to the pseudo-first-order kinetic model is considered to be physical adsorption, and the adsorption rate is proportional to the number of adsorption sites on the surface of the adsorbent. The adsorption process of the pseudo-second-order kinetic model is a chemical adsorption process. The surface of the adsorbent and Cr(VI) achieve a chemical adsorption by sharing electron pairs, charge exchange, or electrostatic attraction. The R^2 of the pseudo-first-order kinetic model is 0.8794, while for the pseudo-second-order kinetic model, R^2 is 0.9973. Hence, the adsorption of Cr(VI) on fibers is better fitted by the pseudo-second-order kinetic model, indicating that the adsorption of Cr(VI) is controlled by chemical phenomena [30]. Hence, the electrostatic attraction between the protonated amine groups and Cr(VI) anions under acidic conditions is the main driving force for Cr(VI) adsorption on the modified fibers.

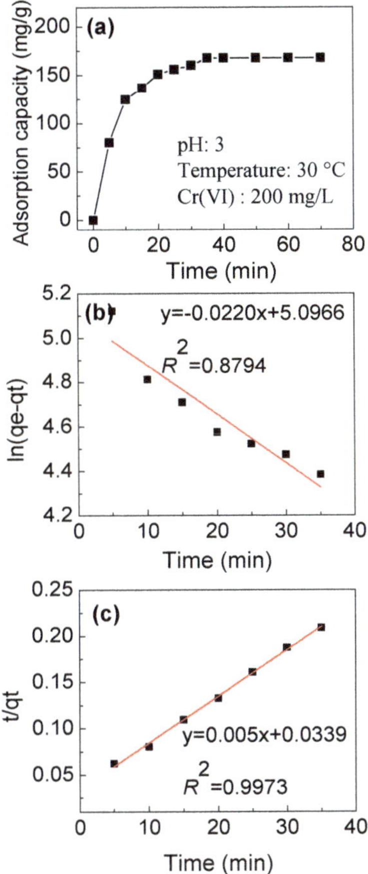

Figure 10. (**a**) The kinetics curve of Cr(VI) adsorption onto IPAT and the linearized (**b**) pseudo-first-order and (**c**) pseudo-second-order kinetics.

3.3.3. Adsorption Isotherm

The adsorption isotherm was drawn based on the results obtained using different initial concentrations of Cr(VI) (30–400 mg/L). As can be seen from Figure 11a, the adsorption capacity of IPAT increased with the increase in initial Cr(VI) concentration. The linearized versions of the two models are illustrated in Figure 11b,c and the values of the kinetic parameters are listed in Table 1. It was found that the R^2 value of Langmuir and Freundlich was 0.993 and 0.996, respectively. Although both models fit the data, the adsorption of Cr(VI) is better fitted by the Freundlich model, showing a multilayer adsorption, and the adsorption process varies at different areas on the surface of the fiber. In the Freundlich model, the value of n is 4.76, indicative of an favorable adsorption of Cr(VI) onto the surface of IPAT fibers [31].

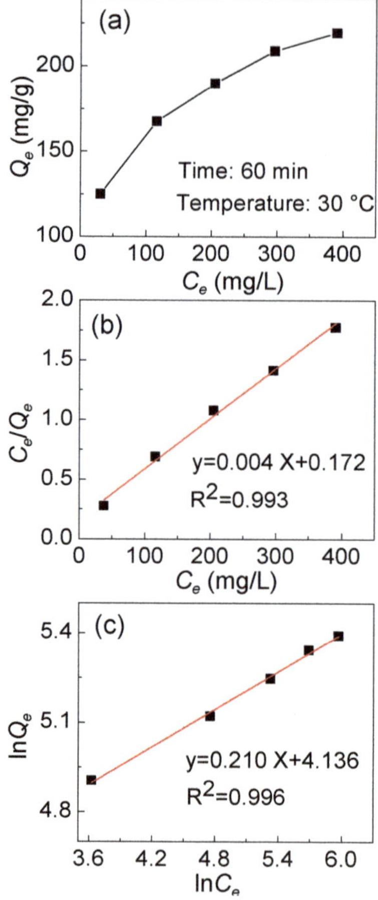

Figure 11. (**a**) The effect of the initial Cr(VI) concentration on the adsorption capacity of Cr(VI) onto IPAT and the linearization of the (**b**) Langmuir and (**c**) Freundlich models.

Table 1. Langmuir and Freundlich isotherm parameters for Cr(VI) adsorption onto IPAT.

	Langmuir			Freundlich	
Q_m (mg/g)	K_L (L/mg)	R^2	n	K_F (mg/g)	R^2
238.1	0.024	0.993	4.76	62.8	0.996

Comparisons between maximum adsorption capacities (q_{max}) of IPAT and other ion imprinted adsorbents for Cr(VI) reported in the literature are presented in Table 2. The results show that IPAT exhibits a reasonable capacity for Cr(VI) adsorption from aqueous solutions.

Table 2. Adsorption capacities of various ion imprinted adsorbents for Cr(VI).

Adsorbents	Morphology	q_{max} (mg/g)	References
styrene/4-VP/EDGMA	Particles 53–90 μm	37.58	[32]
4-VP/HEMA	Particles 75–150 μm	172.12	[33]
Magnetic poly(4-VP)	-	6.20	[34]
4-VP/EDGMA	Particles 50–100 μm	286.56	[35]
SCC/Epichlorohydrin	-	177.62	[36]
Magnetic	Particles ~70 nm	39.3	[29]
Pebax/chitosan/GO/APTES	Nanofibers ~90 nm	204 5	[37]
IPAT	Fiberous ~7 μm	167	This work

4-VP: 4-vinylpyridine; EGDMA: ethylene glycol dimethacrylate; HEMA: hydroxyethyl methacrylate; HEMAH: 2-methacryloylamido histidine; SCC: sodium carboxymethyl cellulose; GO: graphene oxide; APTES: 3-aminopropyltriethoxysilane.

3.3.4. Adsorption Selectivity

It can be seen from Figure 12a that the adsorption capacity of IPAT for Cr(VI) decreases as the concentration of competing ions (i.e., phosphate, sulfate, and carbonate) increases, which demonstrates that the competing ions reduce the adsorption capacity of fibers for Cr(VI) due to the similar affinity for the adsorption sites on the fiber surface. The strength of adsorption of the three competing ions increases in the order of phosphate > sulfate > carbonate, among which phosphate and sulfate have a greater effect on the adsorption of Cr(VI) than that of carbonate. When the concentration of the three competing ions (i.e., phosphate, sulfate, and carbonate) is 2.5 times higher than that of Cr(VI), the adsorption selectivity of IPAT fibers towards Cr(VI) is 32.7, 35.3, and 76.7%, respectively. Compared with IPAT, the adsorption selectivity of NIPAT is significantly lower. When the concentration of three ions (i.e., phosphate, sulfate, and carbonate) is 2.5 times higher than that of Cr(VI), the adsorption selectivity of NIPAT fibers to Cr(VI) is 17.3, 19.3, and 51.2%, respectively.

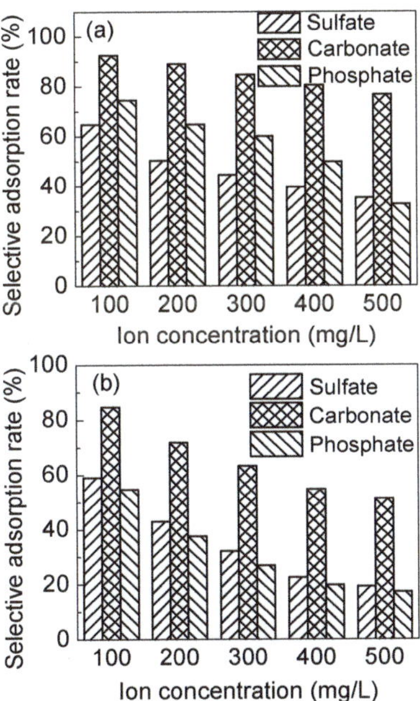

Figure 12. The effects of the competing ions on the adsorption selectivity of (**a**) IPAT and (**b**) NIPAT.

3.3.5. Desorption and Regeneration

It can be seen from Figure 13a that nearly 80% of Cr(VI) can be removed after the first desorption cycle, the adsorbed Cr(VI) being almost completely removed after five cycles, indicating that NaOH solution has a good elution effect on the adsorbed Cr(VI) on the fiber. Under acidic conditions, the positively charged amine groups on the surface of the fibers attract the negatively charged Cr(VI) anions thus promoting the adsorption process. On the other hand, the OH^- in sodium hydroxide solution has a stronger affinity to protons compared with the amine group and can quickly capture the H^+ in the protonated amine groups. As a result, the amine groups loss the ability to bind the Cr(VI) anion and achieve the purpose of regeneration. It can be seen from the SEM images taken for the fibers after elution (insert in Figure 13a) that the fiber surface displays the morphology of the fibers before adsorption, which demonstrates that the adsorption and elution process has no effect on the morphostructural properties of the fibers. Figure 13b shows that the adsorption efficiency of fibers stays above 82% after 10 adsorption and regeneration cycles. This kind of adsorption strength is moderate, neither too small to decrease adsorption quantity, nor too strong to make the elution difficult.

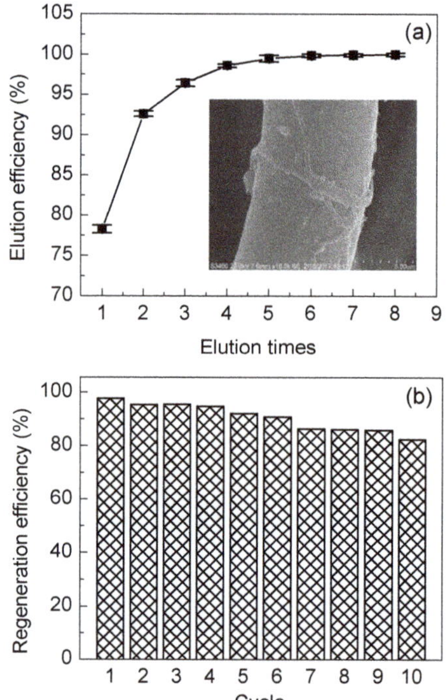

Figure 13. (**a**) The elution efficiency after several elution cycles and (**b**) the regeneration efficiency of IPAT after 10 adsorption–desorption cycles.

3.3.6. The Effect of the Oxidation of Cr(VI) on the IPAT Structural Integrity

After adsorption of Cr(VI) and drying to a constant weight, the morphostructural properties of IPAT were analyzed by SEM and FTIR spectroscopy. As SEM images displayed in Figure 14 show, a large number of fibers were fractured, and deep holes appeared on their surface, which can explain the fiber fracture. In addition, the color of the fibers changed from yellow to dark brown, while the fibers were brittle with almost no hardiness. A large amount of Cr(VI) was found on the surface of the fiber via EDS analysis. The FTIR spectra of IPAT fibers before and after adsorption of Cr(VI) show that the characteristic bands of the original PP fiber remain basically unchanged. However, the bands assigned to an amide group (3322.8, 1403.4, and 1051.6 cm^{-1}) almost disappeared, whereas the vibration band of –C=O at 1556 cm^{-1} was significantly weakened, which reveals the damage to the surface functional amide groups. Meanwhile, the corresponding absorption band for the existing Cr(VI) was found at 1713.4 cm^{-1} in the FTIR spectrum [38]. Therefore, after adsorption of Cr(VI) and drying, the structural integrity of the IPAT fibers formed by amidation and crosslinking was destroyed due to the strong oxidizing effect of Cr(VI). Compared with the morphology of IPAT in Figure 13a, it can be concluded that the oxidation and damage of fibers caused by Cr(VI) only occurred when they were subjected to drying without elution, whereas they exhibited resistance to Cr(VI) oxidation under normal conditions of use, i.e., adsorption, elution, and the regeneration procedure.

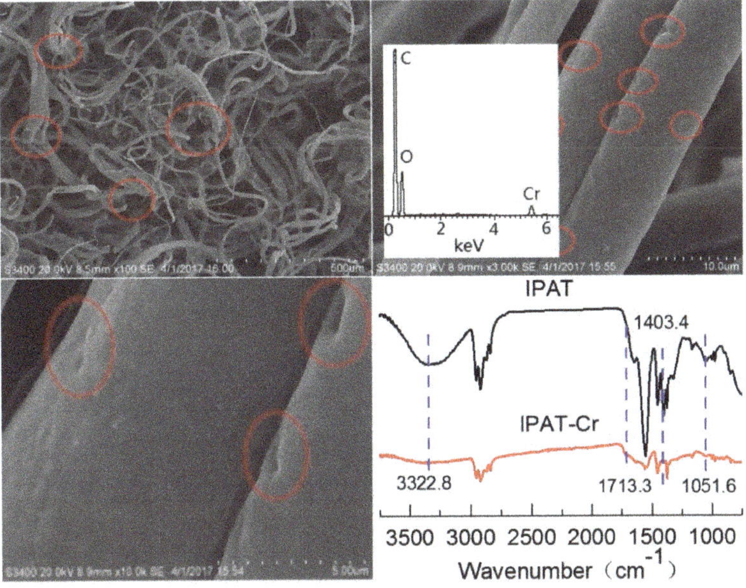

Figure 14. SEM images and FTIR spectra of dried IPAT loaded with Cr(VI).

4. Conclusions

In this work, a non-thermal RF plasma treatment followed by gaseous phase AA grafting was used to prepare ion-imprinted PP fibers. The grafting degree was the highest when AA was used as a plasma atmosphere, and the optimum treatment conditions were 5 min, 120 W and 10 Pa. The output power and processing time had a greater impact on the grafting degree than the monomer pressure. The highest grafting degree of AA was reached under the condition of 150 Pa, 90 min, and 45 °C. When glutaraldehyde (30%) was used as a crosslinking agent, the highest crosslinking degree was obtained at 60 °C for 4 h. SEM and FTIR results indicated that the imprinted fibers were successfully prepared by a step-by-step method. Meanwhile, the hydrophilicity of imprinted fibers was greatly enhanced compared with that of PP fibers, which is conducive to the adsorption process in aqueous solution. The adsorption performance of the obtained IPAT fibers was evaluated in Cr(VI) removal. The highest adsorption capacity of 167 mg/g was obtained at pH = 3. The kinetics of adsorption followed the pseudo-second-order model and the adsorption isotherm data were better fitted by the Freundlich model. The adsorption selectivity of IPAT and NIPAT was significantly affected by phosphate and sulfate, while the carbonate ion had a slight effect. In addition, in the presence of competing ions, IPAT showed higher adsorption selectivity compared with NIPAT. The adsorbed Cr(VI) was eluted rapidly and effectively by a NaOH solution (0.2%). After 10 regenerative adsorption experiments, the adsorption efficiency of IPAT was still higher than 80%, confirming the reusability of the prepared fibers. Under normal conditions of use, IPAT exhibits efficient resistance to Cr(VI) oxidation, which makes it feasible for use in adsorption.

Author Contributions: Conceptualization, Z.L. (Zhengwei Luo), Z.L. (Zhouyang Lian) and W.W.; Methodology, Z.L. (Zhengwei Luo), Z.L. (Zhouyang Lian) and W.W.; Measurement, Z.L. (Zhengwei Luo), J.X. (Jiahuan Xu), D.Z., D.W. and H.J.; Data analysis, Z.L. (Zhengwei Luo), H.J. and Z.L. (Zhouyang Lian); Writing (Original Draft Preparation), Z.L. (Zhengwei Luo), J.X. (Jianjian Xu) and Z.L. (Zhouyang Lian); Writing (Review & Editing), Z.L. (Zhengwei Luo), W.G., Z.L. (Zhouyang Lian) and W.W.

Funding: This work was financially supported by the National Natural Science Foundation of China (Grant Number 21676144).

Acknowledgments: W.W.J. would like to thank the support from the Natural Science Foundation of China (Grant Number 21676144).

Conflicts of Interest: The authors declare no conflict of interest.

References

1.   Kimbrough, D.E.; Cohen, Y. A critical assessment of chromium in the environment. *Crit. Rev. Environ. Control* **1999**, *29*, 1–46. [CrossRef]
2.   Dubey, S.P.; Gopal, K. Adsorption of chromium(VI) on low cost adsorbents derived from agricultural waste material: A comparative study. *J. Hazard. Mater.* **2007**, *145*, 465–470. [CrossRef] [PubMed]
3.   Erdem, M.; Tumen, F. Chromium removal from aqueous solution by the ferrite process. *J. Hazard. Mater.* **2004**, *109*, 71–77. [CrossRef] [PubMed]
4.   Ho, W.S.W.; Poddar, T.K. New membrane technology for removal and recovery of chromium from waste waters. *Environ. Prog. Sustain.* **2000**, *20*, 44–52. [CrossRef]
5.   Luo, S.; Qin, F. Fabrication uniform hollow Bi_2S_3 nanospheres via Kirkendall effect for photocatalytic reduction of Cr(VI) in electroplating industry wastewater. *J. Hazard. Mater.* **2017**, *340*, S508503713X. [CrossRef] [PubMed]
6.   Mohanty, K.; Jha, M. Biosorption of Cr(VI) from aqueous solutions by *Eichhornia crassipes*. *Chem. Eng. J.* **2006**, *117*, 71–77. [CrossRef]
7.   Wang, W.; Li, M. Column adsorption of chromium(VI) by strong alkaline anion-exchange fiber. *J. Appl. Polym. Sci.* **2012**, *126*, 1733–1738. [CrossRef]
8.   Dognani, G.; Hadi, P. Effective chromium removal from water by polyaniline-coated electrospun adsorbent membrane. *Chem. Eng. J.* **2019**, *372*, 341–351. [CrossRef]
9.   Hayashi, N.; Chen, J. Nitrogen-containing fabric adsorbents prepared by radiation grafting for removal of chromium from wastewater. *Polymers* **2018**, *10*, 744. [CrossRef]
10.   Huang, R.; Ma, X. A novel ion-imprinted polymer based on graphene oxide-mesoporous silica nanosheet for fast and efficient removal of chromium (VI) from aqueous solution. *J. Colloid Interface Sci.* **2018**, *514*, 544–553. [CrossRef]
11.   Wei, Q.F.; Mather, R.R. Functional nanostructures generated by plasma-enhanced modification of polypropylene fibre surfaces. *J. Mater. Sci.* **2005**, *40*, 5387–5392. [CrossRef]
12.   Gul Dincmen, M.; Hauser, P.J. Plasma induced graft polymerization of three hydrophilic monomers on nylon 6,6 Fabrics for enhancing antistatic property. *Plasma Chem. Plasma Proc.* **2016**, *36*, 1377–1391. [CrossRef]
13.   Basarir, F.; Choi, E.Y. Electrochemical properties of PP membranes with plasma polymer coatings of acrylic acid. *J. Membrane Sci.* **2005**, *260*, 66–74. [CrossRef]
14.   Chen, H.; Guo, M. Green and efficient synthesis of an adsorbent fiber by plasma-induced grafting of glycidyl methacrylate and its Cd(II) adsorption performance. *Fibers Polym.* **2018**, *19*, 722–733. [CrossRef]
15.   Tseng, C.; Wang, C. Polypropylene fibers modified by plasma treatment for preparation of Ag nanoparticles. *J. Phys. Chem. B* **2006**, *110*, 4020–4029. [CrossRef]
16.   Haji, A.; Shoushtari, A.M. Grafting of poly(propylene imine) dendrimer on polypropylene nonwoven: Preparation optimization, characterization, and application. *Fibers Polym.* **2019**, *20*, 913–921. [CrossRef]
17.   Luo, Z.; Chen, H. Surface grafting of styrene on polypropylene by argon plasma and its adsorption and regeneration of BTX. *J. Appl. Polym. Sci.* **2018**, *135*, 46171. [CrossRef]
18.   Noor, S.; Xiakeer, S. Controlled levofloxacin release and antibacterial properties of β-cyclodextrins-grafted polypropylene mesh devices for hernia repair. *Polymers* **2018**, *10*, 493.

19. Khlyustova, A.; Galmiz, O. Underwater discharge plasma-induced coating of poly(acrylic acid) on polypropylene fiber. *J. Mater. Sci.* **2015**, *50*, 3504–3509. [CrossRef]
20. Buček, A.; Popelka, A. Acrylic acid plasma treatment of polypropylene nonwoven fabric. *Fibres Text. East. Eur.* **2016**, *24*, 161–164. [CrossRef]
21. Sciarratta, V.; Vohrer, U. Plasma functionalization of polypropylene with acrylic acid. *Surf. Coat. Technol.* **2003**, *174–175*, 805–810. [CrossRef]
22. Nishide, H.; Tsuchida, E. Selective adsorption of metal ions on poly(4-vinylpyridine) resins in which the ligand chain is immobilized by crosslinking. *Macromol. Chem. Phys.* **1976**, *177*, 2295–2310. [CrossRef]
23. Li, T.; Chen, S. Preparation of an ion-imprinted fiber for the selective removal of Cu^{2+}. *Langmuir* **2011**, *27*, 6753–6758. [CrossRef] [PubMed]
24. Zhang, L.; Yang, S. Surface ion-imprinted polypropylene nonwoven fabric for potential uranium seawater extraction with high selectivity over vanadium. *Ind. Eng. Chem. Res.* **2017**, *56*, 1860–1867. [CrossRef]
25. Yang, S.; Ji, G. Polypropylene nonwoven fabric modified with oxime and guanidine for antibiofouling and highly selective uranium recovery from seawater. *J. Radioanal. Nucl. Chem.* **2019**, *321*, 323–332. [CrossRef]
26. Guo, M.; Liang, H. Study on melt-blown processing, web structure of polypropylene nonwovens and its BTX adsorption. *Fibers Polym.* **2016**, *17*, 257–265. [CrossRef]
27. Černáková, L.; Kováčik, D. Surface modification of polypropylene non-woven fabrics by atmospheric-pressure plasma activation followed by acrylic acid grafting. *Plasma Chem. Plasma Proc.* **2005**, *25*, 427–437. [CrossRef]
28. Altay, L.; Bozaci, E. The effect of atmospheric plasma treatment of recycled carbon fiber at different plasma powers on recycled carbon fiber and its polypropylene composites. *J. Appl. Polym. Sci.* **2019**, *136*, 47131. [CrossRef]
29. Hassanpour, S.; Taghizadeh, M. Magnetic Cr(VI) ion imprinted polymer for the fast selective adsorption of Cr(VI) from aqueous solution. *J. Polym. Environ.* **2018**, *26*, 101–115. [CrossRef]
30. Wu, S.; Dai, X. Highly sensitive and selective ion-imprinted polymers based on one-step electrodeposition of chitosan-graphene nanocomposites for the determination of Cr(VI). *Carbohydr. Polym.* **2018**, *195*, 199–206. [CrossRef] [PubMed]
31. Xia, L.; Huang, Z. Bagasse cellulose grafted with an amino-terminated hyperbranched polymer for the removal of Cr(VI) from aqueous solution. *Polymers* **2018**, *10*, 931. [CrossRef] [PubMed]
32. Pakade, V.; Cukrowska, E. Selective removal of chromium (VI) from sulphates and other metal anions using an ion-imprinted polymer. *Water Sa* **2011**, *37*, 529–538. [CrossRef]
33. Bayramoglu, G.; Arica, M.Y. Synthesis of Cr(VI)-imprinted poly(4-vinyl pyridine-co-hydroxyethyl methacrylate) particles: Its adsorption propensity to Cr(VI). *J. Hazard. Mater.* **2011**, *187*, 213–221. [CrossRef] [PubMed]
34. Tavengwa, N.T.; Cukrowska, E. Synthesis, adsorption and selectivity studies of N-propyl quaternized magnetic poly(4-vinylpyridine) for hexavalent chromium. *Talanta* **2013**, *116*, 670–677. [CrossRef] [PubMed]
35. Kong, D.; Zhang, F. Fast removal of Cr(VI) from aqueous solution using Cr(VI)-imprinted polymer particles. *Ind. Eng. Chem. Res.* **2014**, *53*, 4434–4441. [CrossRef]
36. Velempini, T.; Pillay, K. Epichlorohydrin crosslinked carboxymethyl cellulose-ethylenediamine imprinted polymer for the selective uptake of Cr(VI). *Int. J. Biol. Macromol.* **2017**, *101*, 837–844. [CrossRef] [PubMed]
37. Etemadi, M.; Samadi, S. Selective adsorption of Cr(VI) ions from aqueous solutions using Cr^{6+}-imprinted Pebax/chitosan/GO/APTES nanofibrous adsorbent. *Int. J. Biol. Macromol.* **2017**, *95*, 725–733. [CrossRef] [PubMed]
38. Ren, Z.; Xu, X. FTIR, Raman, and XPS analysis during phosphate, nitrate and Cr(VI) removal by amine cross-linking biosorbent. *J. Colloid Interf. Sci.* **2016**, *468*, 313–323. [CrossRef] [PubMed]

MDPI
St. Alban-Anlage 66
4052 Basel
Switzerland
Tel. +41 61 683 77 34
Fax +41 61 302 89 18
www.mdpi.com

Polymers Editorial Office
E-mail: polymers@mdpi.com
www.mdpi.com/journal/polymers